# Functions of α-Bounded Type in the Half-Plane

Advances in Complex Analysis and Its Applications
Volume 4

# Functions of $\alpha$-Bounded Type in the Half-Plane

by

ARMEN M. JERBASHIAN
Institute of Mathematics
National Academy of Sciences, Armenia

A C.I.P. Catalogue record for this book is available
from the Library of Congress.

ISBN 1-4899-9989-2 ISBN 0-387-23626-0 (eBook)



Softcover re-print of the Hardcover 1st edition 2005



9 8 7 6 5 4 3 2 1 SPIN 11161851

springeronline.com

# CONTENTS

# PREFACE

This book is related to the theory of functions of $\alpha$-bounded type in the half-plane of the complex plane. I constructed this theory by application of the Liouville integro-differentiation. To some extent, it is similar to M.M.Djrbashian's factorization theory of the classes $N_\alpha$ of functions of $\alpha$-bounded type in the disc, as much as the well known results on different classes and spaces of regular functions in the half-plane are similar to those in the disc.

Besides, the book contains improvements of several results such as the Phragmén-Lindelöf Principle and Nevanlinna Factorization in the Half-Plane and offers a new, equivalent definition of the classical Hardy spaces in the half-plane.

The last chapter of the book presents author's united work with G.M. Gubreev (Odessa). It gives an application of both $\alpha$-theories in the disc and in the half-plane in the spectral theory of linear operators. This is a solution of a problem repeatedly stated by M.G.Krein and being of special interest for a long time.

The book is proposed for a wide range of readers. Some of its parts are comprehensible for graduate students, while the book in the whole is intended for young researchers and qualified specialists in the field.

I am grateful to my father Prof. Mkhitar M. Djrbashian who in 1980's led me into the field of representations of regular functions. I am very glad to express my gratitude to Prof. Cabiria Andreian Cazacu and Prof. Chung-Chun Yang for encouraging me to write this book. Also, I am thankful to my wife Gohar Jerbashian for her valuable assistance in refining the text of the book.

ARMEN M. JERBASHIAN

# INTRODUCTION

*1.* The fundamental role of integral representations and factorizations in complex analysis is well known. In this field, the basic results of Poisson, Cauchy, Weierstrass, and Hadamard where continued by outstanding achievements of Hardy, Blaschke, Fatou, Herglotz, F. and M. Riesz, Segö, R.Nevanlinna, V.I. Privalov and other classics of the contemporary complex analysis. These achievements partially were summarized in the widely known monographs of R. Nevanlinna [84], I.I.Privalov [86, 87], L.De'Branges [14]. A series of further results is due to Paley–Wiener [85], Hille–Tamarkin [47], V.I.Krilov [76], B.Ja. Levin [77], M.Tsuji [106] and others.

The investigated *non-weighted* classes of functions mainly were connected with the following three classical problems, which depend on the specific geometry and technical apparatus of the disc and half-plane of the complex plane.

**Problem I.** Describe some classes of regular functions, the growth of which is restricted by a rotation-invariant condition that does not distinguish the boundary points of the unit disc, i.e. considers *the whole boundary as a single point.* The growth condition

*The unit disc on the Riemann sphere*

• 0

THE GLOBAL PROBLEM IN THE DISC

$$\sup_{0<r<1} \int_0^{2\pi} |f(re^{i\vartheta})|^p d\vartheta < +\infty$$

defining Hardy's spaces $H^p$ $(0 < p < +\infty)$ of functions holomorphic in $|z| < 1$ is an example of such restriction. Another example is the condition defining Nevanlinna's class of meromorphic functions of bounded type in the disc or a similar class of subharmonic functions possessing nonnegative harmonic majorants in $|z| < 1$:

$$\sup_{0<r<1} \int_0^{2\pi} u^+(re^{i\vartheta}) d\vartheta < +\infty \quad (u^+ = \max\{u, 0\}). \tag{1}$$

**Problem II.** Describe some classes of functions which are assumed to be in a sense regular even on the boundary of the half-plane, except the point $\infty$, while the growth of these functions is restricted only near $\infty$. Nevanlinna conditions defining a subclass of subharmonic functions possessing nonnegative harmonic majorants in the upper half-plane (i.e. providing Nevanlinna factorization in

*The upper half-plane on the Riemann sphere*

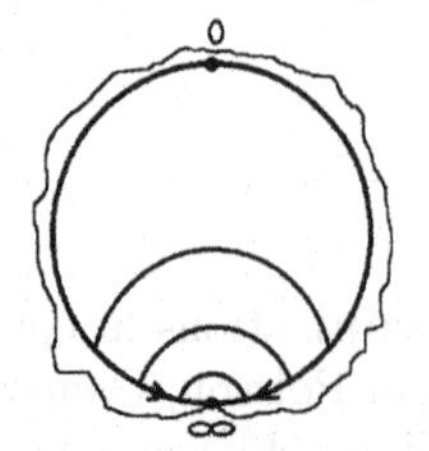


THE LOCAL PROBLEM IN THE HALF-PLANE

the upper half-plane) present an example of such restrictions:

$$\int_{-\infty}^{+\infty} \frac{u^+(x)}{1+x^2} dx < +\infty, \tag{2}$$

$$\liminf_{R\to+\infty} \frac{1}{R} \int_0^\pi u^+(Re^{i\vartheta}) \sin\vartheta d\vartheta < +\infty \tag{3}$$

($u(z)$ is assumed to be subharmonic in a domain containing the closed finite half-plane Im $z \geq 0$). Another example is the Phragmén–Lindelöf principle, where both quantities (2) and (3) are assumed to be zero.

**Problem III.** The growth of functions is restricted by a shift-invariant condition distinguishing only the point $\infty$, while the finite points of the boundary are considered equivalently, i.e. *the boundary is considered as two points, $\infty$ and all finite boundary points.* This is the similarity of the global Problem I in the disc.

*The upper half-plane on the Riemann sphere*

0

∞

THE GLOBAL PROBLEM IN THE HALF-PLANE

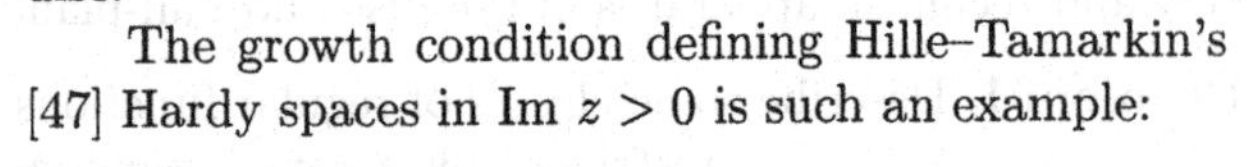

The growth condition defining Hille–Tamarkin's [47] Hardy spaces in Im $z > 0$ is such an example:

$$\sup_{y>0} \int_{-\infty}^{+\infty} |f(x+iy)|^p dx < +\infty.$$

Another condition of such type defines V.I.Krilov[76]–E.D.Solomentsev's [103] classes $\mathfrak{N}$ and $\mathfrak{N}^m$ of subharmonic functions with bounded Tsuji characteristics in Im $z > 0$:

$$\begin{aligned} \mathfrak{N}: &\quad \sup_{y>0} \int_{-\infty}^{+\infty} u^+(x+iy) dx < +\infty, \\ \mathfrak{N}^m: &\quad \sup_{y>0} \int_{-\infty}^{+\infty} |u(x+iy)|\, dx < +\infty. \end{aligned} \tag{4}$$

One can note the differences between the analytic apparatus peculiar to the disc and to the half-plane. Particularly, representation as a Laplace transform in the half-plane corresponds to the Taylor expansion in the disc.

*2.* Further development of representation apparatus was necessary for applications in different fields of science. Particularly, the apparatus was extended to different *weighted* classes of functions regular in different senses. The canonical representations of weighted classes of holomorphic and meromorphic functions, established by M.M.Djrbashian [16] in 1945 were among first results in the field. His results *actually revealed a new way of application of fractional integ-*

*ro-differentiation – in the representation problem of weighted classes of regular functions.* One can find a detailed survey of these and other methods of fractional integro-differentiation in the monograph of S.G.Samko, A.A.Kilbas and O.I.Marichev [92]. The initial work [16] (see also [17] for detailed proofs and additional results) mainly aimed at improving Nevanlinna's result [84, Sec. 216] related to description of the density of zeros and poles of the weighted class of functions meromorphic in the disc $|z| < 1$, the Nevanlinna characteristics of which satisfy the rotation-invariant condition

$$\int_0^1 (1-r)^\alpha T(r,f)dr < +\infty \tag{5}$$

for a given $-1 < \alpha < +\infty$. Namely, Nevanlinna has shown that the zeros and poles of functions from his weighted class (5) satisfy the modular condition

$$\sum_\mu (1-|a_\mu|)^{\alpha+2} < +\infty, \quad \sum_\nu (1-|b_\nu|)^{\alpha+2} < +\infty. \tag{6}$$

In [16] the following canonical factorization was found for meromorphic functions from Nevanlinnna's class (5):

$$\begin{aligned} F(z) =& \frac{k_\alpha}{C_\lambda} \frac{\pi_\alpha(z,\{a_\mu\})}{\pi_\alpha(z,\{b_\nu\})} \\ &\times \exp\left\{ \frac{2(\alpha+1)}{\pi} \iint_{|\zeta|<1} (1-|\zeta|^2)^\alpha \frac{\log|F(\zeta)|}{(1-z\overline{\zeta})^{\alpha+2}} d\sigma(\zeta) \right\}, \quad |z|<1, \end{aligned} \tag{7}$$

where $C_\lambda$ is the first non-zero coefficient of the Laurent expansion of $F(z)$ at $z=0$, $k_\alpha$ is a definite constant, $\{a_\mu\}$ and $\{b_\nu\}$ are zeros and poles of $F(z)$, and $\pi_\alpha$ are some *Blaschke type products* with factors of the form

$$\exp\left\{ -\int_{|\zeta|^2}^1 \frac{(1-t)^{\alpha+1}}{\left(1-\frac{z}{\zeta}t\right)^{\alpha+2}} \frac{dt}{t} \right\}\Bigg|_{\alpha=-1} = \frac{\zeta - z}{1-\overline{\zeta}z} \frac{|\zeta|^2}{\zeta}. \tag{8}$$

Note that later the same products were considered also by Tsuji (see [106, Ch. IV]) for integers $\alpha = 0, 1, 2, \ldots$.

One can easily observe the Riemann–Liouville fractional integro-differentiation in the definition and in the structure of the canonical factorization (7) of Nevanlinna's class (5). Using that factorization apparatus, F.A.Shamoian [94] found some descriptive factorizations of the class (5) and particularly could prove that Nevanlinna's modular condition (6) completely describes the density of zeros and poles (and hence of any $a$-points) of functions from the mentioned class. One can find a series of further results related to these weighted classes and, particularly, a complete modular description of the zero-sets of more general classes in [95, 97, 99].

The papers [16, 17] also contain an investigation of the weighted Hardy type classes $H^p(\alpha)$ (later known as $A^p_\alpha$) of functions holomorphic in $|z|<1$, the growth of which is restricted by the condition

$$\iint_{|\zeta|<1} (1-|\zeta|)^\alpha |f(\zeta)|^p d\sigma(\zeta) < +\infty \quad (d\sigma \text{ is the Lebesgue area measure})$$

for given $\alpha > -1$ and $p \geq 1$. Particularly, the widely known representations of functions $f(z) \in A^p_\alpha$ were first found. The origins of investigations related to $A^2_0$ can be found in a work of L.Biberbach [9] and in other references of [107], where the works [13, 73, 109, 43, 8] relating to $A^2_0$ and even more general spaces are to be especially mentioned. For the later developments in the topics, see the monographs [15], [45] with their reference lists and [95, 97, 98, 100, 101, 90, 91, 68].

*3.* Later the Riemann–Liouville fractional integro-differentiation was used [18], [19, Ch. IX] for construction of the factorization theory of the wider than (5) classes $N_\alpha$ $(-1<\alpha<+\infty)$ of functions meromorphic in the disc, which we call *functions of $\alpha$-bounded type in the disc* along with Nevanlinna's class (5). Namely, the operator

$$D^{-\alpha}u(re^{i\vartheta}) \equiv \frac{r^\alpha}{\Gamma(\alpha)} \int_0^1 (1-t)^{\alpha-1} u(tre^{i\vartheta})dt,$$

$$D^o u(re^{i\vartheta}) \equiv u(re^{i\vartheta}),$$

$$D^\alpha u(re^{i\vartheta}) \equiv \frac{d^p}{dr^p}\left\{D^{-(p-\alpha)}u(re^{i\vartheta})\right\},$$

(where $0<\alpha<+\infty$ and $p$ is the natural number deduced from $p-1<\alpha\leq p$) was applied in $|z|<1$ directly to $u(te^{i\vartheta}) \equiv \log|f(te^{i\vartheta})|$.

It shall be mentioned that later consideration of a more general operator [20], permitted to investigate [21, 22, 23] some classes $N\{\omega\}$ (of functions of $\omega$-bounded type in the disc) which depend on a function-parameter $\omega(x)$ replacing $(1-x)^\alpha$ and exhaust the whole set of functions meromorphic in the disc. Thus, a general theory of factorization and boundary properties of functions meromorphic in the disc was constructed [21, 22]. This theory, together with some of the series of investigations of V.S.Zakarian [110 – 121], was summarized by M.M.Djrbashian and V.S.Zakarian in the monograph [30]. Also, we mention the extension of their theory to functions $\delta$-subharmonic in $|z|<1$ [65, 66]. This extension led to a very explicit understanding of the theory. For $\omega$-generalizations and extension of the results of [16, 17] to $\delta$-subharmonic functions see [68].

Below we restrict ourselves by giving only those results of the above mentioned theory, which relate to functions of $\alpha$-bounded type in the disc, since they, along with the factorizations of functions of $\alpha$-bounded type in the half-plane, will be applied in the spectral theory of linear operators in the last chapter of this book.

In assumption that $f(z)$ is meromorphic and $n(r,f)$ is the quantity of its poles in $|z| \leq r$, consider the following $\alpha$-characteristic functions and $\alpha$-kernels:

$$N_\alpha(r,f) = \frac{r^{-\alpha}}{\Gamma(1+\alpha)} \int_0^r (r-t)^\alpha [n(t,f) - n(0,f)] \frac{dt}{t} + \frac{n(0,f)}{\Gamma(1+\alpha)} [\log r - k_\alpha],$$

$$m_\alpha(r,f) = \frac{1}{2\pi} \int_0^{2\pi} \left[D^{-\alpha} \log |f(re^{i\vartheta})|\right]^+ d\vartheta, \ T_\alpha(r,f) = m_\alpha(r,f) + N_\alpha(r,f),$$

$$S_\alpha(z) = \Gamma(1+\alpha) \left\{ \frac{2}{(1-z)^{1+\alpha}} - 1 \right\}, \quad C_\alpha(z) = \frac{\Gamma(1+\alpha)}{(1-z)^{1+\alpha}},$$

where $k_\alpha = \alpha \sum_{n=1}^\infty [n(n+\alpha)]^{-1}$. The class $N_\alpha$ $(-1 < \alpha < +\infty)$ of functions of $\alpha$-bounded type in $|z| < 1$ is defined by the condition

$$\sup_{0<r<1} T_\alpha(r,f) < +\infty,$$

and the following main theorem is true.

**Theorem 1.** *The class $N_\alpha$ $(-1 < \alpha < +\infty)$ coincides with the set of functions representable in the form*

$$f(z) = e^{i\gamma + \lambda k_\alpha} z^\lambda \frac{B_\alpha(z;a_\mu)}{B_\alpha(z;b_\nu)} \exp\left\{ \frac{1}{2\pi} \int_0^{2\pi} S_\alpha(e^{-i\vartheta} z) d\psi(\vartheta) \right\}, \ |z| < 1, \quad (9)$$

*where $\gamma$ is a real number, $\lambda$ is an integer, $\psi(\vartheta)$ is a function of bounded variation in $[0, 2\pi]$, and*

$$B_\alpha(z;a_\mu) = \prod_\mu A_\alpha(z;a_\mu), \quad B_\alpha(z;b_\nu) = \prod_\nu A_\alpha(z;a_\nu), \quad (10)$$

$$A_\alpha(z,\zeta) = \exp\left\{ -\int_{|\zeta|}^1 \left[ C_\alpha\left(\frac{z}{\zeta}x\right) + C_\alpha\left(\frac{z\bar{\zeta}}{x}\right) - 1 \right] (1-x)^\alpha \frac{dx}{x} \right\}, \ |z|, |\zeta| < 1,$$

*are convergent Blaschke type products vanishing in zeros $\{a_\mu\} \subset \{|z| < 1\}$ and poles $\{b_\nu\} \subset \{|z| < 1\}$ of $f(z)$, which satisfy the density conditions*

$$\sum_\mu (1 - |a_\mu|)^{1+\alpha} < +\infty, \quad \sum_\nu (1 - |b_\nu|)^{1+\alpha} < +\infty.$$

Becoming the classical Nevanlinna factorization for $\alpha = 0$, the above representation describes a wide scale of classes of meromorphic functions, such that $N_\alpha \subset N$ for $-1 < \alpha < 0$, and $N \subset N_\alpha$ for $0 < \alpha < +\infty$.

The classes $N_\alpha$ $(-1 < \alpha < 0)$ which lie in $N$ have good boundary properties. This was shown in [25 – 30], where particularly was established that the exceptional set of functions from $N_\alpha \subset N$, on which no nontangential boundary values $F(e^{i\vartheta}) = \lim_{z\to e^{i\vartheta}} F(z)$ may exist, is *more delicate* than the exceptional set of the class $N$ (i.e. sets of zero Lebesgue measure). Namely, this set is of zero $(1+\alpha)$-capacity in Frostman's [33] sense.

**Theorem 2.** 1°. *If $F(z) \in N_\alpha$ $(-1 < \alpha < 0)$, then the limit $F(e^{i\vartheta}) = \lim\limits_{z\to e^{i\vartheta}} F(z)$ exists and is finite for all $\vartheta \in [0, 2\pi]$ with possible exception of a set $E \subset [0, 2\pi]$ of zero $(1+\alpha)$-capacity.*

2°. *Let $F(z) \in N_\alpha$ $(-1 < \alpha < 0)$ is holomorphic and $F(e^{i\vartheta}) = 0$ on a set $E \subset [0, 2\pi]$ of zero Lebesgue measure. If $E$ is of positive $(1+\alpha)$-capacity, then $F(z) \equiv 0$ in $|z| < 1$.*

*4.* The above results in the disc relate to solution of Problem I for several classes of functions. Whereas, the construction of a factorization theory in the half-plane may relate to solution of Problems II and III for weighted classes of functions. This was done in author's works [49 – 56, 58 – 60, 69, 70]. The papers [65, 66, 67, 68] relate to Problem I while [61] and [71] deal with Problems I and III. This book presents several chapters of the author's Dr. of Sci., Math. Thesis to which only the Sections 1, 2 and 4 of Chapter 9 from author's joint work with G.M.Gubreev [40, 41] are added. The author is thankful to Prof. S.G.Samko for his remarks to Thesis, which are taken into account in this book. The book consists of 9 chapters.

**Chapter 1** ([55, 58]) starts by some main properties of the Liouville fractional integro-differentiation, which are used all over the book. These properties are used for establishing some similarities of the Herglotz [46] – Riesz [88] theorem on the descriptive representation of the class **R** of holomorphic functions having nonnegative real parts. The new theorems describe several classes of functions holomorphic in a half-plane, the real parts of which keep their sign after application of Liouville's integro-differential operator in the imaginary variable. For $\alpha = 0$ the representations of these classes coincide with some well-known results of R.Nevanlinna [81] (see also [102, Ch. II], [2, Ch. III]) and I.S.Kats [72] (see also Addendum I in [4]) on subclasses of **R**, of functions representable by Cauchy integral.

**Chapter 2** ([49, 54, 59]) contains the construction and investigation of a family of Blaschke type products in the lower half-plane $G^- = \{w : \operatorname{Im} w < 0\}$. These products are used as the main tool in the further factorizations.

**Chapter 3** ([55]) is devoted to solutions of II-nd and III-rd Problems for some sets of functions holomorphic in a half-plane, which are larger than the classical ones. Using Liouville's integro-differentiation, a factorization theory of weighted classes of functions holomorphic in a half-plane is constructed. Depending on continuous $\alpha \in (-1, +\infty)$, for $\alpha = 0$ the obtained factorizations in essence coincide with the well known factorizations of R.Nevanlinna [83] and V.I.Krilov [76].

**Chapter 4** ([56]) establishes canonical factorizations of some weighted classes of functions meromorphic in the half-plane, the Tsuji characteristics of which are weighted-summable, similar to Nevanlinna's condition (1). Like [94], descriptive representations of the considered classes are established and the density of zeros and poles of functions of these classes is completely described.

**Chapter 5** ([61]) is devoted to investigation of boundary properties of meromorphic functions which are of $\alpha$-bounded type in the disc or in the half-plane. The main results in the half-plane in a sense are similar to Frostman's [33] theorem on the boundary properties of Blaschke products with "rare" zeros. Besides, a theorem on the boundary properties of Blaschke–M.M.Djrbashian product (8) in the disc improves the results of M.M.Djrbashian–V.S.Zakarian [28, 29].

**Chapter 6** ([60]) mainly is devoted to uniform approximation with arbitrary sharpness in the classes of meromorphic functions of $\alpha$-bounded type in the half-plane. These approximations are done by finite Blaschke type products of Ch. 2 and, in a sense, are similar to the well-known theorem of S.Schur [93], stating that any function bounded and holomorphic in the disc (or in the half-plane) is uniformly approximated by finite Blaschke products. Besides, a necessary and sufficient growth condition is found, under which a function of $\alpha$-bounded type in the half-plane is a Blaschke type product.

**Chapter 7.** It is well known that the classical conditions (2)–(3) and (4) of Problems II and III *describe only the growth of some subclasses of subharmonic functions possessing nonnegative harmonic majorants in* $\operatorname{Im} z > 0$. This was the reason for finding some general conditions [57, 62, 63] improving Nevanlinna factorization in the half-plane by *description of all functions of bounded type in the half-plane.* This led to solution of a new, IV-th problem, which is in essence a "symbiosis" of Problems II and III.

**Problem IV.** Describe some classes of functions regular in $\operatorname{Im} z > 0$, the growth of which is assumed to be restricted by a "global" (shift-invariant) and a "local" conditions simultaneously. It appears that a function $u(z)$ subharmonic in the upper half-plane $\operatorname{Im} z > 0$ has a nonnegative harmonic majorant in $\operatorname{Im} z > 0$ if and only if simultaneously

*The upper half-plane on the Riemann sphere*

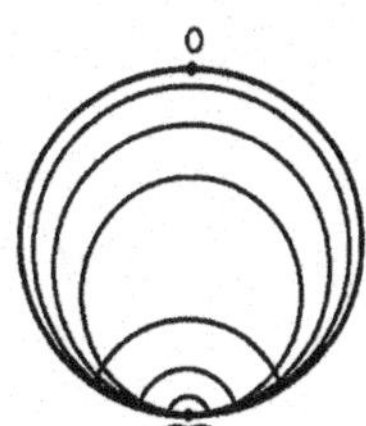


GLOBAL-LOCAL PROBLEM

$$\liminf_{y\to+0}\int_{-\infty}^{+\infty} u^+(x+iy)\frac{dx}{1+x^2} = \lim_{R\to+\infty}\liminf_{y\to+0}\int_{-R}^{R} u^+(x+iy)\frac{dx}{1+x^2} < +\infty$$

and

$$\liminf_{R\to+\infty}\frac{1}{R}\int_0^{\pi} u^+(Re^{i\vartheta})\sin\vartheta d\vartheta < +\infty.$$

In particular, this result contains the classical Nevanlinna factorization [83] in the half-plane (see also [10]). Besides, Nevanlinna's uniqueness theorem and the Phragmén–Lindelöf type theorem holding from a result due to L.Ahlfors – M.Heins [1] are improved by means of replacement of the traditional condition

$$\limsup_{z \to t,\ \mathrm{Im}\ z>0} u(z) \leq 0, \quad -\infty < t < +\infty,$$

by a less restrictive integral condition. In addition a new, equivalent definition of the classical Hardy spaces over the half-plane is found and a theorem on some weighted $H^p$ classes is proved.

**Chapter 8** ([64]): the methods of the previous chapter are used to investigate some weighted classes of subharmonic functions possessing nonnegative harmonic majorants in a half–plane after application of Liouville's fractional integration. Riesz type representations and complete growth characterization are established. In some natural sense, these growth characterization and representations completely describe *all* subharmonic functions, the Liouville primitives of which have nonnegative harmonic majorants in the half-plane. The considered classes of subharmonic functions are generalizations of the classes of functions of $\alpha$-bounded type.

**Chapter 9** gives the author's joint work with G.M.Gubreev [40, 41], which reveal a new way of application of the factorization theory in the disc and some of the results of Chapters 7 and 8 in the spectral theory of linear operators. The application of such results is of long standing interest for operator theory specialists. For example, soon after the monograph [19] was published, M.G.Krein mentioned at Moscow Mathematical Congress the necessity to find spectral interpretations of its results. Later, M.S.Livšic [79] and his pupils (L.Kh.Mehrabyan, Do Kong Khan and others) realized meromorphic functions of classes $N_\alpha$ as transfer functions of some special linear systems in a series of works. However, the applications found in [40, 41] may turn to be more reasonable.

# CHAPTER 1

# THE LIOUVILLE OPERATOR AND HERGLOTZ–RIESZ TYPE THEOREMS

## 1. THE LIOUVILLE INTEGRO–DIFFERENTIATION

*1.1.* Let a generally complex-valued function $U(v)$ be measurable on the ray $-\infty < v < d$ $(d < +\infty)$. Assuming that the below integral is absolutely convergent for almost all $v \in (-\infty, d)$, for $\alpha > 0$ we set

$$W^{-\alpha}U(v) \equiv \frac{1}{\Gamma(\alpha)} \int_{-\infty}^{v} (v-t)^{\alpha-1} U(t)dt = \frac{1}{\Gamma(\alpha)} \int_{0}^{+\infty} \sigma^{\alpha-1} U(v - i\sigma)d\sigma. \tag{1.1}$$

*$U(v)$ is said to be of the class $\mathbf{L}_\alpha(-\infty, d)$ $(0 < \alpha < +\infty)$, if for any $\varepsilon > 0$*

$$(1 + |t|)^{\alpha-1} U(t) \in L_1(-\infty, d - \varepsilon). \tag{1.2}$$

Note that for any $\alpha_1$ and $\alpha_2$ $(0 < \alpha_1 < \alpha_2 < +\infty)$

$$\mathbf{L}_{\alpha_2}(-\infty, d) \subset \mathbf{L}_{\alpha_1}(-\infty, d). \tag{1.3}$$

**Lemma 1.1.** 1°. *If $U(v) \in \mathbf{L}_\alpha(-\infty, d)$ for some $\alpha \in (0, +\infty)$, then the integral $W^{-\alpha}U(v)$ is absolutely convergent for almost all $v \in (-\infty, d)$ and it represents a function summable on any finite interval $(a, b) \subset (-\infty, d)$.*

2°. *If $\alpha_1$, $\alpha_2 \in (0, +\infty)$ and $U(v) \in \mathbf{L}_{\alpha_1+\alpha_2}(-\infty, d)$, then $W^{-\alpha_1}|U(v)| \in \mathbf{L}_{\alpha_2}(-\infty, d)$.*

**Proof.** 1°. Let $-\infty < a < b < d$. Then

$$\int_a^b \left|W^{-\alpha}U(v)\right| dv \le \int_a^b W^{-\alpha}|U(v)| dv$$
$$= \frac{1}{\Gamma(\alpha)} \left( \int_a^b dv \int_{-\infty}^{a} (v-t)^{\alpha-1}|U(v)|dt + \int_a^b dv \int_a^v (v-t)^{\alpha-1}|U(v)|dt \right)$$
$$= \frac{1}{\Gamma(1+\alpha)} \left( \int_{-\infty}^{a} [(b-t)^\alpha - (a-t)^\alpha]\, |U(v)|dt + \int_a^b (b-t)^\alpha |U(v)|dt \right).$$

On the other hand, for enough great $C \equiv C(a,b,\alpha) > 0$

$$0 < (b-t)^\alpha - (a-t)^\alpha \le C(1+|t|)^{\alpha-1} \quad \text{if} \quad -\infty < t < a,$$
$$\text{and} \quad (b-t)^\alpha \le C(1+|t|)^{\alpha-1} \quad \text{if} \quad a < t < b.$$

Therefore

$$\int_a^b \left|W^{-\alpha}U(v)\right| dv \le \frac{C}{\Gamma(1+\alpha)} \int_{-\infty}^b (1+|t|)^{\alpha-1}|U(v)|dt < +\infty.$$

2°. We shall use the well-known formula

$$\int_{t_2}^v (v-t_1)^{\alpha_1-1}(t_1-t_2)^{\alpha_2-1}dt_1 = \frac{\Gamma(\alpha_1)\Gamma(\alpha_2)}{\Gamma(\alpha_1+\alpha_2)}(v-t_2)^{\alpha_1+\alpha_2-1}. \tag{1.4}$$

It is obvious that for any $\varepsilon > 0$

$$\int_{-\infty}^{d-\varepsilon} (1+|t|)^{\alpha_2-1}W^{-\alpha_1}|U(t)|dt$$
$$= \frac{1}{\Gamma(\alpha_1)} \int_{-\infty}^{d-\varepsilon} |U(x)|dx \int_{-d+\varepsilon}^{-x} (-x-t)^{\alpha_1-1}(1+|t|)^{\alpha_2-1}dt.$$

But $(1+|t|)^{\alpha_2-1} \le C_1(t+d)^{\alpha_2-1}$ for enough great $C_1 \equiv C_1(a,d,\varepsilon) > 0$ and any $t \in (-d+\varepsilon, +\infty)$. Therefore, by (1.4)

$$\int_{-d+\varepsilon}^{-x} (-x-t)^{\alpha_1-1}(1+|t|)^{\alpha_2-1}dt$$
$$\le C_1 \int_{-d}^{-x} (-x-t)^{\alpha_1-1}(t+d)^{\alpha_2-1}dt = C_1\frac{\Gamma(\alpha_1)\Gamma(\alpha_2)}{\Gamma(\alpha_1+\alpha_2)}(d-x)^{\alpha_1+\alpha_2-1}.$$

Besides, $(d-x)^{\alpha_1+\alpha_2-1} \le C_2(1+|x|)^{\alpha_1+\alpha_2-1}$ $(-\infty < x < d-\varepsilon)$ for some $C_2 \equiv C_2(\alpha_1+\alpha_2, d, \varepsilon) > 0$. Hence

$$\int_{-\infty}^{d-\varepsilon} (1+|t|)^{\alpha_2-1}W^{-\alpha_1}|U(t)|dt$$
$$= C_1C_2\frac{\Gamma(\alpha_2)}{\Gamma(\alpha_1+\alpha_2)} \int_{-\infty}^{d-\varepsilon} (1+|x|)^{\alpha_1+\alpha_2-1}|U(x)|dx < +\infty.$$

**Lemma 1.2.** *Let $U(v) \in \mathbf{L}_{\alpha_1+\alpha_2+\cdots+\alpha_n}(-\infty,d)$ for some positive numbers $\alpha_1, \alpha_2, \cdots \alpha_n$. Then $W^{-(\alpha_1+\alpha_2+\cdots+\alpha_n)}U(v)$ and $W^{-\alpha_1}W^{-\alpha_2}\ldots W^{-\alpha_n}U(v)$ are absolutely convergent for almost all $v \in (-\infty,d)$, and a.e. in $(-\infty,d)$*

$$W^{-(\alpha_1+\alpha_2+\cdots+\alpha_n)}U(v) = W^{-\alpha_1}W^{-\alpha_2}\ldots W^{-\alpha_n}U(v) \tag{1.5}$$

**Proof.** For $n = 1$ our assertion is trivial. Using (1.4), we come to the following equalities, which prove our lemma for $n = 2$:

$$\begin{aligned} &W^{-\alpha_1}W^{-\alpha_2}|U(v)| \\ &= \frac{1}{\Gamma(\alpha_1)\Gamma(\alpha_1)}\int_{-\infty}^{v}|U(t_2)|dt_2\int_{t_2}^{v}(v-t_1)^{\alpha_1-1}(t_1-t_2)^{\alpha_2-1}dt_1 \\ &= W^{-(\alpha_1+\alpha_2)}|U(v)|. \end{aligned}$$

A similar argument with an induction leads to the proof for any $n \geq 1$.

**Remark 1.1.** If $\alpha \in [1, +\infty)$ and $U(v) \in \mathbf{L}_\alpha(-\infty, d)$, then by Lemmas 1.1 and 1.2 the integral $W^{-\alpha}U(v)$ is absolutely convergent *for all* $v \in (-\infty, d)$ and it represents a function continuous on $(-\infty, d)$. By the same lemmas, if $U(v) \in \mathbf{L}_p(-\infty, d)$ for any natural number $p \geq 1$, then

$$W^{-p}U(v) = \int_{-\infty}^{v} dt_1 \int_{-\infty}^{t_1} dt_2 \cdots \int_{-\infty}^{t_{p-1}} U(t_p)dt_p$$

everywhere in $(-\infty, d)$. *Thus, $W^{-p}U(v)$ is the p-th primitive of $U(v)$.*

*1.2.* By (1.3) and Lemma 1.1, for any $\alpha \in (0, \alpha_0]$ the integral $W^{-\alpha}U(v)$ is absolutely convergent for almost all $v \in (-\infty, d)$, provided $U(v) \in \mathbf{L}_{\alpha_0}(-\infty, d)$ for an $\alpha_0 \in (0, +\infty)$. Besides, the following statement is true.

**Lemma 1.3.** *Let $U(v) \in \mathbf{L}_{\alpha_0}(-\infty, d)$ for some $\alpha_0 \in (0, +\infty)$. Then for any $\alpha \in (0, \alpha_0]$ the integral $W^{-\alpha}U(v)$ is absolutely convergent in any Lebesgue point $v \in (-\infty, d)$ of the function $|U(v)|$.*

**Proof.** For $\alpha \geq 1$ the assertion is obvious by the above Remark 1.1. Let $0 < \alpha < 1$ ($\alpha \in (0, \alpha_0]$). Assuming $v \in (-\infty, d)$ a Lebesgue point of $|U(v)|$, we write

$$W^{-\alpha}|U(v)| = \frac{1}{\Gamma(\alpha)}\left(\int_0^1 + \int_1^{+\infty}\right)\sigma^{\alpha-1}|U(v-\sigma)|d\sigma \equiv I_\alpha^{(1)} + I_\alpha^{(2)}.$$

For evaluation of $I_\alpha^{(2)}$, choose a constant $C \equiv C(v, \alpha_0) \in (0, +\infty)$ such that

$$(v-t)^{\alpha_0-1} \leq C(1+|t|)^{\alpha_0-1} \quad \text{for} \quad t < v-1. \tag{1.6}$$

Then

$$I_\alpha^{(2)} \leq \frac{1}{\Gamma(\alpha)}\int_1^{+\infty}\sigma^{\alpha_0-1}|U(v-\sigma)|d\sigma \leq \frac{C}{\Gamma(\alpha)}\int_{-\infty}^{v-1}(1+|t|)^{\alpha_0-1}|U(t)|dt < +\infty$$

since $U(v) \in \mathbf{L}_{\alpha_0}(-\infty, d)$. Thus, it remains to prove that also $I_\alpha^{(1)}$ is convergent. To this end, it suffices to show that $\lim_{\delta\to+0}\int_\delta^1 \sigma^{\alpha-1}|U(v-\sigma)|d\sigma$ is a finite number. For this, introduce the function

$$\Phi_v(\sigma) \equiv \int_0^\sigma |U(v-t)|dt = \int_{v-\sigma}^{v} |U(\tau)|d\tau,$$

which is continuous in $0 < \sigma < +\infty$, and observe that for any $\sigma > 0$

$$\left| \frac{\Phi_v(\sigma)}{\sigma} - |U(v)| \right| \le \frac{1}{\sigma} \int_0^\sigma ||U(v-t)| - |U(v)|| \, dt.$$

Besides,

$$\lim_{\sigma \to +0} \sigma^{\alpha-1} \Phi_v(\sigma) = |U(v)|$$

since $v \in (-\infty, d)$ is a Lebesgue point of $|U(v)|$. On the other hand, by the definition of $\Phi_v(\sigma)$

$$\int_\delta^1 \sigma^{\alpha-1} |U(v-\sigma)| d\sigma = \int_\delta^1 \sigma^{\alpha-1} d\Phi_v(\sigma)$$
$$= \Phi_v(1) - \delta^{\alpha-1}\Phi_v(\delta) + (1-\alpha) \int_\delta^1 \sigma^{\alpha-2} \Phi_v(\sigma) d\sigma.$$

Therefore, the mentioned limit is finite and the proof is complete.

**Lemma 1.4.** *If $U(v) \in \mathbf{L}_{\alpha_0}(-\infty, d)$ for some $\alpha_0 \in (0, +\infty)$, then*

$$\lim_{\alpha \to +0} W^{-\alpha} U(v) = U(v) \tag{1.7}$$

*in every Lebesgue point $v \in (-\infty, d)$ of $U(v)$.*

**Proof.** Let $\alpha \in (0, \alpha_0]$, $0 < \alpha < 1$, and let $v \in (-\infty, d)$ be a Lebesgue point of $U(v)$. Similar to the proof of the previous lemma,

$$W^{-\alpha} U(v) = \frac{1}{\Gamma(\alpha)} \left( \int_0^1 + \int_1^{+\infty} \right) \sigma^{\alpha-1} U(v-\sigma) d\sigma \equiv J_\alpha^{(1)} + J_\alpha^{(2)}. \tag{1.8}$$

In view of (1.6),

$$|J_\alpha^{(2)}| \le \frac{C}{\Gamma(\alpha)} \int_{-\infty}^{v-1} (1+|t|)^{\alpha_0 - 1} |U(t)| dt \to 0 \quad \text{as} \quad \alpha \to +0. \tag{1.9}$$

Now introduce the function

$$F_v(\sigma) \equiv \int_0^\sigma U(v-\sigma) d\sigma = \int_{v-\sigma}^v U(\tau) d\tau$$

which is continuous for $\sigma > 0$. Since $v \in (-\infty, d)$ is a Lebesgue point of $U$,

$$\left| \frac{F_v(\sigma)}{\sigma} - U(v) \right| \le \frac{1}{\sigma} \int_0^\sigma |U(v-t) - U(v)| \, dt = o(1) \quad \text{as} \quad \sigma \to +0.$$

Hence $\lim_{\sigma\to+0}\sigma^{-1}F_v(\sigma)=U(v)$. Therefore

$$F_v(\sigma)=\sigma\left[U(v)+\omega_v(\sigma)\right],$$

where the function $\omega_v(\sigma)\equiv\sigma^{-1}F_v(\sigma)-U(v)$ is continuous on $0<\sigma<+\infty$ and such that $\omega_v(+0)=0$. Further, for an $\varepsilon>0$ choose $\delta\equiv\delta(\varepsilon)\in(0,1)$ enough small to provide $|\omega_v(\sigma)|<\varepsilon/5$ for $0<\sigma<\delta$. Then

$$\begin{aligned}J_\alpha^{(1)}=&\frac{1}{\Gamma(\alpha)}\int_0^1\sigma^{\alpha-1}dF_v(\sigma)=\frac{F_v(1)}{\Gamma(\alpha)}+\frac{1-\alpha}{\Gamma(1+\alpha)}U(v)\\&+\frac{1-\alpha}{\Gamma(\alpha)}\int_0^\delta\omega_v(\sigma)\sigma^{\alpha-1}d\sigma+\frac{1-\alpha}{\Gamma(\alpha)}\int_\delta^1\omega_v(\sigma)\sigma^{\alpha-1}d\sigma.\end{aligned}$$

But one can verify that

$$\left|\frac{1-\alpha}{\Gamma(\alpha)}\int_0^\delta\omega_v(\sigma)\sigma^{\alpha-1}d\sigma\right|\le\frac{1-\alpha}{\Gamma(\alpha)}\frac{\varepsilon}{5}\int_0^\delta\sigma^{\alpha-1}d\sigma<\frac{\varepsilon}{5}\frac{1-\alpha}{\Gamma(1+\alpha)},$$

$$\begin{aligned}\left|\frac{1-\alpha}{\Gamma(\alpha)}\int_\delta^1\omega_v(\sigma)\sigma^{\alpha-1}d\sigma\right|&\le\delta^{\alpha-1}\frac{1-\alpha}{\Gamma(\alpha)}\int_0^1|\omega_v(\sigma)|d\sigma\\&<\frac{1-\alpha}{\delta\Gamma(\alpha)}\max_{0<\sigma\le1}|\omega_v(\sigma)|.\end{aligned}$$

Hence

$$\begin{aligned}\left|J_\alpha^{(1)}-U(v)\right|\le&\frac{|F_v(1)|}{\Gamma(\alpha)}+\left|\frac{1-\alpha}{\Gamma(1+\alpha)}-1\right||U(v)|\\&+\frac{\varepsilon}{5}\frac{1-\alpha}{\Gamma(1+\alpha)}+\delta^{-1}\frac{1-\alpha}{\Gamma(\alpha)}\max_{0<\sigma\le1}|\omega_v(\sigma)|.\end{aligned}$$

Now choose $\delta'\in(0,\min\{\alpha_0,1\})$ enough small to provide that all terms of the right-hand side of the last inequality be less than $\varepsilon/4$ for $0<\alpha<\delta'$. Then $|J_\alpha^{(1)}-U(v)|<\varepsilon$ for $0<\alpha<\delta'$, and $\lim_{\alpha\to+0}J_\alpha^{(1)}=U(v)$ since $\varepsilon>0$ was arbitrary. Hence our assertion follows by (1.8) and (1.9).

**Remark 1.2.** By Lemmas 1.1 and 1.2, in the requirements of Lemma 1.4

$$\lim_{\alpha\to\alpha_1+0}W^{-\alpha}U(v)=W^{-\alpha_1}U(v)\tag{1.7$'$}$$

for any $\alpha_1\in(0,\alpha_0)$ and almost all $v\in(-\infty,d)$. This relation shows that the operator $W^{-\alpha}$ $(0<\alpha<+\infty)$ is "continuous from the right" by $\alpha$.

In view of (1.7)–(1.7′), the natural extension of the definition (1.1) of $W^{-\alpha}$ $(\alpha>0)$ to $\alpha=0$ is given by the identity

$$W^0U(v)\equiv U(v).\tag{1.10}$$

For defining fractional derivatives, we assume that $\alpha > 0$ and $p \geq 1$ is the natural number deduced from $p-1 < \alpha \leq p$. Then we set

$$W^{\alpha}U(v) \equiv W^{-(p-\alpha)}\left\{\frac{d^p}{dv^p}U(v)\right\}, \tag{1.11}$$

preassuming the existence of the latter function for almost all $v \in (-\infty, d)$. This we call *the $\alpha$-th derivative of $U(v)$*. Observe that by (1.10)

$$W^{p}U(v) = \frac{d^p}{dv^p}U(v), \tag{1.11'}$$

if $p$ is a natural number, i.e. $W^pU(v)$ coincides with the ordinary $p$-th derivative of $U(v)$.

*1.3.* Henceforth we apply Liouville's operator by the variable $v$ to some functions $U(w) \equiv U(u+iv)$ given in several domains which along with inner points $w = u + iv$ contain the rays

$$\Gamma(\infty, w) = \{s = w - i\sigma\ :\ 0 < \sigma < +\infty\} \tag{1.12}$$

directed from $\infty$ to $w$. We call such domains starshaped in respect to $\infty$, or briefly *$\infty$-starshaped.*

Unless otherwise stipulated, we shall assume the function $U(w)$ harmonic in a $\infty$-starshaped domain $D$, with a possible exception of a countable set of points $\{w_m\}$. Besides, we shall assume that any $\Gamma(\infty, w) \subset D$ contains not more than a finite set of points from $\{w_m\}$, and also that on $\Gamma(\infty, w)$ the function $U(w)(w - w_m)$ satisfies Hölder's condition in neighborhoods of the points $\{w_m\} \subset \Gamma(\infty, w)$.

Preassuming absolute convergence in any interval disjoint from the points $\{w_m\} \subset \Gamma(\infty, w)$ and existence of Cauchy's principal value in the rest of the integration contour, we call

$$\begin{aligned} W^{-\alpha}U(w) \equiv & \frac{1}{\Gamma(\alpha)}\int_{-\infty}^{v}(v-t)^{\alpha-1}U(u+it)dt \\ = & \frac{1}{\Gamma(\alpha)}\int_0^{+\infty}\sigma^{\alpha-1}U(w-i\sigma)d\sigma, \quad w = u+iv \in D, \end{aligned} \tag{1.13}$$

*the $\alpha$-th* $(0 < \alpha < +\infty)$ *primitive of $U(w)$*. For $\alpha = 0$ we set

$$W^{0}U(w) \equiv U(w), \quad w \in D. \tag{1.14}$$

For definition of $\alpha$-th $(\alpha > 0)$ derivative of $U(w)$, we assume that the derivative

$$\frac{\partial^p}{\partial v^p}U(w), \quad w = u + iv \in D \quad (p - 1 < \alpha \leq p)$$

satisfies the above requirements. Then we set

$$W^\alpha U(w) \equiv W^{-(p-\alpha)}\left\{\frac{\partial^p}{\partial v^p}U(w)\right\}, \quad w = u + iv \in D. \tag{1.15}$$

**Definition 1.1.** *A function $U(w)$ given in a $\infty$–starshaped domain $D$ is said to be of the class $M_\beta(D)$ $(0 \le \beta < +\infty)$, if there exists an angular domain $\Lambda(\delta_0, R_0) = \{w : |\pi/2 + \arg w| \le \delta_0, |w| \ge R_0\}$ $(0 < \delta_0 \le \pi/2, 0 < R_0 < +\infty)$ such that*

$$\sup_{w\in\mathbf{K}} \int_1^{+\infty} \sigma^{\beta-1}|U(w - i\sigma)|d\sigma < +\infty \tag{1.16}$$

*for any compact $\mathbf{K} \subset D \cap \Lambda(\delta_0, R_0)$. The function $U(w)$ is said to be of the class $K_\gamma(\delta_0, D)$ $(0 < \gamma < +\infty)$, if*

$$|U(w)| \le C|w|^{-\gamma}, \quad w \in D \cap \Lambda(\delta_0, R_0), \tag{1.17}$$

*for some $\delta_0$ $(0 < \delta_0 \le \pi/2)$ and $R_0$ $(0 < R_0 < +\infty)$.*

It is evident that for any $\beta_1$ and $\beta_2$ $(0 \le \beta_1 < \beta_2 < +\infty)$

$$M_{\beta_2}(D) \subset M_{\beta_1}(D).$$

Besides, it is obvious that for any $\gamma_1$, $\gamma_2$ $(0 < \gamma_1 < \gamma_2 < +\infty)$ and $\delta_0$ $(0 < \delta_0 \le \pi/2)$

$$K_{\gamma_2}(\delta_0, D) \subset K_{\gamma_1}(\delta_0, D).$$

Thus, the classes $M_\beta$ and $K_\gamma$ narrow down as $\beta$ and $\gamma$ increase. At the same time, the below lemma is true.

**Lemma 1.5.** *Let $\beta \in [0, +\infty)$ and $\gamma \in (\beta, +\infty)$ be any numbers. Then for any $\delta_0$ $(0 < \delta_0 \le \pi/2)$*

$$K_\gamma(\delta_0, D) \subset M_\beta(D). \tag{1.18}$$

**Proof.** One can verify that $(a + b)^{-\kappa} \le \max\{1, 2^{1-\kappa}\}(a^\kappa + b^\kappa)^{-1}$ for any positive numbers $a$, $b$ and $\kappa$. Consequently,

$$|w - i\sigma|^{-\gamma} \le (|w - i\sigma|^2)^{-\frac{\gamma}{2}} = \left[|w|^2 + \sigma^2(1 + 2\sigma^{-1}|\mathrm{Im}\ w|)\right]^{-\frac{\gamma}{2}} <$$
$$< (|w|^2 + \sigma^2)^{-\frac{\gamma}{2}} \le \max\{1, 2^{1-\gamma/2}\}(|w|^\gamma + \sigma^\gamma)^{-1}$$

if $\mathrm{Im}\ w < 0$ and $\sigma, \gamma > 0$. Thus, for any $w$ $(\mathrm{Im}\ w < 0)$ and $\sigma$, $\gamma > 0$

$$|w - i\sigma|^{-\gamma} < C_\gamma(|w|^\gamma + \sigma^\gamma), \quad \text{where} \quad C_\gamma = \max\{1, 2^{1-\gamma/2}\}. \tag{1.19}$$

Now assume $U(w) \in K_\gamma(\delta_0, D)$ $(0 < \delta_0 \le \pi/2)$, where $\gamma \in (\beta, +\infty)$ $(\beta \ge 0)$ and $\mathbf{K}$ is any compact in $D \cap \Lambda(\delta_0, R_0)$. Then by (1.17) and (1.19)

$$\int_1^{+\infty} \sigma^{\beta-1}|U(w - i\sigma)|d\sigma \le C\int_1^{+\infty} \frac{\sigma^{\beta-1}d\sigma}{|w - i\sigma|^\gamma} \le CC_\gamma \int_1^{+\infty} \frac{\sigma^{\beta-1}d\sigma}{|w|^\gamma + \sigma^\gamma}$$
$$\le CC_\gamma|w|^{\beta-\gamma}\int_{R_0^{-1}}^{+\infty} \frac{x^{\beta-1}dx}{1 + x^\gamma}$$

for any $w \in \mathbf{K}$ (if $w \in \mathbf{K}$, then $\operatorname{Im} w < 0$). Hence (1.16) follows since $|w|^{\beta-\gamma}$ is uniformly bounded for $w \in \mathbf{K}$. Thus, (1.18) is valid.

*1.4.* **Lemma 1.6.** *Let $U(w)$ be a continuous function in a $\infty$-starshaped domain $D$, and let $U(w) \in M_\beta(D)$ for some $\beta > 0$. Then*

$$\sup_{w\in\mathbf{K}} \int_0^{+\infty} \sigma^{\alpha-1}|U(w-i\sigma)|d\sigma < +\infty \tag{1.20}$$

*for any $\alpha \in (0,\beta]$ and any compact $\mathbf{K} \subset D$.*

**Proof.** Choose the number $H_0 \geq 0$ enough great to provide that whole compact $\mathbf{K} - iH_0 = \{\zeta = w - iH_0 : w \in \mathbf{K}\}$ is contained in the angular domain $\Lambda(\delta_0, R_0)$ from the definition of $M_\beta(D)$. Further, assuming $w \in \mathbf{K}$ an arbitrary point, we write

$$\int_0^{+\infty} \sigma^{\alpha-1}|U(w-i\sigma)|d\sigma = \left(\int_0^{H_0+1} + \int_{H_0+1}^{+\infty}\right) \sigma^{\alpha-1}|U(w-i\sigma)|d\sigma$$
$$\equiv I_1(w) + I_2(w).$$

It suffices to show that $I_1(w)$ and $I_2(w)$ are uniformly bounded in $\mathbf{K}$. To this end, observe that

$$I_1(w) = \int_0^{H_0+1} \sigma^{\alpha-1}|U(w-i\sigma)|d\sigma \leq \frac{(H_0+1)^\alpha}{\alpha} \max_{\substack{\zeta\in\mathbf{K}-i\sigma \\ 0\leq\sigma\leq H_0+1}} |U(\zeta)| < +\infty$$

since $U(w)$ is continuous in $D$. For evaluating $I_2(w)$, we use the obvious inequality $(\sigma + H_0)^{\alpha-1} \leq C\sigma^{\beta-1}$ $(1 \leq \sigma < +\infty)$, where $C \equiv C(\alpha, \beta, H_0)$ is a constant. As $\mathbf{K} - iH_0 \subset D \cap \Lambda(\delta_0, R_0)$,

$$I_2(w) = \int_1^{+\infty} (\sigma + H_0)^{\alpha-1}|U(w - i(H_0+\sigma))|d\sigma$$
$$\leq C \int_1^{+\infty} \sigma^{\beta-1}|U(w - i(H_0+\sigma))|d\sigma$$
$$\leq C \sup_{\zeta\in\mathbf{K}-iH_0} \int_1^{+\infty} \sigma^{\beta-1}|U(\zeta - i\sigma)|d\sigma < +\infty, \quad w \in \mathbf{K}.$$

**Lemma 1.7.** *Let $U(w)$ be a continuous function in a $\infty$-starshaped domain $D$, and let $U(w) \in M_\beta(D)$ for some $\beta > 0$. Then:*

*1°. For any $u \in (\inf_{w\in D}\{\operatorname{Re} w\}, \sup_{w\in D}\{\operatorname{Re} w\})$ and $d < \sup_{w\in D}\{\operatorname{Im} w\}$ the function $U_u(v) \equiv U(u+iv)$ belongs to the class $\mathbf{L}_\beta(-\infty, d)$.*

*2°. For any $\alpha \in (0,\beta]$, the functions $W^{-\alpha}U(w)$ and $W^{-\alpha}|U(w)|$ are defined everywhere, finite-valued and measurable in $D$.*

**Proof.** 1°. Let $\varepsilon > 0$ be arbitrary. One can see that if $-\infty < t < d - \varepsilon$, then $(1 + |t|)^{\beta-1} \le C(d - t)^{\beta-1}$, where $C \equiv C(\beta, d, \varepsilon)$ is a constant. Since $U(w)$ is continuous, we conclude

$$\begin{aligned}\int_{-\infty}^{d-\varepsilon} (1+|t|)^{\beta-1}|U(u+it)|dt \le & C \int_{-\infty}^{d-\varepsilon} (d-t)^{\beta-1}|U(u+it)|dt \\ =& C \int_{\varepsilon}^{+\infty} \sigma^{\beta-1}|U(u+id-i\sigma)|d\sigma < +\infty.\end{aligned}$$

2°. It is obvious that along with $U(w)$ for any $N \ge 1$ both functions

$$f_N(w) = \frac{1}{\Gamma(\alpha)} \int_0^N \sigma^{\beta-1} U(w - i\sigma)d\sigma, \quad F_N(w) = \frac{1}{\Gamma(\alpha)} \int_0^N \sigma^{\beta-1}|U(w - i\sigma)|d\sigma$$

are continuous in $D$. Besides, in view of (1.20) the following limits exist and are finite:

$$\lim_{N\to\infty} f_N(w) = W^{-\alpha}U(w), \qquad \lim_{N\to\infty} F_N(w) = W^{-\alpha}|U(w)|.$$

**Lemma 1.8.** *Let $U(w)$ be harmonic in a $\infty$-starshaped domain $D$, and let $U(w) \in M_\beta(D)$ for some $\beta \ge 0$. Then also $W^{-\alpha}U(w)$ is harmonic in $D$ for any $\alpha \in [0, \beta]$.*

**Proof.** Our assertion is trivial for $\alpha = 0$ since $W^0$ is identical. Consider the case $\beta > 0$, $\alpha \in (0, \beta]$. By Lemma 1.7, $W^{-\alpha}U(w)$ is finite-valued and measurable in $D$. Therefore (see, for instance, [11], Addendum, §1, Subsection. 1), $W^{-\alpha}U(w)$ necessarily will be harmonic in $D$, if we prove the equality

$$W^{-\alpha}U(w_0) = \frac{1}{2\pi} \int_0^{2\pi} W^{-\alpha}U(w_0 + re^{i\vartheta})d\vartheta$$

for any $w_0 \in D$ and $r > 0$ such that $\overline{O_r(w_0)} = \{\zeta : |\zeta - w_0| \le r_0\} \subset D$. It follows from Lemma 1.7 and (1.20) that $W^{-\alpha}|U(w)|$ is finite-valued and measurable in $D$. Also, it follows that

$$\int_0^{2\pi} W^{-\alpha} \left|U(w_0 + re^{i\vartheta})\right| d\vartheta < +\infty.$$

Therefore

$$\begin{aligned}\frac{1}{2\pi} \int_0^{2\pi} & W^{-\alpha}U(w_0 + re^{i\vartheta})d\vartheta \\ &= \frac{1}{\Gamma(\alpha)} \int_0^{+\infty} \sigma^{\alpha-1} \left\{ \frac{1}{2\pi} \int_0^{2\pi} U(w_0 + re^{i\vartheta} - i\sigma)d\vartheta \right\} d\sigma \\ &= \frac{1}{\Gamma(\alpha)} \int_0^{+\infty} \sigma^{\alpha-1}U(w_0 - i\sigma)d\sigma = W^{-\alpha}U(w_0),\end{aligned}$$

where all integrals are absolutely convergent. Thus, the proof is complete.

**Remark 1.3.** The requirements of Lemma 1.8 can be weakened. Namely, its assertions remain valid if instead the uniform boundedness of (1.16) in any compact $\mathbf{K}$ lying in $D \cap \Lambda(\delta_0, R_0)$ one require *only its uniform boundedness on a family of continuous curves* $\{l_k\} \subset D$ such that for any $\mathbf{K} \subset D$ there exists $k \geq 1$ for which $\operatorname{Re} \mathbf{K} \subset \operatorname{Re} l_k$. We omit the proof of this statement.

*1.5.* The following lemmas also will be useful.

**Lemma 1.9.** *Let $U(w)$ be continuous in an angular domain $\Lambda(\delta_0, R_0)$ ($0 < \delta_0 \leq \pi/2$, $0 < R_0 < +\infty$), and let*

$$|U(w)| \leq C_1 |w|^{-(\kappa+\alpha)}, \quad w \in \Lambda(\delta_0, R_0),$$

*for some positive $\alpha$ and $\kappa$. Then for a constant $C_1 > 0$*

$$W^{-\alpha}|U(w)| \leq C_1 |w|^{-\kappa}, \quad w \in \Lambda(\delta_0, R_0).$$

**Proof.** If $w \in \Lambda(\delta_0, R_0)$, then by (1.19)

$$\begin{aligned} W^{-\alpha}|U(w)| &\leq \frac{C}{\Gamma(\alpha)} \int_0^{+\infty} \frac{\sigma^{\alpha-1} d\sigma}{|w - i\sigma|^{\alpha+\kappa}} \\ &\leq \frac{CC(\alpha+\kappa)}{\Gamma(\alpha)} |w|^{-\kappa} \int_0^{+\infty} \frac{x^{\alpha-1} dx}{1 + x^{\alpha+\kappa}} \equiv C_1 |w|^{-\kappa}. \end{aligned}$$

**Lemma 1.10.** *Let $U(w)$ be harmonic in a $\infty$–starshaped domain $D$.*

$1°$. *If $U(w) \in M_p(D)$ for some $p = 0, 1, 2, \ldots$, then for any $\alpha \in [0, p]$*

(a) *$W^{-\alpha}U(w)$ is harmonic in $D$;*

(b) $$\frac{\partial^p}{\partial v^p} W^{-(p-\alpha)} \{W^{-\alpha}U(w)\} = U(w), \quad w = u + iv \in D. \tag{1.21}$$

$2°$. *If $\partial/\partial v U(w) \in M_1(D)$, then for any $\alpha \in (-1, 0)$*

(a) *$W^{-\alpha}U(w)$ is harmonic in $D$;*

(b) $$W^{\alpha}W^{-\alpha}U(w) = U(w) - \lim_{t \to +\infty} U(u - it), \quad w = u + iv \in D, \tag{1.22}$$

*besides, for some constants $a_0$, $a_1$*

$$\lim_{t \to +\infty} U(u - it) = a_1 u + a_0, \quad -\infty < u < +\infty. \tag{1.22$'$}$$

**Proof.** $1°$. **(a)** is proved in Lemma 1.8. For proving **(b)**, observe that by Lemmas 1.7($1°$) and 1.2

$$W^{-(p-\alpha)}W^{-\alpha}U(u + it) = W^{-p}U(u + it), \quad w = u + iv \in D,$$

for any $t$ such that $u + it \in D$. Consequently,

$$\frac{\partial^p}{\partial v^p} W^{-(p-\alpha)} \{W^{-\alpha} U(w)\} = \frac{\partial^p}{\partial v^p} W^{-p} U(u+iv)$$
$$= \frac{\partial}{\partial v} \int_{-\infty}^{v} U(u+it)dt = U(u+iv).$$

2°. **(a)** is obvious. For **(b)** observe that

$$W^{\alpha} W^{-\alpha} U(u+iv) = W^{\alpha} W^{-(1+\alpha)} \frac{\partial}{\partial v} U(u+iv) = W^{-1} \frac{\partial}{\partial v} U(u+iv)$$
$$= \lim_{A \to -\infty} \int_A^v \frac{\partial}{\partial t} U(u+it)dt = U(u+iv) - \lim_{t \to +\infty} U(u-it)$$
$$\equiv U(u+iv) - \varphi(u),$$

where the limit exists, is finite and depends solely on $u$. On the other hand, this limit is a function harmonic in $D$, from which (1.22′) follows.

*1.6.* The assertions of the above lemmas can be inverted to similar statements related to an integro-differential operator *with bounded integration contour.* For an arbitrary point $w \in G^-$ (and $z = w^{-1}$), consider the arc

$$L(0,z) = [\Gamma(\infty, w)]^{-1} = \{\zeta = \omega^{-1} : \omega \in \Gamma(\infty, w)\}$$

of the circle centered on the real axis, directed from the origin to $z$. A domain $D$ containing $L(0,z)$ together with any inner point $z \in D$ will be called 0–starshaped. Note that $L(0,z)$ is the inversion of the ray

$$\Gamma(\infty, w) = \{\omega = w - i\sigma : 0 < \sigma < +\infty\}$$

directed from $\infty$ to $w$, and both $z = w^{-1}$ and the arc $L(0,z)$ lie in the upper half-plane $G^+$, since $w \in G^-$ and $\Gamma(\infty, w) \subset G^-$.

Unless otherwise stipulated, everywhere below we shall assume $u(z)$ a function harmonic in a 0–starshaped domain $\mathbf{D} \subset G^+ = \{z : \text{Re } z > 0\}$, with possible exception of a countable set of points $\{z_m\}$. Besides, we shall assume that for any $z \in \mathbf{D}$ there is not more than a finite set of such points on $L(0,z)$, also that $u(z)(z - z_m)$ satisfies Holder's condition in some neighborhoods of all $z_m \in L(0,z)$. Then we set

$$\widetilde{W}^{-\alpha} u(z) \equiv \frac{1}{\Gamma(\alpha)} \int_{L(0,z)} \left(\frac{i}{\zeta} - \frac{i}{z}\right)^{\alpha-1} \zeta^{-2} u(\zeta) d\zeta, \quad 0 < \alpha < +\infty, \tag{1.23}$$

$$\widetilde{W}^{0} u(z) \equiv u(z), \tag{1.24}$$

where $p \geq 1$ is the integer deduced from $p - 1 < \alpha \leq p$. The convergence of the integral of $\widetilde{W}^{-\alpha}$ ($\alpha > 0$) will mean its absolute convergence on the arcs $L(0,\zeta) \subset L(0,z)$, which are disjoint from the points $z_m \in L(0,z)$, and convergence as Cauchy's principal value on the rest of $L(0,z)$. Assuming that the requirements on $u(z)$ are satisfied by the derivative

$$\frac{\partial^p}{\partial(\text{Im}\ \frac{1}{z})^p}u(z) = \frac{\partial^p}{\partial(\text{Im}\ \frac{1}{z})^p}u\left(\left\{\text{Re}\ \frac{1}{z} + i\,\text{Im}\ \frac{1}{z}\right\}^{-1}\right), \quad z \in \mathbf{D},$$

where $p \geq 1$ is the natural number from $p-1 < \alpha \leq p$, we define the fractional derivative $\widetilde{W}^\alpha$ ($\alpha > 0$) as follows:

$$\widetilde{W}^\alpha u(z) \equiv \widetilde{W}^{-(p-\alpha)}\left\{\frac{\partial^p}{\partial\,(\text{Im}\ z^{-1})^p}u(z)\right\}, \quad 0 < \alpha < +\infty, \tag{1.25}$$

If $U(w)$ is defined in a $\infty$-starshaped domain $D$, then the function

$$u(z) \equiv U(w), \quad z^{-1} \in D, \tag{1.26}$$

is defined in the 0-starshaped domain $\mathbf{D} = D^{-1}$. The connection between $\widetilde{W}^\alpha$ and Liouville's operator is given by the following lemma.

**Lemma 1.11.** *Let $U(w)$ and $u(z)$ be defined in a $\infty$-starshaped domain $D \subset G^-$ and in the 0-starshaped domain $\mathbf{D} = D^{-1} \subset G^+$ respectively. If these functions are connected by (1.26), then for any $\alpha \in (-\infty, +\infty)$*

$$W^\alpha U(w) \equiv \widetilde{W}^\alpha u(z), \quad z = w^{-1} \in \mathbf{D} \subset G^+, \tag{1.27}$$

*where both sides exist simultaneously.*

**Proof.** Our assertion holds for $\alpha = 0$, since both $W^0$ and $\widetilde{W}^0$ are identical. Further, for any $p \geq 1$

$$\frac{\partial^p u(z)}{\partial\,(\text{Im}\ z^{-1})^p} \equiv \frac{\partial^p U(w)}{\partial(\text{Im}\ w)^p}, \quad z = w^{-1} \in \mathbf{D}.$$

Therefore, it suffices to prove our lemma for $\alpha \in (-\infty, 0)$. To this end, we assume $w \in D$ an arbitrary fixed point and transform (1.1) by the change of variable $\text{Re}\ w + it = s$:

$$W^\alpha U(w) \equiv -\frac{i}{\Gamma(|\alpha|)}\int_{\Gamma(\infty,w)}[i(s-w)]^{|\alpha|-1}U(s)ds, \quad w \in D.$$

Further, the inversion $s = \zeta^{-1}$ gives

$$W^\alpha U(w) = \frac{i}{\Gamma(|\alpha|)}\int_{L(0,z)}\left(\frac{i}{\zeta} - \frac{i}{z}\right)^{|\alpha|-1}\zeta^{-2}u(\zeta)d\zeta, \quad z = w^{-1},$$

and this change of variable is permitted by the properties of $U(w)$ (see, for instance, [34], Ch. 1, § 3, Subsection 3.5). In the last integral

$$\arg\left[\left(\frac{i}{\zeta}-\frac{i}{z}\right)^{|\alpha|-1}\zeta^{-2}u(\zeta)d\zeta\right]=0.$$

Also, one can verify by a direct calculation that $\arg\left[i(\zeta-z)\right]=\vartheta_0+\vartheta-\pi$,

$$d\zeta=\frac{\rho_0}{|\cos\vartheta_0|}e^{2i\vartheta}|d\vartheta| \text{ and } \arg\left[e^{i(|\alpha|-1)\pi}z^{1-\alpha}\left[i(\zeta-z)\right]^{|\alpha|-1}\zeta^{1-|\alpha|}d\zeta\right]=0$$

for $z=\rho_0e^{i\vartheta_0}\in G^+$ and $\zeta=\rho e^{i\vartheta}\in L(0,z)$. Thus, (1.27) holds, and its sides exist simultaneously because of the invertability of our operations.

The following classes in 0-starshaped domains are similar to the previously defined classes $M_\beta(D)$ and $K_\gamma(\delta_0,D)$.

**Definition 1.2.** *A function $u(z)$ given in a 0-starshaped domain $\mathbf{D}$ is said to be of the class $\widetilde{M}_\beta(\mathbf{D})$ $(0\le\beta<+\infty)$, if there exists an angular domain $\lambda(\delta_0,r_0)=\{z:|\pi/2-\arg z|\le\delta_0,\ |z|\le r_0\}$ $(0<\delta_0\le\pi/2,\ 0<r_0<+\infty)$ such that*

$$\sup_{z\in\mathbf{K}}\int_{L(0,(z^{-1}-i^{-1}))}|z-\zeta|^{\beta-1}|\zeta|^{-1-\beta}|u(\zeta)||d\zeta|<+\infty$$

*for any compact $\mathbf{K}\subset\mathbf{D}\cap\lambda(\delta_0,r_0)$. The function $u(z)$ belongs to $\widetilde{K}_\gamma(\delta_0,\mathbf{D})$ $(0<\gamma<+\infty)$, if*

$$|u(z)|\le C|z|^\gamma,\quad z\in\mathbf{D}\cap\lambda(\delta_0,r_0)$$

*for a constant $C>0$ and some $\delta_0$ $(0<\delta_0\le\pi/2)$ and $r_0>0$.*

One can verify that if $U(w)$ and $u(z)$ are connected by (1.26), then the inclusions $U(w)\in M_\beta(D)$ and $U(w)\in K_\gamma(\delta_0,D)$ are equivalent to $u(z)\in\widetilde{M}_\beta(\mathbf{D})$ and $u(z)\in\widetilde{K}_\gamma(\delta_0,\mathbf{D})$ $(\mathbf{D}=D^{-1})$ respectively, whatever be $\beta\in[0,+\infty)$, $\gamma\in(0,+\infty)$ and $\delta_0\in(0,\pi/2]$, i.e.

$$\begin{aligned}\widetilde{M}_\beta(\mathbf{D})&=\left\{u(z)=U(z^{-1}):\ U(w)\in M_\beta(\mathbf{D}^{-1})\right\},\\ \widetilde{K}_\gamma(\delta_0,\mathbf{D})&=\left\{u(z)=U(z^{-1}):\ U(w)\in K_\gamma(\delta_0,\mathbf{D}^{-1})\right\}.\end{aligned}\tag{1.28}$$

Therefore, all lemmas proved for $M_\beta(D)$ and $K_\gamma(\delta_0,D)$ can be equivalently stated for $\widetilde{M}_\beta(\mathbf{D})$ and $\widetilde{K}_\gamma(\delta_0,\mathbf{D})$ $(\mathbf{D}=D^{-1})$. Omitting these statements, we only note that by (1.27) and Lemma 1.5

$$\widetilde{M}_\beta(\mathbf{D})\subset\widetilde{K}_\gamma(\delta_0,\mathbf{D})\tag{1.29}$$

for any $\beta\in[0,+\infty)$, $\gamma\in(\beta,+\infty)$ and $\delta_0\in(0,\pi/2]$.

## 2. HERGLOTZ–RIESZ TYPE THEOREMS

Below we give some well-known results on descriptive representations of several subclasses of the Herglotz–Riesz class by Cauchy integral. After that, we state the main results of this section.

*2.1.* Everywhere below $\mathbf{R}$ means the Herglotz–Riesz class in the lower half-plane $G^- = \{w : \text{Im } w < 0\}$, i.e. the set of functions holomorphic in $G^-$, having there nonnegative real parts.

**Definition 2.1.** *$\mathbf{R}^\circ$ is the set of functions holomorphic in $G^-$, representable in the form*

$$F(w) = \frac{1}{\pi i} \int_{-\infty}^{+\infty} \frac{d\mu(t)}{w-t}, \quad w \in G^-, \tag{2.1}$$

*where $\mu(t)$ $(-\infty < t < +\infty)$ is a nondecreasing, bounded function. $\mathbf{R}^\gamma$ $(0 < \gamma < 2)$ is the set of those $F(w) \in \mathbf{R}$ for which*

$$\int_1^{+\infty} \frac{\text{Re } F(-it)}{t^\gamma} dt < +\infty.$$

The class $\mathbf{R}^\circ$ was first investigated by R.Nevanlinna [81]. One can find a detailed account of its properties in [4] (Addendum I) containing the below

**Theorem 2.1.** *The following statements are equivalent:*

1) $F(w) \in \mathbf{R}^\circ$;

2) $F(w) \in \mathbf{R}$, $\sup\limits_{t>0} t|F(-it)| < +\infty$;

3) $F(w) \in \mathbf{R}$, $\sup\limits_{t>0} t \,\text{Re } F(-it) < +\infty, \quad \lim\limits_{t\to+\infty} F(-it) = 0.$

The classes $\mathbf{R}^\gamma$ $(0 < \gamma < 2)$ were considered by I.S.Kats [72] (see also [4], Addendum I). The next theorem follows from his results.

**Theorem 2.2.** $\mathbf{R}^\gamma$ $(0 < \gamma \le 1)$ *coincides with the set of functions representable in the form*

$$F(w) = iC + \frac{1}{\pi i} \int_{-\infty}^{+\infty} \frac{d\mu(t)}{w-t}, \quad w \in G^-, \tag{2.2}$$

*where $C$ is a real number and $\mu(t)$ is a nondecreasing function satisfying*

$$\int_{-\infty}^{+\infty} \frac{d\mu(t)}{1+|t|^\gamma} < +\infty. \tag{2.3}$$

**Remark 2.1.** It is well known (see [4], Addendum I), that $\mathbf{R}^\circ \subset \mathbf{R}^\gamma \subset \mathbf{R}$ for $0 < \gamma < 2$. One can see that by Theorem 2.1 formula (2.1), where $\mu(t)$ is a nondecreasing, bounded function, is a descriptive representation for the class of functions defined by any of statements 2), 3) of the same theorem.

*2.2.* For stating our theorems, we need the following definitions.

**Definition 2.2.** $\mathbf{R}^{\circ}_{\alpha}$ $(-1 < \alpha < +\infty)$ *is the set of those holomorphic functions in $G^-$, which are representable in the form*

$$F(w) = \frac{\Gamma(1+\alpha)}{\pi} e^{-i\frac{\pi}{2}(1+\alpha)} \int_{-\infty}^{+\infty} \frac{d\mu(t)}{(w-t)^{1+\alpha}}, \qquad w \in G^-, \tag{2.4}$$

*where $\mu(t)$ $(-\infty < t < +\infty)$ is a nondecreasing, bounded function.*

**Definition 2.3.** $\mathbf{R}_{\alpha}$ $(-1 < \alpha < +\infty)$ *is the set of functions $f(w)$ holomorphic in $G^-$ and such that:*
1°. *If $0 \le \alpha < +\infty$ and $p \ge 0$ is the integer from $p-1 < \alpha \le p$, then*

(a) $\mathrm{Re}\, F(w) \in M_p$;

(b) $W^{-\alpha}\mathrm{Re}\, F(w) \ge 0, \quad w \in G^-$;

(c) $$\int_1^{+\infty} W^{-\alpha}\mathrm{Re}\, f(-it)\frac{dt}{t} < +\infty, \qquad \text{if} \quad \alpha = p. \tag{2.5}$$

2°. *If $-1 < \alpha < 0$, then*

(a) $\mathrm{Im}\, F'(w) \in M_1$;

(b) $W^{-\alpha}\mathrm{Re}\, F(w) \ge 0, \quad w \in G^-$.

**Theorem 2.3.** *Let $F(w)$ be holomorphic in $G^-$.*
1°. *If $0 \le \alpha < +\infty$ and $p \ge 0$ is the integer from $p-1 < \alpha \le p$, then the following statements are equivalent:*

(i) $F(w) \in \mathbf{R}^{\circ}_{\alpha}$;

(ii) $\mathrm{Re}\, F(w) \in M_p$, $W^{-\alpha}\mathrm{Re}\, F(w) \ge 0$ *for* $w \in G^-$,

$$\sup_{t>0} t^{1+\alpha}|F(-it)| < +\infty, \qquad \int_1^{+\infty} W^{-\alpha}\mathrm{Re}\, F(-it)\frac{dt}{t} < +\infty$$

*(if $\alpha \ne p$, then the convergence of the last integral is not required)*;

(iii) $\mathrm{Re}\, F(w) \in M_p$, $W^{-\alpha}\mathrm{Re}\, F(w) \ge 0$ *for* $w \in G^-$,

$$\sup_{t>0} tW^{-\alpha}\mathrm{Re}\, F(-it) < +\infty, \qquad \lim_{t\to+\infty} F(-it) = 0.$$

2°. *If $-1 < \alpha < 0$, then the following statements are equivalent:*

(i) $F(w) \in \mathbf{R}^{\circ}_{\alpha}$;

(ii) $\mathrm{Im}\, F'(w) \in M_1$, $W^{-\alpha}\mathrm{Re}\, F(w) \ge 0$ *for* $w \in G^-$,

$$\sup_{t>0} t^{1+\alpha}|F(-it)| < +\infty, \qquad \sup_{t>0} t^{2+\alpha}|\mathrm{Im}\, F'(-it)| < +\infty;$$

(iii) Im $F'(w) \in M_1$, $W^{-\alpha}\text{Re } F(w) \geq 0$ *for* $w \in G^-$,

$$\sup_{t>0} tW^{-\alpha}\text{Re } F(-it) < +\infty, \qquad \lim_{t\to+\infty} F(-it) = 0.$$

**Theorem 2.4.** $\mathbf{R}_\alpha$ $(-1 < \alpha < +\infty)$ *coincides with the set of functions admitting one of the following representations:*

1°. *For* $0 \leq \alpha < +\infty$ *and* $p \geq 0$ *from* $p - 1 < \alpha \leq p$,

$$F(w) = iC + \frac{\Gamma(1+\alpha)}{\pi} e^{-i\frac{\pi}{2}(1+\alpha)} \int_{-\infty}^{+\infty} \frac{d\mu(t)}{(w-t)^{1+\alpha}}, \quad w \in G^-, \tag{2.6}$$

*where* $C$ *is a real number and* $\mu(t)$ *is a nondecreasing function satisfying*

$$\int_{-\infty}^{+\infty} \frac{d\mu(t)}{1+|t|^{1+\alpha-p}} < +\infty. \tag{2.7}$$

2°. *For* $-1 < \alpha < 0$

$$F(w) = a_1 w + a_0 + iC + \frac{\Gamma(1+\alpha)}{\pi} e^{-i\frac{\pi}{2}(1+\alpha)} \int_{-\infty}^{+\infty} \frac{d\mu(t)}{(w-t)^{1+\alpha}} \tag{2.8}$$

$(w \in G^-)$, *where* $a_1$, $a_0$ *and* $C$ *are real numbers and* $\mu(t)$ *is a nondecreasing function such that*

$$\int_{-\infty}^{+\infty} \frac{d\mu(t)}{1+|t|^{1+\alpha}} < +\infty. \tag{2.9}$$

**Remark 2.2.** 1°. If a function $F(w)$ holomorphic in $G^-$ admits any of representations (2.4), (2.6)–(2.7) or (2.8)–(2.9), then

$$W^{-\alpha}\text{Re } F(w) = \frac{|v|}{\pi} \int_{-\infty}^{+\infty} \frac{d\mu(t)}{(u-t)^2+v^2}, \qquad w = u + iv \in G^-. \tag{2.10}$$

2°. If a function $F(w)$ is representable in the form (2.8)–(2.9), then

$$a_1 u + a_0 = \lim_{t\to+\infty} \text{Re } F(u - it), \quad -\infty < u < \infty. \tag{2.11}$$

These assertions will be established in Sec. 3, in Proof of Theorem 2.4.

**Remark 2.3.** It is obvious that strictly

$$\mathbf{R}_\alpha^\circ \subset \mathbf{R}_\alpha, \quad -1 < \alpha < +\infty. \tag{2.12}$$

## 3. PROOFS OF THEOREMS 2.3 AND 2.4

*3.1.* First we prove some lemmas which will be used in this and some other sections.

**Lemma 3.1.** *Let $\mu(t)$ be a function of locally bounded variation on the real axis, and let*

$$\varphi(w) \equiv \frac{h}{iw} + \frac{1}{\pi i}\int_{-\infty}^{+\infty} \frac{d\mu(t)}{w-t}, \quad w \in G^-,$$

*where $h$ is a real number. Then:* 1°. *If*

$$\int_{-\infty}^{+\infty} \frac{|d\mu(t)|}{1+|t|^{1+\alpha-p}} < +\infty \tag{3.1}$$

*for a given $\alpha \in [0,+\infty)$ and the integer $p \geq 0$ $(p-1 < \alpha \leq p)$, then for any $w = u+iv \in G^-$*

$$\frac{\partial^p}{\partial v^p} W^{-(p-\alpha)}\varphi(w) = \Gamma(1+\alpha)\left\{\frac{h}{(iw)^{1+\alpha}} + \frac{1}{\pi}\int_{-\infty}^{+\infty} \frac{d\mu(t)}{[i(w-t)]^{1+\alpha}}\right\}. \tag{3.2}$$

2°. *If*

$$\int_{-\infty}^{+\infty} \frac{|d\mu(t)|}{1+|t|^{1+\alpha}} < +\infty \tag{3.1'}$$

*for some $\alpha \in (-1,0)$, then*

$$W^{\alpha}\varphi(w) = \Gamma(1+\alpha)\left\{\frac{h}{(iw)^{1+\alpha}} + \frac{1}{\pi}\int_{-\infty}^{+\infty} \frac{d\mu(t)}{[i(w-t)]^{1+\alpha}}\right\}, \quad w \in G^-. \tag{3.2'}$$

**Proof.** We shall use the well-known formula

$$\int_{0}^{+\infty} \frac{\sigma^{\gamma-1}d\sigma}{\sigma+iz} = \Gamma(\gamma)\Gamma(1-\gamma)(iz)^{\gamma-1}, \quad \gamma \in (0,1), \quad z \in G^-, \tag{3.3}$$

and the following estimate which is true for any compact $\mathbf{K} \subset G^-$:

$$\sup_{\substack{-\infty<t<+\infty \\ w\in\mathbf{K}}} \left\{\frac{1+|t|^{1+\gamma}}{|w-t|^{1+\gamma}}\right\} = C_\gamma(\mathbf{K}) < +\infty, \quad \lambda \in (-1,+\infty), \tag{3.4}$$

where $C_\gamma(\mathbf{K}) > 0$ is a constant depending only on $\lambda$ and $\mathbf{K}$.

1°. If $\alpha = p \geq 0$ is an integer, then (3.2) is trivial since $W^{-(p-\alpha)} \equiv W^0$ is identical and (3.1) is true. If $\alpha > 0$ is not an integer (i.e. $0 < p-\alpha < 1$), then it suffices to show that

$$W^{-(p-\alpha)}\varphi(w) = \Gamma(1+\alpha-p)\left\{\frac{h}{(iw)^{1+\alpha-p}} + \frac{1}{\pi}\int_{-\infty}^{+\infty}\frac{d\mu(t)}{[i(w-t)]^{1+\alpha-p}}\right\}. \tag{3.5}$$

To this end, first we prove that the integral $W^{-(p-\alpha)}\varphi(w)$ is absolutely convergent. Indeed, one can verify that by (3.4), (1.19) and (3.1)

$$W^{-(p-\alpha)}\int_{-\infty}^{+\infty}\frac{|d\mu(t)|}{|w-t|}\le\frac{\sqrt{2}C_{\alpha-p}(w)}{\Gamma(p-\alpha)}\int_{0}^{+\infty}\frac{x^{p-\alpha-1}dx}{1+x}\int_{-\infty}^{+\infty}\frac{|d\mu(t)|}{1+|t|^{1+\alpha-p}},$$

where the last integral is convergent. Hence, by (3.3) a change of integration order gives (3.5). The assertion 2° is proved similarly.

**Lemma 3.2.** *Let $\beta\ge 0$ be not a natural number, and let $p\ge 0$ be the integer deduced from $p-1<\beta\le p$. Further, let a function $U(w)$ be harmonic in $G^-$ and such that $U(w)\in M_p\{G^-\}$. Then*

$$\int_{1}^{+\infty}W^{-\beta}|U(-it)|\frac{dt}{t^{1+\beta-p}}<+\infty. \tag{3.6}$$

**Proof.** If $\beta=0$, then $p=\beta=0$ and $W^{-\beta}$ is identical, and $\int_1^\infty|U(-it)|\frac{dt}{t}<+\infty$ since $U(w)\in M_0\{G^-\}$. Now let $\beta>0$ be not an integer (i.e. $p-1<\beta<p$). Then the change of integration order gives

$$\begin{aligned}\int_{1}^{+\infty}W^{-\beta}|U(-it)|\frac{dt}{t^{1+\beta-p}}&=\frac{1}{\Gamma(\beta)}\int_{1}^{+\infty}\frac{dt}{t^{1+\beta-p}}\int_{t}^{+\infty}(\tau-t)^{\beta-1}|U(-i\tau)|d\tau\\&<\frac{1}{\Gamma(\beta)}\int_{0}^{1}\frac{(1-x)^{\beta-1}}{x^{1+\beta-p}}dx\int_{1}^{+\infty}\tau^{\beta-1}|U(-i\tau)|d\tau<+\infty.\end{aligned}$$

**Lemma 3.3.** *Let $\nu(t)$ be a function of locally bounded variation on the real axis, and let*

$$\varphi_\alpha(w)\equiv\frac{\Gamma(1+\alpha)}{\pi}e^{-i\frac{\pi}{2}(1+\alpha)}\int_{-\infty}^{+\infty}\frac{d\nu(t)}{(w-t)^{1+\alpha}}.$$

1°. *If $\alpha\in[0,+\infty)$, $p\ge 0$ is the integer from $p-1<\alpha\le p$ and*

$$\int_{-\infty}^{+\infty}\frac{|d\nu(t)|}{1+|t|^{1+\alpha-p}}<+\infty,$$

*then* Re $\varphi_\alpha(w)\in M_0$ *for $\alpha=0$, and $\varphi_\alpha(w)\in M_p$ for $\alpha>0$. Besides,*

$$\frac{\partial^p}{\partial v^p}W^{-(p-\alpha)}\varphi_0(w)=\varphi_\alpha(w),\quad W^{-\alpha}\varphi_\alpha(w)=\varphi_0(w),\quad w=u+iv\in G^-.$$

2°. *If $\alpha\in(-1,0)$ and*

$$\int_{-\infty}^{+\infty}\frac{|d\nu(t)|}{1+|t|^{1+\alpha}}<+\infty,$$

*then* $\varphi'_\alpha(w) \in M_1$. *Besides,*

$$W^\alpha \varphi_0(w) = \varphi_\alpha(w) \quad and \quad W^{-\alpha}\varphi_\alpha(w) = \varphi_0(w), \quad w \in G^-.$$

**Proof** is obvious.

*3.2.* **Proof of Theorem 2.4.** 1°. Let $F(w) \in \mathbf{R}_\alpha$ for some $\alpha > 0$. Then the function $W^{-\alpha}\mathrm{Re}\, F(w)$ is harmonic in $G^-$ by Lemma 1.10(a) with $D = G^-$. Besides, this function is nonnegative, and by Lemma 3.2

$$\int_1^{+\infty} W^{-\alpha}\mathrm{Re}\, F(-it)\frac{dt}{t^{1+\alpha-p}} < +\infty.$$

Thus, the function $\Phi(w) = W^{-\alpha}\mathrm{Re}\, F(w) + iV(w)$ (where $V(w)$ is the conjugate harmonic of $W^{-\alpha}\mathrm{Re}\, F(w)$) belongs to $\mathbf{R}^{1+\alpha-p}$. By Theorem 2.2, $\Phi(w)$ is representable in the form (2.2)–(2.3) for $\gamma = 1 + \alpha - p$. Therefore, the representation (2.6)–(2.7) follows by the assertions 1° of Lemmas 3.1 and 3.3.

Conversely, let $F(w)$ admit the representation (2.6)–(2.7) for some $\alpha > 0$. Then, by Lemma 3.3(1°) $\mathrm{Re}\, F(w) \in M_p$. By the same lemma, also (2.10) is valid, and hence $W^{-\alpha}F(w) \geq 0$ for $w \in G^-$. On the other hand, by (2.10) the function $\Phi(w)$ is representable in the form (2.2)–(2.3), where $\gamma = 1 + \alpha - p$. Hence (2.5) holds for integers $\alpha = p \geq 0$ by Theorem 2.2.

2°. For $-1 < \alpha < 0$ our assertion and formulas (2.10) and (2.11) are proved similarly, using Theorem 2.2 and Lemmas 1.10 and 3.1, 3.2, 3.3.

**Proof of Theorem 2.3.** In both cases 1° $(0 < \alpha < +\infty)$ and 2° $(-1 < \alpha < 0)$ we prove the implications (i) $\Rightarrow$ (ii) $\Rightarrow$ (iii) $\Rightarrow$ (i).

1°. Let $\alpha \geq 0$ and (i) is true, i.e. $F(w)$ admits (2.4). Then, by Lemma 3.3 $\mathrm{Re}\, F(w) \in M_p$ and (2.10) is valid. Hence $W^{-\alpha}\mathrm{Re}\, F(w) \geq 0$ for $w \in G^-$. Besides, (2.5) follows from (2.12) and Theorem 2.4. Now observe that by (2.4)

$$t^{1+\alpha}|F(-it)| \leq \frac{\Gamma(1+\alpha)}{\pi}\int_{-\infty}^{+\infty} d\mu(t) < +\infty, \quad t > 0.$$

Hence (i) $\Rightarrow$ (ii) is true. Let (ii) be valid. Then $|F(-it)| \leq C_0 t^{-1-\alpha}$ $(0 < t < +\infty)$ for some $C_0 > 0$. Hence $F(-it) = o(1)$ as $t \to +\infty$ and for any $t > 0$

$$tW^{-\alpha}\mathrm{Re}\, F(-it) \leq \frac{C_0}{\Gamma(\alpha)}\int_0^{+\infty} \frac{x^{\alpha-1}dx}{(1+x)^{1+\alpha}} < +\infty.$$

This proves (iii). Now let (iii) be true. Then, by Lemma 1.10(a) the function $W^{-\alpha}\mathrm{Re}\, F(w) \geq 0$ is harmonic in $G^-$ and $W^{-\alpha}\mathrm{Re}\, F(-it) \leq C_1 t^{-1}$ $(t > 0)$ for some $C_1 > 0$. Consequently, for any $\gamma > 0$ the above function $\Phi(w)$ is of $\mathbf{R}^\gamma$. Using Lemmas 3.1 and 3.3 we conclude that $F(w)$ is representable in the form (2.6). As it follows from (2.10),

$$\frac{1}{\pi}\int_{-\infty}^{+\infty} d\mu(\sigma) = \sup_{t>0}\, tW^{-\alpha}\mathrm{Re}\, F(-it) < +\infty.$$

Thus, the function $\mu(t)$ is bounded. On the other hand, it is obvious that

$$|C| \leq |F(-it)| + \frac{\Gamma(1+\alpha)}{\pi} \int_{-\infty}^{+\infty} \frac{d\mu(\sigma)}{|t - i\sigma|^{1+\alpha}} = o(1) \quad \text{as} \quad t \to +\infty.$$

Hence $C = 0$ and the representation (2.4) is true, i.e. (i) is valid.

2°. The proofs of the case $-1 < \alpha < 0$ (including formulas (2.10) and (2.11)) are quite similar.

NOTES. Some of the auxiliary statements on Liouville's fractional integro-differentiation, which are given in Section 1, are well known. One can find them, for instance, in [92].

# CHAPTER 2

# BLASCHKE TYPE PRODUCTS

We shall consider a family of Blaschke type products in the lower half-plane $G^- = \{w : \text{Im } w < 0\}$ and investigate some of the properties of these products. The factors of our products depend on a parameter $-1 < \alpha < +\infty$ and are chosen to be of the form

$$b_\alpha(w, \zeta) = \exp\left\{\int_0^{|\eta|} \left( [\tau + i(w - \zeta)]^{-1-\alpha} + [i(w - \overline{\zeta}) - \tau]^{-1-\alpha} \right) \tau^\alpha d\tau \right\} \quad (1)$$

For $\alpha = 0$ this expression becomes $\frac{w-\zeta}{w-\overline{\zeta}}$, and the inversion $z = w^{-1}$ ($z_k = w_k^{-1}$, $k = 1, 2, \ldots$) transfers the properties of (1) into ones for several products in the upper half-plane $G^+ = \{z : \text{Im } z > 0\}$, which for $\alpha = 0$ coincide with the widely known Blaschke product

$$\widetilde{B}(z, \{z_k\}) = \prod_k \frac{1 - z/z_k}{1 - z/\overline{z}_k}, \quad \{z_k\} \subset G^+, \quad \sum_k \left| \text{Im } \frac{1}{z_k} \right| < +\infty. \quad (2)$$

## 1. ELEMENTARY PROPERTIES OF FACTORS

*1.1.* We start by another representation of $b_\alpha(w, \zeta)$. Assuming $\zeta = \xi + i\eta \in G^-$ a fixed point, for any $\alpha \in (-1, +\infty)$ we introduce the function

$$\Omega_\alpha(w, \zeta) = \int_0^{|\eta|} \frac{\tau^\alpha d\tau}{[\tau + i(w - \zeta)]^{1+\alpha}} + \int_0^{|\eta|} \frac{\tau^\alpha d\tau}{[i(w - \overline{\zeta}) - \tau]^{1+\alpha}}. \quad (1.1)$$

The changes of variables $t = -(\tau + \eta)$ and $t = \tau + \eta$ in the above integrals give

$$\Omega_\alpha(w, \zeta) = \int_{-|\eta|}^{|\eta|} \frac{(|\eta| - |t|)^\alpha dt}{[i(w - \xi) - t]^{1+\alpha}}, \quad -1 < \alpha < +\infty. \quad (1.1')$$

By (1) and (1.1)

$$b_\alpha(w, \zeta) = \exp\{-\Omega_\alpha(w, \zeta)\}, \quad -1 < \alpha < +\infty. \quad (1.2)$$

Sometimes we shall write $\Omega_\alpha(w,\zeta)$ in the form

$$\Omega_\alpha(w,\zeta) = U_\alpha(w,\zeta) + V_\alpha(w,\zeta), \quad -1 < \alpha < +\infty, \tag{1.3}$$

where

$$U_\alpha(w,\zeta) = \int_{-|\eta|}^{0} \frac{(|\eta|+t)^\alpha dt}{[i(w-\xi)-t]^{1+\alpha}}, \quad V_\alpha(w,\zeta) = \int_0^{|\eta|} \frac{(|\eta|-t)^\alpha dt}{[i(w-\xi)-t]^{1+\alpha}}. \tag{1.3'}$$

**Theorem 1.1.** *The function $b_\alpha(w,\zeta)$ $(-1 < \alpha < +\infty)$ is holomorphic in the region $\mathbb{C}\backslash\{\xi + ih : 0 \le h < +\infty\}$ and it vanishes only at the point $\zeta \in G^-$, which is a first order zero.*

**Proof.** By (1.2), (1.3), (1.4)

$$b_\alpha(w,\zeta) = \frac{w-\zeta}{w-\bar{\zeta}} \exp\Big\{U_0(w,\zeta) - U_\alpha(w,\zeta) + V_0(w,\zeta) - V_\alpha(w,\zeta)\Big\}.$$

The functions $U_\alpha(w,\zeta)$ and $U_0(w,\zeta)$ obviously are holomorphic in $\mathbb{C}\backslash\{\xi + ih : 0 \le h < +\infty\}$. Thus, it suffices to show that

$$\begin{aligned} F_\alpha(w,\zeta) \equiv & V_0(w,\zeta) - V_\alpha(w,\zeta) \\ = & -\int_0^{|\eta|} \left\{ \frac{(|\eta|-t)^\alpha}{[i(w-\xi)-t]^{1+\alpha}} - \frac{1}{i(w-\xi)-t} \right\} dt \end{aligned}$$

is holomorphic in the same domain. To this end, we write

$$\begin{aligned} F_\alpha(w,\zeta) = & \left(\int_{-\infty}^{0} - \int_{-\infty}^{|\eta|}\right) \left\{ \frac{(|\eta|-t)^\alpha}{[i(w-\xi)-t]^{1+\alpha}} - \frac{1}{i(w-\xi)-t} \right\} dt \\ \equiv & I_\alpha^{(1)}(w,\zeta) - I_\alpha^{(2)}(w,\zeta) \end{aligned}$$

and separately prove the holomorphity of $I_\alpha^{(1)}$ and $I_\alpha^{(2)}$. First observe that

$$\varphi_\alpha(w,\zeta,t) \equiv \frac{(|\eta|-t)^\alpha}{[i(w-\xi)-t]^{1+\alpha}} - \frac{1}{i(w-\xi)-t} \quad (-\infty < t < 0)$$

is holomorphic in $\mathbb{C}\backslash\{\xi + ih : 0 \le h < +\infty\}$, and for $-\infty < t < |\eta|$ the same function is holomorphic in the smaller domain $\mathbb{C}\backslash\{\xi + ih : -|\eta| \le h < +\infty\}$. Next, for any compact $\mathbf{K} \subset \mathbb{C}$

$$\varphi_\alpha(w,\zeta,t) = O(t^{-2}) \quad \text{as} \quad t \to -\infty$$

uniformly in respect to $w \in \mathbf{K}$. Therefore, by uniform convergence, the integrals $I_\alpha^{(1)}(w,\zeta)$ and $I_\alpha^{(2)}(w,\zeta)$ represent holomorphic functions in the domains

$\mathbb{C}\backslash\{\xi+ih : 0 \le h < +\infty\}$ and $\mathbb{C}\backslash\{\xi+ih : -|\eta| \le h < +\infty\}$ respectively. But $I_\alpha^{(2)}(w,\zeta)$ is a constant. Indeed, considering this function on the ray $w = \zeta - ih$ $(0 < h < +\infty)$ and taking $t = -\sigma h + |\eta|$ we get

$$I_\alpha^{(2)}(\zeta - ih, \zeta) = \int_0^{+\infty} \left\{ \frac{\sigma^\alpha}{(1+\sigma)^{1+\alpha}} - \frac{1}{1+\sigma} \right\} d\sigma \equiv \text{const} \neq \infty.$$

*1.2.* The below recurrent formulas for $U_\alpha(w,\zeta)$ and $V_\alpha(w,\zeta)$ lead to some representations of $\Omega_\alpha(w,\zeta)$. We shall assume that $\alpha > 0$ and $p \ge 1$ is the natural number deduced from $p-1 < \alpha \le p$. First observe that integration by parts in (1.3′) gives

$$U_\alpha(w,\zeta) = \frac{1}{\alpha} \int_{t=-|\eta|}^{0} (|\eta| + t)^\alpha d[i(w-\zeta) - t]^{-\alpha} = \frac{1}{\alpha} \left( \frac{i\eta}{w-\xi} \right)^\alpha - U_{\alpha-1}(w,\zeta).$$

Hence

$$U_\alpha(w,\zeta) = \sum_{j=1}^{p} \frac{(-1)^{p-j}}{\alpha - p + j} \left( \frac{i\eta}{w-\xi} \right)^{\alpha-p+j} + (-1)^p U_{\alpha-p}(w,\zeta). \tag{1.4}$$

Similar to this,

$$V_\alpha(w,\zeta) = -\sum_{j=1}^{p} \frac{1}{\alpha - p + j} \left( \frac{i\eta}{w-\xi} \right)^{\alpha-p+j} + V_{\alpha-p}(w,\zeta). \tag{1.4′}$$

By these recurrent formulas and (1.2), (1.3), (1.4)

$$\begin{aligned} b_\alpha(w,\zeta) = \exp&\Big\{(-1)^{p+1} U_{\alpha-p}(w,\zeta) - V_{\alpha-p}(w,\zeta)\Big\} \\ &\times \exp\left\{ \sum_{j=1}^{p} \frac{1-(-1)^{p-j}}{\alpha-p+j} \left( \frac{i\eta}{w-\xi} \right)^{\alpha-p+j} \right\}. \end{aligned} \tag{1.5}$$

*1.3.* **Lemma 1.1.** *If $\alpha \in (-1, +\infty)$ and $\zeta = \xi + i\eta \in G^-$ are arbitrary, and $w = u + iv$ is an arbitrary point from the cut complex plane $\mathbb{C}\backslash\{\zeta + ih : 0 \le h < +\infty\}$, such that $|w - \xi| > |\eta|$, then:*

$$|\Omega_\alpha(w,\zeta)| \le \frac{2}{(|w-\xi| - |\eta|)^{1+\alpha}} \frac{|\eta|^{1+\alpha}}{1+\alpha}, \tag{1.6}$$

$$\left| \frac{\partial}{\partial v} \Omega_\alpha(w,\zeta) \right| \le \frac{2}{(|w-\xi| - |\eta|)^{2+\alpha}} |\eta|^{1+\alpha}. \tag{1.6′}$$

**Proof.** Evidently $|i(w-\xi) - t|^{1+\alpha} \ge (|w-\xi| - |\eta|)^{1+\alpha}$ for any $t \in [-|\eta|, |\eta|]$. Hence by (1.1′) we come to (1.6). For proving (1.6′), one have to repeat the same argument after a differentiation of (1.1′) by $v = \text{Im } w$.

**Lemma 1.2.** *If $\alpha \in (-1, +\infty)$, then for any compact $\mathbf{K} \subset G^-$*

$$\Omega_\alpha(w, \xi + i\eta) = \frac{2e^{-i\frac{\pi}{2}(1+\alpha)}}{(w-\xi)^{1+\alpha}} \frac{|\eta|^{1+\alpha}}{1+\alpha} + O(|\eta|^{2+\alpha}) \quad as \quad \eta \to -0 \tag{1.7}$$

*uniformly in respect to $w \in \mathbf{K}$ and $\xi \in (-\infty, +\infty)$.*

**Proof.** Integration by parts in (1.3′) gives

$$U_\alpha(w, \zeta) = \frac{1}{1+\alpha}\left(\frac{i\eta}{w-\xi}\right)^{1+\alpha} - U_{1+\alpha}(w, \zeta),$$

$$V_\alpha(w, \zeta) = \frac{1}{1+\alpha}\left(\frac{i\eta}{w-\xi}\right)^{1+\alpha} + V_{1+\alpha}(w, \zeta).$$

Hence by (1.3)

$$\Omega_\alpha(w, \xi + i\eta) = \frac{2e^{-i\frac{\pi}{2}(1+\alpha)}}{(w-\xi)^{1+\alpha}} \frac{|\eta|^{1+\alpha}}{1+\alpha} + R_\alpha(w, \zeta),$$

where $R_\alpha(w, \zeta) = -U_{1+\alpha}(w, \zeta) + V_{1+\alpha}(w, \zeta)$. We shall prove that there exists a constant $C$ depending on $\alpha$ and on the compact $\mathbf{K}$, for which

$$|R_\alpha(w, \zeta)| \leq C|\eta|^{2+\alpha} \quad \text{as} \quad \eta \to -0$$

uniformly in respect to $w \in \mathbf{K}$ and $\xi \in (-\infty, +\infty)$. To this end, we denote $\rho = \min_{w \in \mathbf{K}} |\mathrm{Im}\, w|$ and observe that if $|\eta| < \rho/2$, then $|w - \xi| - |\eta| \geq \rho/2$ for any $w \in \mathbf{K}$ and $\xi \in (-\infty, +\infty)$. The desired estimate for $R_\alpha$ follows from (1.3′).

**Remark 1.1.** If $|w| \geq R_0 \geq 4|\zeta|$, then $|w - \xi| - |\eta| \geq |w| - 2|\zeta| \geq |w|/2$ since $2|\zeta| > |\xi| + |\eta|$. Hence by (1.2), (1.6) and (1.6′)

$$|\log |b_\alpha(w, \zeta)|| \leq |\Omega_\alpha(w, \zeta)| \leq \frac{2^{2+\alpha}}{1+\alpha}|\eta|^{1+\alpha}|w|^{-(1+\alpha)}, \tag{1.8}$$

$$\left|\frac{\partial}{\partial v} \log |b_\alpha(w, \zeta)|\right| \leq \left|\frac{\partial}{\partial v}\Omega_\alpha(w, \zeta)\right| \leq 2^{3+\alpha}|\eta|^{1+\alpha}|w|^{-(2+\alpha)} \tag{1.8′}$$

for any $w \in \mathbb{C}\backslash\{\zeta + ih : 0 \leq h < +\infty\}$ ($|w| \geq R_0$). More simple representations than (1.5) are true for $b_\alpha(w, \zeta)$ for $\alpha = p$, where $p$ is an even or odd number (see [54], formulas (1.6)–(1.8′)).

*1.4.* Assuming that $s = \delta + i\lambda$ is an arbitrary fixed point from $G^+$ and $\zeta = \xi + i\eta \in G^- = s^{-1}$ ($\in G^-$), consider the ray

$$\Gamma^*[\xi, \infty] = \{w = \xi + ih : 0 \leq h \leq +\infty\} \tag{1.9}$$

and the arc

$$L^*[\xi^{-1},0] = \{\Gamma^*[\xi,\infty)\}^{-1} = \{z = w^{-1} : w \in \Gamma^*[\xi,\infty)\}. \tag{1.10}$$

One can see that $L^*[\xi^{-1},0]$ is contained in the closed half-plane $\overline{G^-} = \{w : \text{Im } w \le 0\}$ and for $\xi \ne 0$ it becomes a semi-circle connecting $\xi^{-1} = |s|^2/\delta$ with the origin, and the center is on the real axis. Now introduce the functions

$$\widetilde{b}_\alpha(z,s) \equiv b_\alpha(w,\zeta)\Big|_{w=z^{-1},\,\zeta=s^{-1}}, \quad -1<\alpha<+\infty, \tag{1.11}$$

and observe that their properties are simple restatement of those of $b_\alpha(w,\zeta)$. Particularly,

$$\widetilde{b}_\alpha(z,s) = \frac{1-z/s}{1-z/\overline{s}}. \tag{1.12}$$

Further, in view of (1.2), (1.2′) and (1.5)

$$\widetilde{b}_\alpha(z,s) = \exp\left\{-\int_{-|\text{Im } s^{-1}|}^{|\text{Im } s^{-1}|} \frac{\left(|\text{Im } s^{-1}| - |t|\right)^\alpha dt}{\left[i(z^{-1} - \text{Re } s^{-1}) - t\right]^{1+\alpha}}\right\} \tag{1.13}$$

for $-1<\alpha<+\infty$, and the following similarity of Theorem 1.1 is true.

**Theorem 1.1*.** *The function $\widetilde{b}_\alpha(z,s)$ $(-1<\alpha<+\infty)$ is holomorphic in $\mathbb{C}\backslash L^*\left[|s|^2/\delta,0\right]$ and vanishes only at the point $s \in G^+$, which is a zero of first order.*

## 2. INTEGRO-DIFFERENTIAL PROPERTIES OF FACTORS

Assuming $\zeta = \xi + i\eta$ a fixed point in $G^-$, we shall apply Liouville's integro-differential operator $W^{-\alpha}$ (formulas (1.13)-(1.15) in Ch. 1) to $\log b_\alpha(w,\zeta) \equiv -\Omega_\alpha(w,z)$ and consider the properties of the function

$$W^{-\alpha}\log|b_\alpha(w,\zeta)| = -W^{-\alpha}\text{Re } \Omega_\alpha(w,\zeta), \quad -1<\alpha<+\infty.$$

*2.1.* Introduce the following line intercepts:

$$\begin{aligned} (\zeta,\overline{\zeta}) &= \{\xi + ih : -|\eta| < h < |\eta|\}, \\ [\zeta,\overline{\zeta}] &= \{\xi + ih : -|\eta| \le h \le |\eta|\}, \\ (\zeta,\xi) &= \{\xi + ih : -|\eta| < h < 0\}. \end{aligned} \tag{2.1}$$

**Lemma 2.1.** *The function $W^{-\alpha}\log b_\alpha(w,\zeta)$ $(-1<\alpha<+\infty)$ is holomorphic in the domain $\mathbb{C}\backslash[\zeta,\overline{\zeta}]$, where*

$$W^{-\alpha}\log b_\alpha(w,\zeta) = -W^{-\alpha}\Omega_\alpha(w,\zeta) = \frac{1}{\Gamma(1+\alpha)}\int_{-|\eta|}^{|\eta|} \frac{(|\eta|-|t|)^\alpha dt}{t - i(w-\xi)}. \tag{2.2}$$

**Proof.** As $W^0$ is identical, for $\alpha = 0$ our assertion is trivial by (1.1′) and (1.2). The cases $\alpha > 0$ and $\alpha < 0$ have to be considered separately.

**(a)** $0 < \alpha < +\infty$. By (1.1′), (1.2) and (1.13) of Ch. 1

$$W^{-\alpha}\log b_\alpha(w,\zeta) \equiv -W^{-\alpha}\Omega_\alpha(w,\zeta)$$
$$= -\frac{1}{\Gamma(\alpha)}\int_0^{+\infty}\sigma^{\alpha-1}d\sigma\int_{-|\eta|}^{|\eta|}\frac{(|\eta|-|t|)^\alpha dt}{[i(w-\xi)-t+\sigma]^{1+\alpha}}$$

for $w \in \mathbb{C}\backslash\{\zeta + ih;\, 0 \le h < +\infty\}$. It is obvious that the right-hand side integral in this equality is absolutely convergent. Therefore, using the well-known formula

$$\int_0^{+\infty}\frac{\sigma^{\alpha-1}d\sigma}{(z+\sigma)^{1+\alpha}} = \frac{1}{\alpha z}\left(\frac{\sigma}{z+\sigma}\right)^\alpha\Bigg|_{\alpha=0}^{+\infty} = \frac{1}{\alpha z}, \quad \alpha > 0, \tag{2.3}$$

we come to (2.2) for $w \in \mathbb{C}\backslash\{\zeta + ih : 0 \le h < +\infty\}$. From (2.2) it follows that $W^{-\alpha}\Omega_\alpha(w,\zeta)$ permits a holomorphic continuation onto $\mathbb{C}\backslash[\zeta,\overline{\zeta}]$. Therefore, (2.2) is true in whole $\mathbb{C}\backslash[\zeta,\overline{\zeta}]$.

**(b)** $-1 < \alpha < 0$. By (1.1′), (1.2) and (1.15) of Ch. 1

$$W^{-\alpha}\log b_\alpha(w,\zeta) \equiv -W^{-\alpha}\Omega_\alpha(w,\zeta)$$
$$= -\frac{1+\alpha}{\Gamma(1+\alpha)}\int_0^{+\infty}\sigma^{\alpha}d\sigma\int_{-|\eta|}^{|\eta|}\frac{(|\eta|-|t|)^\alpha dt}{[i(w-\xi)-t+\sigma]^{2+\alpha}}$$

for $w \in \mathbb{C}\backslash\{\zeta + ih : 0 \le h < +\infty\}$. Hence we again come to (2.2).

**Remark 2.1.** From (2.2) it follows that

$$\left|W^{-\alpha}\log|b_\alpha(w,\zeta)|\right| \le \frac{2}{|w-\xi|-|\eta|}\frac{|\eta|^{1+\alpha}}{\Gamma(2+\alpha)}, \quad -1 < \alpha < +\infty, \tag{2.4}$$

for any $w \in \mathbb{C}$ such that $|w-\xi| > |\eta|$. The proof of this estimate is similar to that of (1.6).

**Lemma 2.2.** 1°. *For any* $\alpha \in (-1,+\infty)$ *and any* $w \in \mathbb{C}\backslash[\zeta,\overline{\zeta}]$

$$W^{-\alpha}\log|b_\alpha(w,\zeta)| = -\mathrm{Re}\, W^{-\alpha}\Omega_\alpha(w,\zeta)$$
$$= \frac{2\mathrm{Im}\, w}{\Gamma(1+\alpha)}\int_0^{|\eta|}\frac{|w-\xi|^2-t^2}{|(w-\xi)^2+t^2|^2}(|\eta|-t)^\alpha dt. \tag{2.5}$$

2°. *For any* $\alpha \in (0,+\infty)$ *and any* $w \in \mathbb{C}$

$$W^{-\alpha}\log|b_\alpha(w,\zeta)| = -\mathrm{Re}\, W^{-\alpha}\Omega_\alpha(w,\zeta)$$
$$= \frac{1}{\Gamma(\alpha)}\int_0^{|\eta|}\log\left|\frac{\xi-it-w}{\xi+it-w}\right|(|\eta|-t)^\alpha dt$$
$$= \frac{1}{\Gamma(\alpha)}\iint_{G^-}\log\left|\frac{w-\overline{s}}{w-s}\right||\xi-s|^{\alpha-1}\mathfrak{X}_\zeta(s)d\sigma(s), \tag{2.6}$$

*where $\mathfrak{X}_\zeta(s)$ is the characteristic function of the interval $(\zeta, \xi)$ and $d\sigma(s)$ is the area element.*

**Proof.** 1°. Denoting $z = i(w - \xi)$, from (2.2) we obtain

$$W^{-\alpha} \log |b_\alpha(w, \zeta)| = \frac{1}{\Gamma(1+\alpha)} \int_0^{|\eta|} \mathrm{Re} \left\{ \frac{1}{t-z} - \frac{1}{t+z} \right\} (|\eta| - t)^\alpha dt.$$

Hence (2.5) follows.

2°. Observe that $\mathrm{Re}\,\{dt/[t - i(w - \xi)]\} = d_t \log |t - i(w - \xi)|$ for $w \in \mathbb{C}$ and $t \in (-\infty, +\infty)$. Therefore, integration by parts in (2.2) gives

$$\begin{aligned} W^{-\alpha} \log |b_\alpha(w, \zeta)| &= \frac{1}{\Gamma(1+\alpha)} \int_{-|\eta|}^{|\eta|} \log \frac{1}{|t - i(w-\xi)|} d(|\eta| - t)^\alpha \\ &= \frac{1}{\Gamma(\alpha)} \int_0^{|\eta|} \log \left| \frac{\xi - it - w}{\xi + it - w} \right| (|\eta| - t)^\alpha dt \\ &= \frac{1}{\Gamma(\alpha)} \iint_{G^-} \log \left| \frac{w - \overline{s}}{w - s} \right| |\xi - s|^{\alpha - 1} \mathfrak{X}_\zeta(s) d\sigma(s), \end{aligned}$$

and the proof is complete.

From (2.5) it follows that $W^{-\alpha} \log |b_\alpha(w, \zeta)|$ is harmonic outside the intercept $[\zeta, \overline{\zeta}]$, and also that this function vanishes on the real axis, i.e.

$$W^{-\alpha} \log |b_\alpha(u, \zeta)| = 0, \quad -\infty < u < +\infty, \quad u \neq \xi \quad \alpha > -1. \tag{2.7}$$

Further, for $|w - \xi| \geq |\eta|$ the integrand in (2.5) is nonnegative. Hence, for $|w - \xi| > |\eta|$

$$W^{-\alpha} \log |b_\alpha(w, \zeta)| \begin{cases} < 0, & \text{if} \quad w \in G^-, \\ > 0, & \text{if} \quad w \in G^+, \end{cases} \quad -1 < \alpha < +\infty. \tag{2.8}$$

*2.2.* For further analysis of $W^{-\alpha} \log |b_\alpha(w, \zeta)|$ we use some well-known properties of Cauchy type integrals. From (2.2) it follows that for $z \notin \mathbb{C} \backslash [-|\eta|, |\eta|]$

$$W^{-\alpha} \log |b_\alpha(-iz + \xi, \zeta)| = -\frac{2}{\Gamma(1+\alpha)} \mathrm{Im}\ \Phi_\alpha(z), \quad -1 < \alpha < +\infty, \tag{2.9}$$

where

$$\Phi_\alpha(z) = \frac{1}{2\pi} \int_{-|\eta|}^{|\eta|} \frac{(|\eta| - |t|)^\alpha}{t - z} dt. \tag{2.9'}$$

In view of these formulas

$$W^{-\alpha} \log |b_\alpha(-iz + \xi, \zeta)| \equiv -W^{-\alpha} \log |b_\alpha(iz + \xi, \zeta)|, \quad z \in \mathbb{C}. \tag{2.10}$$

The function $\Phi_\alpha(z)$ is holomorphic in $\mathbb{C}\backslash[-|\eta|,|\eta|]$. Besides, $\varphi_\alpha(t)=(|\eta|-|t|)^\alpha$ satisfies the Lipschitz condition (for $\alpha=0$ and $\alpha\geq 1$ of order 1 on $[-|\eta|,|\eta|]$, for $0<\alpha<1$ of order $\alpha$ on $[-|\eta|,|\eta|]$ and for $-1<\alpha<0$ of order 1 on $(-|\eta|,|\eta|)$). Therefore, the below statements are true by the well-known theory of Cauchy type integrals (see, for instance, [34], Ch. 1).

1°. The Cauchy type integral

$$\Phi_\alpha(x)=\frac{1}{2\pi}\int_{-|\eta|}^{|\eta|}\frac{(|\eta|-|t|)^\alpha}{t-x}dt,\quad -1<\alpha<+\infty,$$

is a continuous function of Lip$\lambda$ in $(-|\eta|,|\eta|)$ for some $\lambda\in(0,1)$.

2°. For any $\alpha\in(-1,+\infty)$ and any $x\in(-|\eta|,|\eta|)$ the limits

$$\lim_{z\to x,\, z\in G+}\Phi_\alpha(z)=\Phi_\alpha^+(x),\qquad \lim_{z\to x,\, z\in G-}\Phi_\alpha(z)=\Phi_\alpha^-(x)$$

exist, are finite and connected by the following formulas:

$$\Phi_\alpha^+(x)-\Phi_\alpha^-(x)=(|\eta|-|t|)^\alpha,\quad \Phi_\alpha^+(x)+\Phi_\alpha^-(x)=2\Phi_\alpha(x).$$

3°. For $\alpha>-1$ these limits are continuous functions of Lip $\lambda$ on $(-|\eta|,|\eta|)$ with some $\lambda\in(0,1)$.

4°. If $\alpha>0$, then $\Phi_\alpha(z)$ is continuous at the points $-|\eta|$ and $|\eta|$, as a function of complex variable.

5°. If $-1<\alpha<0$, then the following representations are true in enough small neighborhoods of the points $-|\eta|$ and $|\eta|$:

$$\Phi_\alpha(z)=-\frac{e^{-i\pi\alpha}}{2i\sin\pi\alpha}(z+|\eta|)^\alpha+\psi_\alpha^0(z),\ \ \Phi_\alpha(z)=\frac{e^{-i\pi\alpha}}{2i\sin\pi\alpha}(-z+|\eta|)^\alpha+\psi_\alpha^1(z),$$

where $\psi_\alpha^0(z)$ and $\psi_\alpha^1(z)$ are holomorphic in the same neighborhoods.

In view of (2.9) and (2.9′) we come to the following

**Lemma 2.3.** *The function $W^{-\alpha}\log|b_\alpha(w,\zeta)|$ $(-1<\alpha<+\infty)$, which is harmonic in $\mathbb{C}\backslash[\zeta,\overline{\zeta}]$, is continuous on $[\zeta,\overline{\zeta}]$ for $\alpha>0$, and is continuous on $(\zeta,\overline{\zeta})$ for $-1<\alpha<0$. Besides, for $-1<\alpha<0$*

$$W^{-\alpha}\log|b_\alpha(w,\zeta)| = \begin{cases}\dfrac{\Gamma(1-\alpha)}{\alpha}|w-\zeta|^\alpha\cos\left[\alpha\arg i(w-\zeta)\right]+u_\alpha^0(w),\\[2mm] \dfrac{\Gamma(1-\alpha)}{\alpha}|w-\overline{\zeta}|^\alpha\cos\left[\alpha\arg i(\overline{\zeta}-w)\right]+u_\alpha^1(w),\end{cases}\tag{2.11}$$

*in enough small neighborhoods of* $\zeta \in G^-$ *and* $\overline{\zeta} \in G^+$, *where* $u^1_\alpha(w)$ *and* $u^1_\alpha(w)$ *are harmonic.*

In contrast to the properties of $\log|b_0(w,\zeta)|$, it appears that the entire intercept $[\zeta,\overline{\zeta}]$ is the support of the mass of $W^{-\alpha}\log|b_\alpha(w,\zeta)|$.

**Theorem 2.1.** $1^\circ$. *For* $\alpha \in (0,+\infty)$ *the function* $W^{-\alpha}\log|b_\alpha(w,\zeta)|$ *is continuous in the closed* $w$*-plane, harmonic out of* $[\zeta,\overline{\zeta}]$, *subharmonic in* $G^-$ *and superharmonic in* $G^+$. *Besides,*

$$W^{-\alpha}\log|b_\alpha(w,\zeta)|\begin{cases} <0 & \text{for} \quad w\in G^-,\\ >0 & \text{for} \quad w\in G^+,\end{cases} \qquad 0<\alpha<+\infty. \tag{2.12}$$

$2^\circ$. *For* $\alpha\in(-1,0)$ *the function* $W^{-\alpha}\log|b_\alpha(w,\zeta)|$ *is continuous in the closed* $w$*-plane, except the points* $\zeta\in G^-$ *and* $\overline{\zeta}\in G^+$, *is harmonic out of* $[\zeta,\overline{\zeta}]$, *superharmonic in* $G^-\backslash\zeta$ *and subharmonic in* $G^+\backslash\overline{\zeta}$. *Besides, for* $-1<\alpha<0$.

$$W^{-\alpha}\log|b_\alpha(w,\zeta)|\begin{cases} <0 & \text{for} \quad w\in G^-, \quad |w-\xi|>|\eta|,\\ >0 & \text{for} \quad w\in G^+, \quad |w-\xi|>|\eta|.\end{cases} \tag{2.13}$$

$3^\circ$. *For any* $\alpha\in(-1,+\infty)$

$$W^{-\alpha}\log|b_\alpha(u,\zeta)|=0, \quad -\infty<u<+\infty. \tag{2.14}$$

**Proof.** Whatever be $\alpha\in(-1,+\infty)$, by (2.2) $W^{-\alpha}\log|b_\alpha(w,\zeta)|$ is continuous in the closure of $w$-plane (by (2.2) $W^{-\alpha}\log b_\alpha(\infty,\zeta)=0$), with possible exception of points $\zeta$ and $\overline{\zeta}$. Therefore, $3^\circ$ follows from (2.7).

$1^\circ$. Let $\alpha>0$. Then in view of (2.9) and (2.9′) $W^{-\alpha}\log|b_\alpha(w,\zeta)|$ is harmonic in $w$-plane, except $[\zeta,\overline{\zeta}]$. Because of continuity of this function, it suffices to prove that for any $s\in[\zeta,\overline{\zeta}]$ and enough small $\rho>0$

$$\frac{1}{2\pi}\int_0^{2\pi} W^{-\alpha}\log|b_\alpha(s+\rho e^{i\vartheta},\zeta)|d\vartheta$$
$$\begin{cases} >W^{-\alpha}\log|b_\alpha(s,\zeta)|, & s\in[\zeta,\xi),\\ <W^{-\alpha}\log|b_\alpha(s,\zeta)|, & s\in(\xi,\overline{\zeta}].\end{cases} \tag{2.15}$$

Hence our assertion will follow by the maximum and the minimum principles of subharmonic and superharmonic functions. Before proving (2.15), observe that by (2.10) and (2.14)

$$\frac{1}{2\pi}\int_0^{2\pi} W^{-\alpha}\log|b_\alpha(\xi+\rho e^{i\vartheta},\zeta)|d\vartheta = W^{-\alpha}\log|b_\alpha(\xi,\zeta)|=0 \tag{2.16}$$

for any $\alpha\in(-1,+\infty)$ and $\rho>0$. Now assume that $0<h<|\eta|$ and $0<\rho<\max\{h,|\eta|-h\}$. Then by (2.2)

$$\frac{1}{2\pi}\int_0^{2\pi} W^{-\alpha}\log|b_\alpha(\xi-ih+\rho e^{i\vartheta},\zeta)|d\vartheta$$
$$=\frac{1}{\Gamma(1+\alpha)}\int_{-|\eta|}^{|\eta|}(|\eta|-|t|)^\alpha \operatorname{Re}\left\{\frac{1}{2\pi}\int_{|s|=\rho}\frac{ds}{s(t-h-is)}\right\}dt$$

for any $1 < \alpha < +\infty$. The last inner integral vanishes for $|t-h| < \rho$ and equals $(t-h)^{-1}$ for $|t-h| > \rho$. Hence

$$\frac{1}{2\pi}\int_0^{2\pi} W^{-\alpha}\log|b_\alpha(\xi - ih + \rho e^{i\vartheta}, \zeta)|d\vartheta \\ = \frac{1}{\Gamma(1+\alpha)}\left(\int_{-|\eta|}^{h-\rho} + \int_{h+\rho}^{|\eta|}\right)\frac{(|\eta|-|t|)^\alpha}{t-h}dt.$$

Consequently, by (2.2) we obtain that

$$\frac{1}{2\pi}\int_0^{2\pi} W^{-\alpha}\log|b_\alpha(\xi - ih + \rho e^{i\vartheta}, \zeta)|d\vartheta - W^{-\alpha}\log|b_\alpha(\xi - ih, \zeta)| \\ = -\frac{1}{\Gamma(1+\alpha)}\int_{h-\rho}^{h+\rho}\frac{(|\eta|-|t|)^\alpha}{t-h}dt \tag{2.17}$$

for $\alpha \in (-1, +\infty)$, $0 < h < |\eta|$ and $0 < \rho < \max\{h, |\eta| - h\}$. One can be convinced that also

$$\frac{1}{2\pi}\int_0^{2\pi} W^{-\alpha}\log|b_\alpha(\zeta + \rho e^{i\vartheta}, \zeta)|d\vartheta - W^{-\alpha}\log|b_\alpha(\zeta, \zeta)| \\ = \frac{1}{\Gamma(1+\alpha)}\frac{\rho^\alpha}{\alpha} > 0. \tag{2.17'}$$

for any $\alpha > 0$ and $\rho \in (0, |\eta|)$. Now observe that

$$\int_{h-\rho}^{h+\rho}\frac{(|\eta|-|t|)^\alpha}{t-h}dt = \int_0^\rho \frac{(|\eta| - x - h)^\alpha - (|\eta| + x - h)^\alpha}{x}dx.$$

Consequently,

$$\frac{1}{2\pi}\int_0^{2\pi} W^{-\alpha}\log|b_\alpha(\xi - ih + \rho e^{i\vartheta}, \zeta)|d\vartheta - W^{-\alpha}\log|b_\alpha(\xi - ih, \zeta)| \\ = \frac{1}{\Gamma(1+\alpha)}\int_0^\rho \frac{(|\eta| + x - h)^\alpha - (|\eta| - x - h)^\alpha}{x}dx \\ \begin{cases} > 0, & 0 < \alpha < +\infty \\ < 0, & -1 < \alpha < 0 \end{cases} \tag{2.18}$$

for any $\alpha \in (-1, +\infty)$, $0 < h < |\eta|$ and $0 < \rho < \max\{h, |\eta| - h\}$. Hence the estimates (2.15) follow by (2.17′) and (2.10).

2°. As we have shown, the function $W^{-\alpha}\log|b_\alpha(w, \zeta)|$ $(-1 < \alpha < 0)$ is continuous in the closure of $w$-plane, except the points $\zeta$ and $\overline{\zeta}$. Besides, this function is harmonic out of the intercept $[\zeta, \overline{\zeta}]$ and the estimates (2.13) (see (2.8)) are true. Therefore, it suffices to see that by the second inequality in (2.18) and (2.10) $W^{-\alpha}\log|b_\alpha(w,\zeta)|$ is superharmonic in $G^-\backslash\zeta$ and subharmonic in $G^+\backslash\overline{\zeta}$.

*2.3.* **Lemma 2.4.** *For any* $\alpha\in(-1,+\infty)$

$$\lim_{v\to 0}\int_{-\infty}^{+\infty}\left|W^{-\alpha}\log|b_\alpha(u+iv,\zeta)|\right|du=0. \tag{2.19}$$

**Proof.** By (2.2)

$$W^{-\alpha}\log|b_\alpha(u+iv,\zeta)|=\frac{1}{\Gamma(1+\alpha)}\int_{-|\eta|}^{|\eta|}\frac{t+v}{(u-\xi)^2+(t+v)^2}(|\eta|-|t|)^\alpha dt.$$

for any $u$, $v\in(-\infty,+\infty)$. Hence, changing the order of integration we get

$$\int_{-\infty}^{+\infty}W^{-\alpha}\log|b_\alpha(u+iv,\zeta)|du$$
$$=\frac{\pi}{\Gamma(1+\alpha)}\int_{-|\eta|}^{|\eta|}(|\eta|-|t|)^\alpha\text{sign }(t+v)dt\equiv I_\alpha(|\eta|,v).$$

For calculation of the last integral, observe that sign $(t+v)=$ sign $v$ for $|v|\geq|\eta|$ and $-|\eta|<t<|\eta|$. Hence, for $|v|\geq|\eta|$

$$I_\alpha(|\eta|,v)=(\text{sign }v)\frac{\pi}{\Gamma(1+\alpha)}\int_{-|\eta|}^{|\eta|}(|\eta|-|t|)^\alpha dt=(\text{sign }v)\frac{2\pi}{\Gamma(2+\alpha)}|\eta|^{1+\alpha}.$$

On the other hand,

$$I_\alpha(|\eta|,v)=(\text{sign }v)\frac{2\pi}{\Gamma(2+\alpha)}\left[|\eta|^{1+\alpha}-(|\eta|-|v|)^{1+\alpha}\right]$$

for $|v|<|\eta|$. Consequently, for any $v\in(-\infty,+\infty)$

$$\int_{-\infty}^{+\infty}W^{-\alpha}\log|b_\alpha(u+iv,\zeta)|du$$
$$=\begin{cases}(\text{sign }v)\dfrac{2\pi}{\Gamma(2+\alpha)}|\eta|^{1+\alpha} & \text{if }\ |v|\geq|\eta|,\\ (\text{sign }v)\dfrac{2\pi}{\Gamma(2+\alpha)}\left[|\eta|^{1+\alpha}-(|\eta|-|v|)^{1+\alpha}\right] & \text{if }\ |v|<|\eta|.\end{cases} \tag{2.20}$$

As the function $W^{-\alpha}\log|b_\alpha(w,\zeta)|$ $(\alpha\geq 0)$ keeps its sign in $G^-$ and $G^+$, (2.20) implies (2.19). For proving (2.19) with $\alpha\in(-1,0)$, note that in view of (2.10) it suffices to prove (2.19) as $v\to -0$. To this end, suppose $v<0$ and denote by $S_v^+$ the part of the line $\{w = u + iv : -\infty < u < +\infty\}$, where $W^{-\alpha}\log|b_\alpha(w,\zeta)| \geq 0$, and by $S_v^-$ the remaining part of the same line, where $W^{-\alpha}\log|b_\alpha(w,\zeta)| < 0$. Then

$$\int_{-\infty}^{+\infty} |W^{-\alpha}\log|b_\alpha(u+iv,\zeta)||du = \left(\int_{S_v^+} - \int_{S_v^-}\right) W^{-\alpha}\log|b_\alpha(u+iv,\zeta)|du$$
$$= \left(2\int_{S_v^+} - \int_{-\infty}^{+\infty}\right) W^{-\alpha}\log|b_\alpha(u+iv,\zeta)|du. \qquad (2.21)$$

But the last integral in the right-hand side of this formula vanishes as $v \to -0$. Therefore, it remains to show that

$$\lim_{v\to -0}\int_{S_v^+} W^{-\alpha}\log|b_\alpha(u+iv,\zeta)|du = 0.$$

To this end, observe that in virtue of first inequality in (2.13) $S_v^+$ is contained in the semidisc $\{w : |w-\xi| \leq |\eta|,\ w \in G^-\}$ for $|v| < |\eta|$ $(v < 0)$. Hence

$$0 \leq \int_{S_v^+} W^{-\alpha}\log|b_\alpha(u+iv,\zeta)|du < \int_{\xi-|\eta|}^{\xi+|\eta|} \left|W^{-\alpha}\log|b_\alpha(u+iv,\zeta)|\right| du. \quad (2.22)$$

Further, by Theorem 2.1 $W^{-\alpha}\log|b_\alpha(w,\zeta)|$ is continuous in the closed quadrat $\{w = u+iv : |u-\xi| \leq |\eta|,\ \eta/2 \leq v \leq 0\}$, and therefore is bounded there. Thus, by Fatou's lemma and (2.14)

$$\limsup_{v\to -0}\int_{\xi-|\eta|}^{\xi+|\eta|} \left|W^{-\alpha}\log|b_\alpha(u+iv,\zeta)|\right| du$$
$$\leq \int_{\xi-|\eta|}^{\xi+|\eta|} \lim_{v\to -0}\left|W^{-\alpha}\log|b_\alpha(u+iv,\zeta)|\right| du = 0.$$

*2.4.* Note that by (2.20) and (2.21)

$$\int_{-\infty}^{+\infty} \left|W^{-\alpha}\log|b_\alpha(u+iv,\zeta)|\right| du \leq \frac{6\pi}{\Gamma(2+\alpha)}|\eta|^{1+\alpha} \qquad (2.23)$$

for any $-1 < \alpha < +\infty$ and $-\infty < v < +\infty$. Further, note that the results of this section can be inverted to similar results on properties of $\widetilde{W}^{-\alpha}\log|\widetilde{b}_\alpha(z,s)|$, where $\widetilde{W}^{-\alpha}$ is the integro-differential operator mentioned in Ch. 1 (formulas (1.21)–(1.23)) since by our notation (1.11)

$$\widetilde{W}^{-\alpha}\log|\widetilde{b}_\alpha(z,s)|\Big|_{z=w^{-1},\, s=\zeta^{-1}} \equiv W^{-\alpha}\log|b_\alpha(w,\zeta)|, \quad z \in \overline{\mathbb{C}}. \qquad (2.24)$$

## 3. BLASCHKE TYPE PRODUCTS

Now we proceed to the convergence and main properties of the products

$$B_\alpha(w, \{w_k\}) \equiv \prod_k b_\alpha(w, w_k), \quad -1 < \alpha < +\infty, \tag{3.1}$$

with zeros $\{w_k\}$ in the lower half-plane $G^-$. These will result in similar properties of the product

$$\widetilde{B}_\alpha(z, \{z_k\}) \equiv \prod_k \widetilde{b}_\alpha(z, z_k), \quad -1 < \alpha < +\infty, \tag{3.2}$$

the zeros $z_k = w_k^{-1}$ $(k = 1, 2, \dots)$ of which lie in the upper half-plane $G^+$.

*3.1.* For $\rho < 0$ we set $G_\rho^- = \{w : \text{Im } w < \rho\}$ and $\overline{G_\rho^-} = \{w : \text{Im } w \le \rho\}$.

**Theorem 3.1.** 1°. *Let $\{w_k\}_1^\infty \subset G^-$ be any sequence of complex numbers satisfying*

$$\sum_{k=1}^\infty |\text{Im } w_k|^{1+\alpha} < +\infty \tag{3.3}$$

*for a given $\alpha \in (-1, +\infty)$. Then for any $\rho < 0$ the infinite product*

$$B_\alpha(w, \{w_k\}_1^\infty) \equiv \prod_{k=1}^\infty b_\alpha(w, w_k) = \exp\Big\{ -\sum_{k=1}^\infty \Omega_\alpha(w, w_k) \Big\} \tag{3.4}$$

*is absolutely and uniformly convergent in the closed half-plane $\overline{G_\rho^-}$, and the function $B_\alpha(w, \{w_k\}_1^\infty)$ is holomorphic in $G^-$, where $\{w_k\}_1^\infty$ are its zeros.*

2°. *If $\{w_k\}_1^\infty \subset G^-$ is a bounded sequence, and for a given $\alpha \in (-1, +\infty)$ the product (3.4) is absolutely and uniformly convergent inside $G^-$, then $\{w_k\}_1^\infty$ satisfies (3.3).*

**Poof.** 1°. Let $\rho < 0$ be a fixed number. Since by (3.3) $\text{Im } w_k \to -0$ as $k \to \infty$, one can choose a natural number $N_\rho \ge 1$ such that $w_k \notin \overline{G_\rho^-}$ for $k \ge N_\rho + 1$. Besides, $w_k \in \overline{G_\rho^-}$, then $|w - \text{Re } w_k| \ge |\text{Im } w| \ge |\rho| > |\text{Im } w_k|$ for $k \ge N_\rho + 1$. Therefore, by (1.6)

$$|\Omega_\alpha(w, w_k)| \equiv |\log b_\alpha(w, w_k)| \le \frac{1}{(|\rho| - |\text{Im } w_k|)^{1+\alpha}} \frac{|\text{Im } w_k|^{1+\alpha}}{1+\alpha}$$

for any $w_k \in \overline{G_\rho^-}$ and $k \ge N_\rho + 1$. Together with (3.3), this implies the absolute and uniform convergence of the series

$$\sum_{k=N_\rho+1}^\infty \log b_\alpha(w, w_k) = -\sum_{k=N_\rho+1}^\infty \Omega_\alpha(w, w_k) \tag{3.5}$$

in $\overline{G_\rho^-}$. Hence, also the product (3.4) is absolutely and uniformly convergent in $\overline{G_\rho^-}$. Consequently, $B_\alpha(w,\{w_k\}_1^\infty)$ is holomorphic in $G^-$ and $\{w_k\}_1^\infty$ are its zeros.

2°. The function $B_\alpha(w,\{w_k\}_1^\infty)\not\equiv 0$ is holomorphic in $G^-$ since (3.4) is assumed uniform convergent inside $G^-$. Besides, Im $w_k\to 0$ as $k\to\infty$ since the sequence $\{w_k\}_1^\infty$ is bounded. Observe that the absolute convergence of the product (3.4) in any fixed point $a\in G^-$ is equivalent to the absolute convergence of the series (3.5) (where $\rho=\text{Im } a$ and $N_\rho$ is chosen by $\rho$ as above) in the same point. Besides, by our requirements $\sup_k|\text{Re } w_k|\le\sup_k|w_k|=M<+\infty$. Hence

$$\left|\frac{2e^{-i\frac{\pi}{2}(1+\alpha)}|\text{Im } w_k|^{1+\alpha}}{(a-\text{Re } w_k)^{1+\alpha}(1+\alpha)}\right|\ge(|a+M|+|a-M|)^{-1-\alpha}\,\frac{2|\text{Im } w_k|^{1+\alpha}}{1+\alpha}.$$

Since Im $w_k\to 0$ as $k\to\infty$, (1.7) and the last inequality lead to

$$|\Omega_\alpha(a,w_k)|\ge C|\text{Im } w_k|^{1+\alpha},\quad k\ge k_0,$$

for enough great $k_0\ge 1$, where $C\equiv C(\alpha,a,M)>0$ is a constant independent of $k$. Therefore, the absolute convergence of (3.5) in $w=a$ implies (3.3), and the proof is complete.

**Remark 3.1.** Considering the convergence of $B_\alpha(w,\{w_k\}_1^\infty)$ on compacts $\mathbf{K}\subset\mathbb{C}$, one can show that under (3.3) $B_\alpha(w,\{w_k\}_1^\infty)$ $(\alpha>-1)$ admits holomorphic continuation into the strip $\{w: a<\text{Re } w<b,\ 0<\text{Im } w<+\infty\}\subset G^+$, through any interval $(a,b)$ disjoint from the points $u_k=\text{Re } w_k$ $(k=1,2,\dots)$.

Let $\{w_k\}\subset G^-$ be a bounded sequence (finite or infinite) satisfying (3.3) for a given $\alpha\in(-1,+\infty)$. Denote $\sup_k|w_k|=M$ $(<+\infty)$ and assume $w\in G^-$ a fixed point such that $|w|>4M$. Then, in view of absolute convergence of (3.5) and (1.8)

$$|\log|B_\alpha(w,\{w_k\})||\le\sum_k|\log|b_\alpha(w,w_k)||$$
$$\le\left\{\frac{2^{2+\alpha}}{1+\alpha}\sum_k|\text{Im } w_k|^{1+\alpha}\right\}|w|^{-(1+\alpha)}. \tag{3.6}$$

*3.2.* The below lemmas are to be used in investigation of the properties of $W^{-\alpha}\log|B_\alpha(w,\{w_k\})|$. Let the sequence $\{w_k\}_1^\infty\subset G^-$ satisfy (3.3) for a given $\alpha\in(-1,+\infty)$. Then Im $w_k\to 0$ as $k\to\infty$ and, as it was mentioned above, for any closed half-plane $\overline{G_\rho^-}=\{w: \text{Im } w\le\rho<0\}$ one can choose $N\equiv N_\rho+1$ such that $w_k\notin\overline{G_\rho^-}$ for $k\ge N+1$. Introduce the function

$$\psi_\alpha(w)\equiv\log|B_\alpha(w,\{w_k\}_1^\infty)|-\sum_{k=1}^N\log|b_\alpha(w,w_k)|$$
$$\equiv\sum_{k=N+1}^\infty\log|b_\alpha(w,w_k)|. \tag{3.7}$$

For $k \geq N+1$ the summands in the last series are harmonic functions in $\overline{G_\rho^-}$. On the other hand, from the proof of Theorem 3.1 it follows that this series is absolutely and uniformly convergent in $\overline{G_\rho^-}$. Therefore $\psi_\alpha(w)$ is harmonic in $G_\rho^-$. Using the inequalities (1.6′) one can show that in $\overline{G_\rho^-}$ there is also another absolutely and uniformly convergent representation:

$$\frac{\partial}{\partial(\text{Im } w)}\psi_\alpha(w) = \sum_{k=N+1}^{\infty} \frac{\partial}{\partial(\text{Im } w)} \log|b_\alpha(w, w_k)|. \tag{3.7′}$$

**Lemma 3.2.** *If (3.3) is true for a given $\alpha \in (-1, +\infty)$, then the series*

$$\sum_{k=N+1}^{\infty} W^{-\alpha} \log|b_\alpha(w, w_k)|$$

*is absolutely and uniformly convergent in the closed half-plane $\overline{G_\rho^-}$.*

**Proof.** We have $w_k \notin \overline{G_\rho^-}$ for $k \geq N+1$ by the choice of $N \equiv N_\rho$. Besides, for $k \geq N+1$

$$\left|W^{-\alpha} \log|b_\alpha(w, w_k)|\right| \leq \frac{2}{\Gamma(2+\alpha)} \frac{|\text{Im } w_k|^{1+\alpha}}{|\rho| - \delta}, \quad w_k \in \overline{G_\rho^-}, \tag{3.8}$$

where $\delta = \max_{k \geq N+1} |\text{Im } w_k| < |\rho|$. Hence the desired assertion follows.

**Lemma 3.3.** *If (3.3) is true for a given $\alpha \in (-1, +\infty)$, then*

$$W^{-\alpha}\psi_\alpha(w) \equiv \sum_{k=N+1}^{\infty} W^{-\alpha} \log|b_\alpha(w, w_k)|, \quad w \in \overline{G_\rho^-}, \tag{3.9}$$

*where the series is absolutely and uniformly convergent in $\overline{G_\rho^-}$.*

**Proof.** Setting $\delta = \max_{k \geq N+1} |\text{Im } w_k|$, preliminarily we give two estimates which are consequences of (1.6) and (1.6 ′) and are true for $u \in (-\infty, +\infty)$, $t \in (-\infty, \rho)$ and $k \geq N+1$:

$$|\log|b_\alpha(u+it, w_k)|| \leq \frac{2}{(|t| - \delta)^{1+\alpha}} \frac{|\text{Im } w_k|^{1+\alpha}}{1+\alpha}, \tag{3.10}$$

$$\left|\frac{\partial}{\partial t} \log|b_\alpha(u+it, w_k)|\right| \leq \frac{2}{(|t| - \delta)^{2+\alpha}} |\text{Im } w_k|^{1+\alpha}. \tag{3.10′}$$

(a) Let $\alpha > 0$, then from (3.10) it follows that the integral

$$I(w) \equiv \frac{1}{\Gamma(\alpha)} \int_{-\infty}^{v} (v-t)^{\alpha-1} \sum_{k=N+1}^{\infty} |\log|b_\alpha(u+it, w_k)|| \, dt$$

is convergent for $w = u + iv \in \overline{G_\rho^-}$. Indeed, since $-\infty < t < v < \rho$

$$I(w) \leq \frac{2}{1+\alpha}\frac{1}{\Gamma(\alpha)}\left(\sum_{k=N+1}^{\infty} |\text{Im } w_k|^{1+\alpha}\right)\int_{-\infty}^{v}\frac{(v-t)^{\alpha-1}}{(|t|-\delta)^{1+\alpha}}dt < +\infty$$

by (3.10) and the convergence of the series (3.3). Hence, for $w = u + iv \in \overline{G_\rho^-}$

$$\begin{aligned} W^{-\alpha}\psi_\alpha(w) &\equiv \frac{1}{\Gamma(\alpha)}\int_{-\infty}^{v}(v-t)^{\alpha-1}\sum_{k=N+1}^{\infty}\log|b_\alpha(u+it, w_k)|dt \\ &= \sum_{k=N+1}^{\infty}\frac{1}{\Gamma(\alpha)}\int_{-\infty}^{v}(v-t)^{\alpha-1}\log|b_\alpha(u+it, w_k)|dt \\ &= \sum_{k=N+1}^{\infty} W^{-\alpha}\log|b_\alpha(w, w_k)| \end{aligned}$$

in view of the definition (3.7) of $\psi_\alpha(w)$ and the convergence of $I(w)$.

(b) Let $-1 < \alpha < 0$. Similar to previous case, (3.10′) implies the convergence of

$$J(w) \equiv \frac{1}{\Gamma(1+\alpha)}\int_{-\infty}^{v}(v-t)^{\alpha}\sum_{k=N+1}^{\infty}\left|\frac{\partial}{\partial t}\log|b_\alpha(u+it, w_k)|\right|dt$$

for $w = u + iv \in \overline{G_\rho^-}$. Hence, at any $w = u + iv \in \overline{G_\rho^-}$

$$\begin{aligned} W^{-\alpha}\psi_\alpha(w) &\equiv \frac{1}{\Gamma(1+\alpha)}\int_{-\infty}^{v}(v-t)^{\alpha}\sum_{k=N+1}^{\infty}\frac{\partial}{\partial t}\log|b_\alpha(u+it, w_k)|dt \\ &= \sum_{k=N+1}^{\infty}\frac{1}{\Gamma(1+\alpha)}\int_{-\infty}^{v}(v-t)^{\alpha}\frac{\partial}{\partial t}\log|b_\alpha(u+it, w_k)|dt \\ &= \sum_{k=N+1}^{\infty} W^{-\alpha}\log|b_\alpha(w, w_k)|. \end{aligned}$$

*3.3.* The following theorem is a consequence of formula (3.7), Lemmas 3.2 and 3.3 and the properties of $W^{-\alpha}\log|b_\alpha(w, w_k)|$ stated in Theorem 2.1.

**Theorem 3.2.** 1°. *If a sequence $\{w_k\} \subset G^-$ satisfies (3.3) for a given $\alpha \in (0, +\infty)$, then the function $W^{-\alpha}\log|B_\alpha(w, \{w_k\})|$ is continuous and subharmonic in $G^-$ and harmonic in the domain $G^- \setminus \cup_k [w_k, \text{Re } w_k)$, and*

$$W^{-\alpha}\log|B_\alpha(w, \{w_k\})| \leq 0, \quad w \in G^-.$$

2°. *If* $\{w_k\} \subset G^-$ *satisfies (3.3) for an* $\alpha \in (-1, 0)$, *then the function* $W^{-\alpha} \log |B_\alpha(w, \{w_k\})|$ *is continuous and superharmonic in* $G^- \setminus \{w_k\}$ *and is harmonic in the domain* $G^- \setminus \cup_k [w_k, \operatorname{Re} w_k)$. *Besides,*

$$W^{-\alpha} \log |B_\alpha(w, \{w_k\})| \leq 0$$

*for any* $w \in G^-$ *such that* $|w - \operatorname{Re} w_k| > |\operatorname{Im} w_k|$ $(k = 1, 2, \ldots)$.

**Remark 3.2.** By a similar argument, one can also show that under (3.3) the series

$$W^{-\alpha} \log |B_\alpha(w, \{w_k\})| = \sum_k W^{-\alpha} \log |b_\alpha(w, w_k)|, \quad \alpha \in (-1, +\infty), \tag{3.11}$$

is absolutely and uniformly convergent in any compact $\mathbf{K} \subset \mathbb{C}$ disjoint from condensation points of the sequence $\{\operatorname{Re} w_k\}$. Therefore, by Theorem 2.1 $W^{-\alpha} \log |B_\alpha(w, \{w_k\})|$ permits continuous extension from $G^-$ through any interval $(a, b)$ containing not more than finite number of points $u_k = \operatorname{Re} w_k$. Besides,

$$W^{-\alpha} \log |B_\alpha(u, \{w_k\})| = 0, \quad a < u < b. \tag{3.11'}$$

**Theorem 3.3.** *If a sequence* $\{w_k\} \subset G^-$ *satisfies (3.3) for a given* $\alpha \in (-1, +\infty)$, *then*

$$\lim_{v \to -0} \int_{-\infty}^{+\infty} \left| W^{-\alpha} \log |B_\alpha(u + iv, \{w_k\})| \right| du = 0. \tag{3.12}$$

**Proof.** By (3.7) for $w \in G^- \setminus \{w_k\}$

$$\left| W^{-\alpha} \log |B_\alpha(w, \{w_k\})| \right| \leq \sum_k \left| W^{-\alpha} \log |b_\alpha(w, w_k)| \right|, \tag{3.13}$$

where the series is convergent. Therefore, by (2.23)

$$\int_{-\infty}^{+\infty} \left| W^{-\alpha} \log |B_\alpha(u + iv, \{w_k\})| \right| du$$
$$\leq \frac{6\pi}{\Gamma(2 + \alpha)} \sum_k |\operatorname{Im} w_k|^{1+\alpha} < +\infty \tag{3.14}$$

for any $v < 0$. Hence, by (2.19) and Fatou's lemma

$$0 \leq \limsup_{v \to -0} \int_{-\infty}^{+\infty} \left| W^{-\alpha} \log |B_\alpha(u + iv, \{w_k\})| \right| du$$
$$\leq \sum_k \limsup_{v \to -0} \int_{-\infty}^{+\infty} \left| W^{-\alpha} \log |b_\alpha(u + iv, \{w_k\})| \right| du = 0.$$

**Remark 3.3.** Since $W^{-\alpha}\log|B_\alpha(w,\{w_k\})|$ $(\alpha\ge 0)$ is subharmonic in $G^-$ and satisfies (3.12), one can use the well-known Littlewood's theorem on boundary values of subharmonic functions (see, for instance, [105], Ch. IV, Sec. 10, Theorem IV.34) to prove that for almost all $u\in(-\infty,+\infty)$

$$\lim_{w\to u_0,\, w\in l(u_0)} W^{-\alpha}\log|B_\alpha(w,\{w_k\})|=0, \tag{3.15}$$

where $l(u_0)$ is the image of a radius for any conformal mapping of $|z|<1$ onto $G^-$.

**Lemma 3.4.** *If $\{w_k\}\subset G^-$ is a bounded sequence (i.e. $\sup_k|w_k|=r_0<+\infty$) satisfying (3.3) for a given $\alpha\in(-1,+\infty)$, then for $w\in G^-$ $(|w|>4r_0)$*

$$\left|W^{-\alpha}\log|B_\alpha(w,\{w_k\})|\right|\le\left\{\frac{4}{\Gamma(2+\alpha)}\sum_k|\mathrm{Im}\ w_k|^{1+\alpha}\right\}|w|^{-1}. \tag{3.16}$$

**Proof.** Assume $w\in G^-$ and $|w|>4r_0$. Since $|w-\mathrm{Re}\ w_k|-|\mathrm{Im}\ w_k|\ge |w|-(|\mathrm{Re}\ w_k|+|\mathrm{Im}\ w_k|)\ge|w|-2|w_k|>|w|/2+2(r_0-|w_k|)\ge|w_k|/2$, (2.4) implies

$$\left|W^{-\alpha}\log|b_\alpha(w,\{w_k\})|\right|\le\frac{4}{\Gamma(2+\alpha)}|\mathrm{Im}\ w_k|^{1+\alpha}|w|^{-1}.$$

Therefore, (3.16) follows from (3.11).

*3.4.* Now we pass to the properties of the products

$$\widetilde{B}_\alpha(z,\{z_k\})=\prod_k\widetilde{b}_\alpha(z,z_k),\quad -1<\alpha<+\infty, \tag{3.17}$$

with zeros in the upper half-plane $G^+$. Some of these properties are described by means of the integro-differential operator $\widetilde{W}^{-\alpha}$ (formulas (1.23)–(1.25) in Ch. 1). One can verify that the inversion $w=z^{-1}$ $(w_k=z_k^{-1},\ k=1,2,\dots)$ transforms Theorems 3.1, 3.2 and 3.3 into the following equivalent assertions.

**Theorem 3.1*.** 1°. *Let $\{z_k\}_1^\infty$ be a sequence of complex numbers in the upper half-plane $G^+$ and let*

$$\sum_{k=1}^\infty\left|\mathrm{Im}\ \frac{1}{z_k}\right|^{1+\alpha}<+\infty \tag{3.18}$$

*for a given $\alpha\in(-1,+\infty)$. Then the infinite product*

$$\widetilde{B}_\alpha(z,\{z_k\}_1^\infty)\equiv\prod_{k=1}^\infty\widetilde{b}_\alpha(z,z_k) \tag{3.19}$$

*is absolutely and uniformly convergent inside $G^+$, and $\widetilde{B}_\alpha(z,\{z_k\})$ is a holomorphic function in $G^+$, with zeros $\{z_k\}$.*

2°. *If an infinite sequence* $\{z_k\}_1^\infty \subset G^+$ *lies out of some neighborhood of the origin, and for an* $\alpha \in (-1,+\infty)$ *the product (3.4) is absolutely and uniformly convergent inside* $G^+$, *then* $\{z_k\}_1^\infty$ *satisfies (3.18).*

In the next theorem $L[s, \mathrm{Re}\, s^{-1})$ ($s \in G^+$) means an arc with the endpoints $z_k$ and $\mathrm{Re}\, z_k{}^{-1}$, which is the part of a tangential to the imaginary axis circle centered on the real axis.

**Theorem 3.2***. 1°. *If a sequence* $\{z_k\} \subset G^+$ *satisfies (3.18) for a given* $\alpha \in (0,+\infty)$, *then* $\widetilde{W}^{-\alpha} \log|\widetilde{B}_\alpha(z,\{z_k\})|$ *is continuous and subharmonic in* $G^+$ *and this function is harmonic in* $G^+ \setminus \cup_k L[z_k, \mathrm{Re}\, z_k)$. *Besides*

$$\widetilde{W}^{-\alpha} \log|\widetilde{B}_\alpha(z,\{z_k\})| \le 0, \quad z \in G^+.$$

2°. *If a sequence* $\{z_k\} \subset G^+$ *satisfies (3.18) for some* $\alpha \in (-1,0)$, *then* $\widetilde{W}^{-\alpha} \log|\widetilde{B}_\alpha(z,\{z_k\})|$ *is continuous and superharmonic in* $G^+ \setminus \{z_k\}$ *and is harmonic in* $G^+ \setminus \cup_k L[z_k, \mathrm{Re}\, z_k)$. *Besides,*

$$\widetilde{W}^{-\alpha} \log|\widetilde{B}_\alpha(z,\{z_k\})| \le 0$$

*for any* $z \in G^+$ *such that* $|1/z - \mathrm{Re}\, 1/z_k| > |\mathrm{Im}\, 1/z_k|$ $(k = 1,2,\dots)$.

**Theorem 3.3***. *If* $\{z_k\} \subset G^+$ *satisfies (3.18) for an* $\alpha \in (-1,+\infty)$, *then*

$$\sup_{0<R<+\infty} \int_{\pi}^{\pi} \left|\widetilde{W}^{-\alpha} \log|\widetilde{B}_\alpha(R \sin\vartheta e^{i\vartheta},\{z_k\})|\right| \frac{d\vartheta}{R\sin^2\vartheta}$$
$$\le \frac{6\pi}{\Gamma(2+\alpha)} \sum_k \left|\mathrm{Im}\, \frac{1}{z_k}\right|^{1+\alpha} < +\infty, \tag{3.20}$$

$$\lim_{R\to+\infty} \int_{-\pi}^{\pi} \left|\widetilde{W}^{-\alpha} \log|\widetilde{B}_\alpha(R \sin\vartheta e^{i\vartheta},\{z_k\})|\right| \frac{d\vartheta}{R\sin^2\vartheta} = 0. \tag{3.21}$$

## 4. A REPRESENTATION OF THE PRODUCT

*4.1.* Below we shall prove a representation which will be used later, in Ch. 5, for investigation of boundary properties of our Blaschke type products.

**Theorem 4.1.** *Let* $\alpha \in (-1,1)$ *be any number, and let* $\{w_k\} \subset G^-$ *be a sequence of complex numbers such that simultaneously*

$$\sum_k |\mathrm{Im}\, w_k|^{1+\alpha} < +\infty \quad \textit{and} \quad \sum_k |\mathrm{Im}\, w_k| < +\infty. \tag{4.1}$$

*Then for any* $w \in G^-$

$$B_\alpha(w,\{w_k\}) = B_0(w,\{w_k\})$$
$$\times \exp\left\{\frac{\Gamma(1+\alpha)}{\pi} e^{-i\frac{\pi}{2}(1+\alpha)} \int_{-\infty}^{+\infty} \frac{d\mu(t)}{(w-t)^{1+\alpha}}\right\}, \tag{4.2}$$

*where* $\mu(t)$ *is a nonincreasing, bounded function in* $(-\infty,+\infty)$ *for* $-1<\alpha<0$, *and for* $0<\alpha<1$ *this function is nondecreasing in* $(-\infty,+\infty)$ *but such that*

$$\int_{-\infty}^{+\infty} \frac{d\mu(t)}{1+|t|^{\alpha+\varepsilon}} < +\infty \tag{4.3}$$

*for any* $\varepsilon>0$. *Besides, whatever be* $\alpha\in(-1,1)$ *there exists a sequence* $\delta_n\downarrow 0$ *by which*

$$\mu(t)=\lim_{n\to\infty}\int_0^t W^{-\alpha}\log\left|\frac{B_\alpha(u-i\delta_n,\{w_k\})}{B_0(u-i\delta_n,\{w_k\})}\right|du, \quad -\infty<t<+\infty. \tag{4.4}$$

*4.2.* Before establishing some necessary lemmas, note that for $\alpha=0$

$$\log b_\alpha(w,\zeta)\equiv-\Omega_\alpha(w,\zeta), \quad w,\zeta\in G^-, \quad \alpha>-1,$$

becomes the main branch of the logarithm.

**Lemma 4.1.** *Let* $\alpha\in(-1,1)$ *and* $\zeta=\xi+i\eta\in G^-$ *be any fixed numbers. Then the function*

$$\varphi_\alpha(w,\zeta)=W^{-\alpha}\log\frac{b_\alpha(w,\zeta)}{b_0(w,\zeta)} \tag{4.5}$$

is holomorphic in $\mathbb{C}\backslash\{\xi+ih:\,0\le h<+\infty\}$ where

$$\begin{aligned}\Gamma(1+\alpha)\varphi_\alpha(w,\zeta)&=\int_{|\eta|}^{+\infty}\left\{[\sigma+i(w-\zeta)]^{-1}-\left[\sigma+i(w-\overline{\zeta})\right]^{-1}\right\}\sigma^\alpha d\sigma\\&+\int_0^{|\eta|}\left\{\left[\sigma-i(w-\overline{\zeta})\right]^{-1}-\left[\sigma+i(w-\overline{\zeta})\right]^{-1}\right\}\sigma^\alpha d\sigma.\end{aligned} \tag{4.6}$$

**Proof.** Preliminarily we shall show that

$$\begin{aligned}&W^{-\alpha}\log b_0(w,\zeta)\\&=-\frac{1}{\Gamma(1+\alpha)}\int_0^{+\infty}\left\{[\sigma+i(w-\zeta)]^{-1}-[\sigma+i(w-\zeta)]^{-1}\right\}\sigma^\alpha d\sigma\end{aligned} \tag{4.7}$$

for $w\notin\{\zeta+ih:\,0\le h<+\infty\}$. Indeed, for $\alpha=0$ (4.7) is trivial since $W^0$ is identical operator. For $0<\alpha<1$ integration by parts gives

$$W^{-\alpha}\log b_0(w,\zeta)=\frac{1}{\Gamma(1+\alpha)}\int_0^{+\infty}\sigma^\alpha d\log\frac{\sigma+i(w-\overline{\zeta})}{\sigma+i(w-\zeta)}$$

which implies (4.7). If $-1<\alpha<0$, then

$$W^{-\alpha}\log b_0(w,\zeta)=W^{-(1+\alpha)}\left\{\left[i(w-\overline{\zeta})\right]^{-1}-[i(w-\zeta)]^{-1}\right\}.$$

Hence (4.7) follows. The representation (4.6) providing the holomorphity of $\varphi_\alpha(w,\zeta)$ in the required domain follows from (4.7) and (2.2).

**Lemma 4.2.** *Let $\alpha \in (-1,1)$ and $\zeta = \xi + i\eta \in G^-$ be fixed. Then the function* Re $\varphi_\alpha(w,\zeta)$ *is harmonic in $G^-$ and continuous in $\overline{G^-}$.*

**Proof.** By Lemma 4.1, it suffices to prove the continuity of Re $\varphi_\alpha(w,\zeta)$ at the point $w = \xi$ or, which is the same, the continuity of

$$\varphi^*_\alpha(w,\zeta) = \Gamma(1+\alpha)\varphi_\alpha(w+\xi,\zeta) \tag{4.9}$$

at the origin. To this end, we subtract from (4.6) the similar formula for $\varphi_0(w,\zeta)$ ($\equiv 0$) multiplied by $|\eta|^\alpha$. This gives

$$\varphi^*_\alpha(w,\zeta) = \int_{|\eta|}^{+\infty} \left\{[\sigma - |\eta| + iw]^{-1} - [\sigma + |\eta| + iw]^{-1}\right\}(\sigma^\alpha - |\eta|^\alpha)d\sigma$$
$$+ \int_0^{|\eta|} \left\{[\sigma - |\eta| - iw]^{-1} - [\sigma - |\eta| + iw]^{-1}\right\}(\sigma^\alpha - |\eta|^\alpha)d\sigma.$$

But $|\sigma^\alpha - |\eta|^\alpha| < C_\alpha|\sigma - |\eta||$ $(0 < \sigma < +\infty)$, where $C_\alpha$ depends only on $\alpha$ and $|\eta|$. Therefore, estimating the real parts of integrands and using the Lebesgue theorem on dominated convergence, we conclude that Re $\varphi^*_\alpha(w,\zeta)$ is continuous in $\overline{G^-}$ and

$$\lim_{\substack{w\to 0\\ \mathrm{Im}\ w\le 0}} \mathrm{Re}\ \varphi^*_\alpha(w,\zeta) = 2|\eta|\int_0^{+\infty} \frac{\sigma^\alpha - |\eta|^\alpha}{\sigma^2 - |\eta|^2}d\sigma.$$

*4.3.* **Lemma 4.3.** *For any $\zeta \in G^-$ and $w \in \overline{G^-}$*

$$\mathrm{Re}\ \varphi^*_\alpha(w,\zeta) \begin{cases} < 0, & \text{if } -1 < \alpha < 0, \\ > 0, & \text{if } 0 < \alpha < 1. \end{cases} \tag{4.10}$$

**Proof.** We shall initially prove these estimates for $w = u \in (-\infty,+\infty)$. By (4.9) and (4.6)

$$\mathrm{Re}\ \varphi^*_\alpha(u,\zeta) = \left(\int_0^{|\eta|} + \int_{|\eta|}^{+\infty}\right)\sigma^\alpha d\log\left|\frac{\sigma - |\eta| + iu}{\sigma + |\eta| + iu}\right| \equiv I_1(u) + I_2(u).$$

Besides, if $0 < \sigma < |\eta|$, then

$$\frac{\partial}{\partial\sigma}\log\left|\frac{\sigma - |\eta| + iu}{\sigma + |\eta| + iu}\right| = -2|\eta|\frac{u^2 + \eta^2 - \sigma^2}{|(\sigma + iu)^2 - \eta^2|^2} < 0.$$

But $\sigma^\alpha < |\eta|^\alpha$ for $\alpha > 0$, and $\sigma^\alpha > |\eta|^\alpha$ for $\alpha < 0$. Consequently,

$$-|\eta|^\alpha \log\left|\frac{2|\eta| + iu}{u}\right| \begin{cases} > I_1(u) & \text{if } -1 < \alpha < 0, \\ < I_1(u) & \text{if } 0 < \alpha < 1, \end{cases} \qquad -\infty < u < +\infty.$$

On the other hand, integration by parts gives

$$I_2(u) = |\eta|^\alpha \log\left|\frac{2|\eta|+iu}{u}\right| - \alpha\int_{|\eta|}^{+\infty} \log\left|\frac{\sigma-|\eta|+iu}{\sigma+|\eta|+iu}\right| \sigma^{\alpha-1}d\sigma.$$

The above estimates of $I_1(u)$ imply (4.10) for Re $\varphi_\alpha^*(w,\zeta)$ (Im $w=0$). For extending the inequalities (4.10) to all $w \in \overline{G^-}$, it suffices to use the Phragmén–Lindelöf principle and the estimate

$$|\varphi_\alpha^*(w,\zeta)| \le \frac{8}{3}\left[\frac{|\eta|}{|w|^{1-\alpha}} + \frac{1}{1+\alpha}\frac{|\eta|^{1+\alpha}}{|w|}\right], \quad w \in G^-, \ |w| > 2|\eta|,$$

which follows from the representation (4.7), (4.9) of $\varphi_\alpha^*(w,\zeta)$ by evaluation of modules of integrands.

**Remark 4.1.** The last estimate of $\varphi_\alpha^*(w,\zeta)$ implies that

$$|\varphi_\alpha(u+iv,\zeta)| \le \frac{8}{3\Gamma(1+\alpha)}\left[\frac{|\eta|}{|v|^{1-\alpha}} + \frac{1}{1+\alpha}\frac{|\eta|^{1+\alpha}}{|v|}\right], \quad v < -2|\eta|. \tag{4.11}$$

On the other hand, using (4.6) and (4.10) one can verify that if $v<0$ is enough small, then for any $\eta \in (-1,0)$

$$|\mathrm{Re}\ \varphi_\alpha(iv,i\eta)| > \begin{cases} C_\alpha|\eta||v|^{-1+\alpha} & \text{if } \ 0<\alpha<1, \\ C'_\alpha|\eta|^{1+\alpha}|v|^{-1} & \text{if } \ -1<\alpha<0, \end{cases} \tag{4.12}$$

where $C_\alpha$ and $C'_\alpha$ are positive constants depending solely on $\alpha$.

*4.4.* Assuming $\{w_k\} \subset G^-$ an arbitrary sequence satisfying the conditions (4.1) for an $\alpha \in (-1,1)$, consider the following function which is holomorphic in $G^-$:

$$\begin{aligned}\Phi_\alpha(w,\{w_k\}) &\equiv W^{-\alpha}\log\frac{B_\alpha(w,\{w_k\})}{B_0(w,\{w_k\})} \\ &= \sum_k W^{-\alpha}\log\frac{b_\alpha(w,w_k)}{b_0(w,w_k)} \equiv \sum_k \varphi_\alpha(w,w_k).\end{aligned} \tag{4.13}$$

Note that the last two series are absolutely and uniformly convergent inside $G^-$ (this follows by (4.11) and by the convergence of the series of $W^{-\alpha}|\log b_\alpha|$). On the other hand, the first two equalities in (4.13) are true by Fubini's theorem (see (1.8) and (2.4)).

**Lemma 4.4.** *Let $\alpha \in (-1,1)$ be arbitrary, and let a sequence $\{w_k\} \subset G^-$ satisfy (4.1). Then*

$$\Phi_\alpha(w,\{w_k\}) = \frac{1}{\pi i}\int_{-\infty}^{+\infty}\frac{d\mu(t)}{w-t}, \quad w \in G^-, \tag{4.14}$$

*where $\mu(t)$ is a nonincreasing and bounded function $(-\infty,+\infty)$ for $-1<\alpha<0$, and for $0<\alpha<1$ this function is nondecreasing in $(-\infty,+\infty)$ but satisfying (4.3) for any $\varepsilon>0$. Besides, for any $\alpha\in(-1,1)$ there exists a sequence $\delta_n\downarrow 0$ by which (4.4) holds.*

**Proof.** By (4.13), (4.10) and the Herglotz-Riesz' theorem, in both cases $-1<\alpha<0$ and $0<\alpha<1$

$$\Phi_\alpha(w,\{w_k\})=ipw+\frac{1}{\pi i}\int_{-\infty}^{+\infty}\left\{\frac{1}{w-t}+\frac{t}{1+t^2}\right\}d\mu(t)+iC,\quad w\in G^-,$$

where $p$ and $C$ are some real numbers and the function $\mu(t)$ is nonincreasing for $-1<\alpha<0$, nondecreasing for $0<\alpha<1$ and such that

$$\int_{-\infty}^{+\infty}\frac{|d\mu(t)|}{1+t^2}<+\infty,\quad -1<\alpha<+\infty.$$

Besides (see, for instance, [4], Ch. I, Sec. 4), there exists a sequence $\delta_n\downarrow 0$ by which (4.4) holds. Now observe that by (4.11) and (4.13)

$$|\Phi_\alpha(u+iv,\{w_k\})| \le \frac{8}{3\Gamma(1+\alpha)}\left[\frac{1}{|v|^{1-\alpha}}\sum_k|\mathrm{Im}\,w_k|+\frac{1}{(1+\alpha)|v|}\sum_k|\mathrm{Im}\,w_k|^{1+\alpha}\right]$$

for $v<-2\max_k|\mathrm{Im}\,w_k|$. Consequently, $\sup_{v<0}|v\Phi_\alpha(iv,\{w_k\})|<+\infty$ for $-1<\alpha<0$. Hence (4.14) and the boundedness of $\mu(t)$ follows (see, for instance, [4], Addendum I, Sec. 4). As to the case $0<\alpha<1$, it is obvious that

$$\lim_{v\to-\infty}\Phi_\alpha(iv,\{w_k\})=0\quad\text{and}\quad\int_1^{+\infty}|\Phi_\alpha(-it,\{w_k\})|\frac{dt}{t^{\alpha+\varepsilon}}<+\infty$$

for any $\varepsilon>0$ (we cannot get rid of $\varepsilon$ in view of the below estimate for Re $\Phi_\alpha$, which follows from (4.12)). Hence we come to the representation (4.14) and to the relation (4.3) with obligatory $\varepsilon>0$ (see [4], Addendum I, Sec. 3).

*4.5.* **Lemma 4.5.** *Under the conditions of Lemma* 4.4,

$$W^\alpha\Phi_\alpha(w,\{w_k\})\equiv\log\frac{B_\alpha(w,\{w_k\})}{B_0(w,\{w_k\})},\qquad \mathrm{Im}\,w<-\max_k|\mathrm{Im}\,w_k|.$$

**Proof.** We start by the case when the sequence $\{w_k\}$ consists of a single term $\zeta=\xi+i\eta\in G^-$ and use the formula

$$\int_0^{+\infty}\frac{\sigma^{-\gamma}d\sigma}{(z+\sigma)^{n+1}}=\frac{\Gamma(\gamma+n)\Gamma(1-\gamma)}{n!}\frac{1}{z^{\gamma+n}},\quad z\notin(-\infty,0],\tag{4.15}$$

which particularly is true for any integer $n \geq 0$ and any $\gamma \in (-n, 1)$.

Let $-1 < \alpha < 0$ and $\operatorname{Im} w < \eta$. Then by formulas (2.2), (1.15) of Ch. 1 and (1.1′)–(1.2)

$$\begin{aligned} &W^{\alpha}W^{-\alpha}\log b_\alpha(w,\zeta) \\ &\quad = -\frac{1}{\Gamma(1+\alpha)\Gamma(1-\alpha)}\int_{-|\eta|}^{|\eta|}(|\eta|-|t|)^{\alpha}dt\int_0^{+\infty}\frac{\sigma^{-\alpha}d\sigma}{[i(w-\xi)-t+\sigma]^2} \\ &\quad = \log b_\alpha(w,\zeta). \end{aligned}$$

Besides,

$$\begin{aligned} &W^{\alpha}W^{-\alpha}\log b_0(w,\zeta) \\ &\quad = -(1-\alpha)\int_{-|\eta|}^{|\eta|}dt\int_0^{+\infty}\frac{\sigma^{-\alpha}d\sigma}{[i(w-\xi)-t+\sigma]^{2-\alpha}} = \log b_0(w,\zeta) \end{aligned}$$

since by formula (1.13) of Ch. 1 and (4.15)

$$W^{-\alpha}\log b_0(w,\zeta) = -\Gamma(1-\alpha)\int_{-|\eta|}^{|\eta|}\frac{dt}{[i(w-\xi)-t]^{1-\alpha}}.$$

For $0 < \alpha < 1$ and $\operatorname{Im} w < \eta$ one can obtain the same equalities in a similar way. Thus

$$W^{\alpha}\varphi_\alpha(w,\zeta) = \log\frac{b_\alpha(w,\zeta)}{b_0(w,\zeta)}, \quad \operatorname{Im} w < \eta, \quad \alpha \in (-1,1).$$

Hence the desired formula follows by (4.5) and the absolute and uniform convergence of the series

$$\log\frac{B_\alpha(w,\{w_k\})}{b_0(w,\{w_k\})} = \sum_k \log\frac{b_\alpha(w,w_k)}{b_0(w,w_k)}$$

for $\operatorname{Im} w < -\max_k|\operatorname{Im} w_k|$ (see. (3.5)).

**Proof of Theorem 4.1.** In view of Lemmas 4.4 and 4.5, it is sufficient to prove the equality

$$W^{\alpha}\left\{\frac{1}{\pi i}\int_{-\infty}^{+\infty}\frac{d\mu(t)}{w-t}\right\} = \frac{\Gamma(1+\alpha)}{\pi}\int_{-\infty}^{+\infty}\frac{d\mu(t)}{[i(w-t)]^{1+\alpha}}, \quad w \in G^-,$$

under the assumption that $\mu(t)$ is as required. One can directly verify this equality using (4.15).

# CHAPTER 3

# EQUILIBRIUM RELATIONS AND FACTORIZATIONS

## 1. F. AND R. NEVANLINNA, CARLEMAN AND LEVIN FORMULAS

In this section F. and R. Nevanlinna type formulas are established in two starshaped domains: in the semidisc $G^+(R) = \{z : \text{Im } z > 0, |z| < R\}$ $(0 < R < +\infty)$ and the half-plane $G^-_\rho = \{w : \text{Im } w < \rho\}$ $(-\infty < \rho < 0)$. Then some limit passages lead to a Carleman type formula for the semidisc and a B.Ja.Levin type formula for a disc that is tangential to the real axis.

*1.1.* For $w \in \mathbb{C}$ we set

$$\Gamma^*[w, \infty) = \{\zeta = w + i\sigma : 0 \leq \sigma < +\infty\}.$$

Further, for any domain $D$ we denote by $D^*$ its $\infty$-starshaped continuation

$$D^* = \{\zeta = w - i\sigma : \ w \in D, 0 \leq \sigma < +\infty\}.$$

We shall use the following estimates for the products of Ch. 2:

$$|\log |B_\alpha(w)|| \leq \frac{2^{2+\alpha}}{1+\alpha} \left( \sum_k |\text{Im } w_k|^{1+\alpha} \right) |\text{Im } w|^{-1-\alpha}, \qquad (1.1)$$

$$\left| \frac{\partial}{\partial(\text{Im } w)} \log |B_\alpha(w)| \right| \leq 2^{3+\alpha} \left( \sum_k |\text{Im } w_k|^{1+\alpha} \right) |\text{Im } w|^{-2-\alpha} \qquad (1.1')$$

which are true for $\text{Im } w < -2 \max_k |\text{Im } w_k|$ and

$$|\log |B_\alpha(w)|| \leq \frac{2^{2+\alpha}}{1+\alpha} \left( \sum_k |\text{Im } w_k|^{1+\alpha} \right) |w|^{-1-\alpha}, \qquad (1.2)$$

$$\left| \frac{\partial}{\partial(\text{Im } w)} \log |B_\alpha(w)| \right| \leq 2^{3+\alpha} \left( \sum_k |\text{Im } w_k|^{1+\alpha} \right) |w|^{-2-\alpha} \qquad (1.2')$$

which are true for $\sup_k |w_k| = M < +\infty$ and $|w| \geq 4M$ $(w \in \overline{G^-})$. The proofs of these estimates are similar to that of inequality (3.6) in Ch. 2.

*1.2* **Lemma 1.1.** *Let $F(w)$ be meromorphic in a $\infty$-starshaped domain $D$, and let $F(w)$ have finite sets of zeros $a_k$ and poles $b_n$ in the $\infty$-starshaped continuation of any domain $D_1 \subset D$ for which $\overline{D_1}\backslash\{\infty\} \subset D$. Then:*

*1°. If $\log|F(w)| \in M_\alpha(D)$ for some $\alpha > 0$, then $W^{-\alpha}\log|F(w)|$ is continuous in $D$ and harmonic in $D\backslash\{[\cup_k\Gamma^*[a_k,\infty)] \cup [\cup_k\Gamma^*[b_n,\infty)]\}$.*

*2°. If $\partial/\partial(\text{Im } w)\log|F(w)| \in M_{1+\alpha}(D)$ for some $\alpha \in (-1,0)$, then the function $W^{-\alpha}\log|F(w)|$ is continuous in $D\backslash[\{a_k\}\cup\{b_n\}]$ and harmonic in $D\backslash\{[\cup_k\Gamma^*[a_k,\infty)] \cup [\cup_k\Gamma^*[b_n,\infty)]\}$. Besides, in a neighborhood of each point $A \in \{a_k\}\cup\{b_n\}$*

$$W^{-\alpha}\log|F(w)| = d_A\frac{\Gamma(1-\alpha)}{\alpha}|w-A|^\alpha\cos\{\alpha\arg[i(w-A)]\} + \psi_\alpha(w,A),$$

*where $d_a$ is the order of zero (then $d_a \geq 1$) or pole (then $d_a \leq -1$) in $A$, and $\psi_\alpha(w,A)$ is a harmonic function.*

**Proof.** Assuming $\varepsilon > 0$ and $\rho > 0$ arbitrary numbers, introduce the domains $D_\varepsilon = \{w \in D : \overline{O_\varepsilon(w)} = \{\zeta : |\zeta - w| \leq \varepsilon\} \subset D\}$ and $D_{\varepsilon,\rho} = D_\varepsilon \cap G_\rho^-$. Since $F(w)$ has a finite set of zeros and poles in $D_{\varepsilon,\rho}$, the function

$$U_\alpha(w) \equiv \log\left|\frac{\prod\limits_{b_n\in D_{\varepsilon,\rho}} b_\alpha(w-i\rho, b_n-i\rho)}{\prod\limits_{a_k\in D_{\varepsilon,\rho}} b_\alpha(w-i\rho, a_k-i\rho)}F(w)\right|, \quad -1<\alpha<+\infty,$$

is harmonic in $D_{\varepsilon,\rho}$, and

$$W^{-\alpha}\log|F(w)| \equiv W^{-\alpha}U_\alpha(w) + \sum_{a_k\in D_{\varepsilon,\rho}} W^{-\alpha}\log|b_\alpha(w-i\rho, a_k-i\rho)|$$

$$- \sum_{b_n\in D_{\varepsilon,\rho}} W^{-\alpha}\log|b_\alpha(w-i\rho, b_n-i\rho)|, \quad -1<\alpha<+\infty. \tag{1.3}$$

For proving 1°, observe that by (1.2)

$$|\log|b_\alpha(w-i\rho,\zeta-i\rho)|| < C(1+\alpha)\frac{2^{2+\alpha}}{1+\alpha}|\eta-\rho|^{1+\alpha}|w|^{-(1+a)}$$

($|w| > R_0$, $\text{Im } w < 0$) for enough great $R_0 > 0$ and any $\zeta = \xi + i\eta \in G_\rho^-$. Hence

$$\log|b_\alpha(w-i\rho,\zeta-i\rho)| \in M_\alpha(D_{\varepsilon,\rho}) \quad \text{and} \quad U_\alpha(w) \in M_\alpha(D_{\varepsilon,\rho}).$$

Consequently, by Lemma 1.8 (Ch. 1) $W^{-\alpha}U_\alpha(w)$ is harmonic in $D_{\varepsilon,\rho}$, and by and Theorem 3.2 of Ch. 2 $W^{-\alpha}\log|F(w)|$ is continuous in $D_{\varepsilon,\rho}$ and harmonic in $D_{\varepsilon,\rho}\backslash\{[\cup_k\Gamma^*[a_k,\infty)]\cup[\cup_n\Gamma^*[b_n,\infty)]\}$. Hence our assertion follows since $\varepsilon$ and $\rho$ are arbitrary. The proof of 2° is similar. It is based on the inclusions

$$\frac{\partial}{\partial \mathrm{Im}\ w}\log|b_\alpha(w-i\rho,\zeta-i\rho)|\in K_{2+\alpha}\left(\frac{\pi}{2,}D_{\varepsilon,\rho}\right)\subset M_{1+\alpha}(D_{\varepsilon,\rho})$$

which follow from (1.2′).

**Remark 1.1.** If the zeros and the poles of $F(w)$ from Lemma 1.1 are such that the sets $\{\cup_k\Gamma^*[a_k,\infty)\}$ and $\{\cup_n\Gamma^*[b_n,\infty)\}$ are disjoint, then (1.3) and the properties of $W^{-\alpha}\log|b_\alpha|$ (see Theorem 3.2 in Ch. 2) provide the validity of the following additional assertions:

**(a)** if $\alpha\in(0,+\infty)$, then $W^{-\alpha}\log|F(w)|$ is subharmonic in the points $w\in\{\cup_k\Gamma^*[a_k,\infty)\}$ and superharmonic in the points $w\in\{\cup_n\Gamma^*[b_n,\infty)\}$, i.e. for enough small $r>0$

$$\frac{1}{2\pi}\int_0^{2\pi}W^{-\alpha}\log|F(w+re^{i\vartheta})|d\vartheta\begin{cases}>W^{-\alpha}\log|F(w)|, & w\in\{\cup_k\Gamma^*[a_k,\infty)\},\\ >W^{-\alpha}\log|F(w)|, & w\in\{\cup_n\Gamma^*[b_n,\infty)\};\end{cases}$$

**(b)** if $\alpha\in(-1,0)$, then in the same sense the function $W^{-\alpha}\log|F(w)|$ is superharmonic in the points $w\in\{\cup_k\Gamma^*[a_k,\infty)\backslash\{a_k\}\}$ and subharmonic in the points $w\in\{\cup_n\Gamma^*[b_n,\infty)\backslash\{b_n\}\}$.

*1.3.* The following two lemmas will be necessary.

**Lemma 1.2.** *Let a function $u(z)$ be harmonic in the semidisc $G^+(R)=\{z:\mathrm{Im}\ z>0,\ |z|<R\}$ $(R>0)$ and be continuous in its closure, with possible exception of a finite set of points $\{c_m\}_1^q\subset\partial G^+(R)$ $(c_m\neq 0,\ 1\le m\le q)$ in neighborhoods of which*

$$\begin{aligned}&u(z)=d_m\log|z-c_m|+\Phi_m(z)\qquad or\\ &u(z)=E_m|z-c_m|^{-\gamma}|c_m z|^{\gamma}\cos\left\{\gamma\arg\left[i(z^{-1}-c_m^{-1}]\right\}+\Psi_m(z)\end{aligned}\tag{1.4}$$

*$(0\le\gamma<1)$, where $d_m$ and $E_m$ are real constants, and $\Phi_m(z)$, $\Psi_m(z)$ are harmonic functions. Then for $z\in G^+(R)$*

$$\begin{aligned}u(z)&=\frac{1}{2\pi}\int_{\partial G^+(R)}u(\zeta)\frac{\partial G(\zeta,z)}{\partial n}ds\\ &=\frac{1}{2\pi}\int_{\partial G^+(R)}u(\zeta)\left[\frac{1}{\zeta-z}-\frac{1}{\zeta-\overline{z}}-\left(\frac{z}{R^2-\zeta z}-\frac{\overline{z}}{R^2-\zeta\overline{z}}\right)\right]d\zeta,\end{aligned}\tag{1.5}$$

*where $G(\zeta,z)$ is the Green function of $G^+(R)$, $\partial/\partial n$ is the differentiation along inner normal, and $ds$ is the element of the curve length.*

**Proof.** $u(z)$ can be written as a Poisson integral in $\partial G^+(R)$ since it can have only integrable singularities (1.4) on $\partial G^+(R)$. For a similar situation, one can find a detailed proof in [38] (Theorem 1.1, Sec. 1, Ch. 1).

**Lemma 1.3.** *Let $U(w)$ be harmonic in the half-plane $G_\rho^-$ $(\rho < 0)$ and continuous in its closure $\overline{G_\rho^-} = \{w : \text{Im } w \le \rho\}$, with possible exception of a finite set of points $\{A_m\}_1^q \subset \partial G_\rho^-$ $(A_m \neq \infty, 1 \le m \le q)$ in neighborhoods of which*

$$U(w) = d_m \log |w - A_m| + \varphi_m(w) \qquad \textit{or}$$
$$U(w) = E_m |w - A_m|^{-\gamma} \cos\{\gamma[i(w - A_m)]\} + \psi_m(w) \quad (0 \le \gamma < 1) \qquad (1.4')$$

*where $d_m$ and $E_m$ are real constants, and $\varphi_m(w)$, $\psi_m(w)$ are harmonic functions. If*

$$\sup_{v<\rho} \int_{-\infty}^{+\infty} |U(u+iv)| du < +\infty,$$

*then*

$$\int_{-\infty}^{+\infty} |U(u+i\rho)| du < +\infty \tag{1.6}$$

*and*

$$U(w) = \frac{|v-\rho|}{\pi} \int_{-\infty}^{+\infty} \frac{U(u+i\rho)du}{(u-t)^2 + (v-\rho)^2} < +\infty, \;\; w = u + iv \in G_\rho^-. \tag{1.7}$$

**Proof.** The function

$$V(w) = U(w + i\rho), \quad w \in G^-, \tag{1.8}$$

is harmonic in $G^-$. Besides, by the properties of $U(w)$

$$V(w) = \frac{|v|}{\pi} \int_{-\infty}^{+\infty} \frac{d\mu(t)}{(u-t)^2 + v^2}, \qquad w = u + iv \in G^-, \tag{1.9}$$

where $\mu(t)$ is of bounded variation in $(-\infty, +\infty)$ and

$$\lim_{v \to -\infty} \int_{-\infty}^{+\infty} f(u) V(u+iv) du = \int_{-\infty}^{+\infty} f(u) d\mu(t)$$

for any $f(u)$ continuous in $(-\infty, +\infty)$ and such that $\lim_{u\to\pm\infty} f(u) = 0$ (see, for instance, [36], Ch. 1, Theorems 5.3 and 3.1(c)). It suffices to prove that in our case the measure $d\mu(t)$ is absolutely continuous and equals $V(t)dt$. To this end, for an arbitrary segment $[a, b] \in (-\infty, +\infty)$ introduce a sequence $\{f_n(u)\}_1^\infty$ of finite, continuous functions in the following way: $f_n(u) \equiv 1$ for $u \in [a, b]$, $f_n(u) \equiv 0$ for $u \in (-\infty, a - 1/n]$ and $u \in [b + 1/n, +\infty)$, and $f_n(u)$

to be continued to the remaining intervals as a linear function. Then, by the properties of $V(w)$

$$\int_{-\infty}^{+\infty} f_n(u)V(u)du = \lim_{v\to -0} \int_{-\infty}^{+\infty} f_n(u)V(u+iv)du = \int_{-\infty}^{+\infty} f_n(u)d\mu(u)$$

for $n \geq 1$. On the other hand,

$$\int_a^b V(u)du = \lim_{n\to\infty} \int_{-\infty}^{+\infty} f_n(u)V(u)du = \int_a^b d\mu(u).$$

Therefore, in (1.9) $d\mu(t) \equiv V(t)dt$ and $V(t) \in L^1(-\infty, +\infty)$. Hence our assertion follows by the definition of $V(t)$.

*1.4.* Now we turn to F. and R. Nevanlinna type formulas.

**Theorem 1.1.** *Let $f(z)$ be meromorphic in the closed half-plane $\overline{G^+}\backslash\{0\} = \{z : \operatorname{Im} z \geq 0,\ z \neq 0\}$, and let for some $R_0 > 0$ this function has no zeros and poles in the semidisc $\overline{G^+(R_0)}\backslash\{0\} = \{z : \operatorname{Im} z \geq 0,\ |z| \leq R_0,\ z \neq 0\}$. Further, let $\alpha > -1$ and $\kappa > 0$ be any numbers for which one of the following conditions is satisfied:*

1°. $\log|f(z)| \in \widetilde{K}_{\alpha+\kappa}\{\pi/2, G^+\}$, *if* $\alpha \in [0, +\infty)$,

2°. $\partial/\partial(\operatorname{Im} 1/z) \log|f(z)| \in \widetilde{K}_{1+\alpha+\kappa}\{\pi/2, G^+\}$, *if* $\alpha \in (-1, 0)$.

*Then for any $-1 < \alpha < +\infty$, any $R > 0$ and any $z \in G^+(R)$ which does not belong to the set of zeros $\{a_k\}$ or to the set of poles $\{b_n\}$ of $f(z)$,*

$$\sum_{|a_k|<R} \left\{ \frac{1}{2\pi} \int_{\partial G^+(R)} \widetilde{W}^{-\alpha} \log|\widetilde{b}_\alpha(\zeta, a_k)| \frac{\partial G(\zeta, z)}{\partial n} ds - \widetilde{W}^{-\alpha} \log|\widetilde{b}_\alpha(\zeta, a_k)| \right\}$$
$$- \sum_{|b_n|<R} \left\{ \frac{1}{2\pi} \int_{\partial G^+(R)} \widetilde{W}^{-\alpha} \log|\widetilde{b}_\alpha(\zeta, b_n)| \frac{\partial G(\zeta, z)}{\partial n} ds - \widetilde{W}^{-\alpha} \log|\widetilde{b}_\alpha(\zeta, b_n)| \right\}$$
$$= \frac{1}{2\pi} \int_{\partial G^+(R)} \widetilde{W}^{-\alpha} \log|f(\zeta)| \frac{\partial G(\zeta, z)}{\partial n} ds - \widetilde{W}^{-\alpha} \log|f(\zeta)|, \qquad (1.10)$$

*where $G(\zeta, z)$ is the Green function of $G^+(R)$.*

**Proof.** The function

$$f_\alpha(z) = \frac{\prod_{|b_n|<R} \widetilde{b}_\alpha(z, b_n)}{\prod_{|a_k|<R} \widetilde{b}_\alpha(z, a_k)} f(z) \quad (-1 < \alpha < +\infty) \qquad (1.11)$$

is meromorphic in $\overline{G^+(R)}\backslash\{0\}$ and it can have only a finite set of zeros and poles $\{A_m\}_1^q \subset \partial G^+(R)$. Therefore, by Lemma 1.1 $\widetilde{W}^{-\alpha} \log|f_\alpha(z)|$ is harmonic in

$\overline{G^+(R)}\backslash\{0\}$, except the points $\{A_m\}_1^q$, where one of representations (1.4) holds (the first one for $\alpha = 0$, the second one with $\gamma = |\alpha|$ for $-1 < \alpha < 0$, and the second one with $\gamma = 0$ for $0 < \alpha < +\infty$). On the other hand, by (1.2)–(1.2′) for $s \in G^+$

$$\log|\widetilde{b}_\alpha(z,s)| \in \widetilde{K}_{1+\alpha}\{\pi/2, G^+\} \quad \text{if} \quad 0 \leq \alpha < +\infty$$
$$\partial/\partial(\text{Im } 1/z) \log|\widetilde{b}_\alpha(z,s)| \in \widetilde{K}_{2+\alpha}\{\pi/2, G^+\} \quad \text{if} \quad -1 < \alpha < 0.$$

Hence, by Lemma 1.9 of Ch. 1 and our requirements in both cases 1° and 2°, $\widetilde{W}^{-\alpha} \log|f_\alpha(z)| \to 0$ as $z \to 0$ remaining in $\overline{G^+}$, i.e. $\widetilde{W}^{-\alpha} \log|f_\alpha(z)|$ is continuous at $z = 0$. Thus, this function satisfies the conditions of Lemma 1.2, it admits the representation (1.5) which is easily transformed to (1.10).

Note that for $\alpha = 0$ (1.10) coincides with R. and F. Nevanlinna formula (see, for instance, [38], Ch. 1, Theorem 2.1) for the semidisc $G^+(R)$. Besides, if $\alpha > 0$, then (1.10) is true in all points $z \in G^+(R)$. Also the following useful remark is true.

**Remark 1.2.** For $\alpha \geq 0$ the function $\widetilde{W}^{-\alpha} \log|\widetilde{b}_\alpha(z,s)|$ is subharmonic in $G^+$. Hence, *for $\alpha \geq 0$ the terms of left-hand side sums in (1.10) are nonnegative.*

**Theorem 1.2.** *Let $F(w)$ be meromorphic in the lower half-plane $G^-$, and let for some $\rho < 0$ this function have not more than a finite set of zeros and poles in the closed half-plane $\overline{G_\rho^-} = \{w : \text{Im } w \leq \rho\}$. Further, let $\alpha \in (-1, +\infty)$ be any number and*

1°. $\log|F(w)| \in M_\alpha\{G^-\}$ *if* $\alpha \in [0, +\infty)$,

2°. $\partial/\partial(\text{Im } w) \log|F(w)| \in M_{1+\alpha}\{G^-\}$ *if* $\alpha \in (-1, 0)$.

*If*

$$\sup_{v<\rho} \int_{-\infty}^{+\infty} \left|W^{-\alpha} \log|F(u+iv)|\right| du < +\infty,$$

*then $W^{-\alpha} \log|F(u+i\rho)| \in L^1(-\infty, +\infty)$, and for any $w = u + iv \in G_\rho^-$, which does not belong the set of zeros $\{a_k\}$ or to the set of poles $\{b_n\}$ of $F(w)$*

$$- \sum_{\text{Im } a_k < \rho} W^{-\alpha} \log|b_\alpha(w - i\rho, a_k - i\rho)| + \sum_{\text{Im } b_n < \rho} W^{-\alpha} \log|b_\alpha(w - i\rho, b_n - i\rho)|$$
$$= \frac{|v - \rho|}{\pi} \int_{-\infty}^{+\infty} \frac{W^{-\alpha} \log|F(t + i\rho)|}{(u - t)^2 + (v - \rho)^2} dt - W^{-\alpha} \log|F(w)|, \ \alpha > -1. \tag{1.12}$$

**Proof.** The function

$$F_\alpha(w) = \frac{\prod_{\text{Im } b_n < \rho} b_\alpha(w - i\rho, b_n - i\rho)}{\prod_{\text{Im } a_k < \rho} b_\alpha(w - i\rho, a_k - i\rho)} F(w) \quad (-1 < \alpha < +\infty) \tag{1.13}$$

is meromorphic in $\overline{G_\rho^-}$, except the sets $\{\text{Re } a_k + i\rho : \text{Im } a_k < \rho\} \subset \partial G_\rho^-$ and $\{\text{Re } b_n + i\rho : \text{Im } b_n < \rho\} \subset \partial G_\rho^-$. And this function can have not more than a finite set of zeros and poles $\{q_m\}_1^N \subset \partial G_\rho^-$ ($q_m \neq \infty$, $1 \leq m \leq N$) which are some of zeros and poles of $F(w)$. By Lemma 1.1 and Theorem 3.2 in Ch. 2, $W^{-\alpha} \log |F_\alpha(w)|$ is harmonic in $\overline{G_\rho^-}$, except a finite set of points on $\partial G_\rho^-$, where one of the representations (1.4′) is valid. Besides, by (3.14) of Ch. 2

$$\sup_{v<\rho} \int_{-\infty}^{+\infty} \left| W^{-\alpha} \log |F_\alpha(u+iv)| \right| du < +\infty.$$

Thus, $W^{-\alpha} \log |F_\alpha(w)|$ satisfies the conditions of Lemma 1.3. Consequently, $W^{-\alpha} \log |F(u+i\rho)| \in L^1(-\infty, +\infty)$ and (1.12) holds. This completes the proof.

Note that for $\alpha = 0$ (1.12) becomes the formula of F. and R. Nevanlinna (see, for instance, [38], Ch. 1, Theorem 1.1) in $G_\rho^-$. Besides, for $\alpha > 0$ formula (1.12) is true in all points $w \in G_\rho^-$. Besides, formula (1.10) of Theorem 1.1 as well as formula (1.12) of Theorem 1.2 can be written in equivalent forms by means of inversion $w = z^{-1}$ and similar changes of variables.

*1.5.* The following two lemmas will be used later for proving a Carleman type formula. For any $\alpha \in (-1, +\infty)$, $r > 0$ and $w, \zeta \in G^-(r) = \{w : \text{Im } w < 0, |w| > r\}$ we set

$$\gamma_r = \{s = re^{i\vartheta} : -\pi \leq \vartheta \leq 0\}, \quad (\gamma_r = [\Gamma_{r^{-1}}]^{-1}) \tag{1.14}$$

(we assume that this arc is directed clockwise) and

$$I_\alpha(w, \zeta, r) \equiv \frac{1}{2\pi i} \int_{\gamma_r} W^{-\alpha} \log |b_\alpha(s, \zeta)| \times \left[ \frac{1}{s - w} - \frac{1}{s - \overline{w}} - \left( \frac{w}{r^2 - sw} - \frac{\overline{w}}{r^2 - \overline{w}} \right) \right] ds. \tag{1.15}$$

**Lemma 1.4.** *For any $\alpha \in (-1, +\infty)$, $r > 0$ and $\zeta = \xi + i\eta \in G^-(r)$*

$$I_\alpha(\zeta, r) \equiv \frac{1}{2} \lim_{v \to -\infty} |v| I_\alpha(iv, \zeta, r) = \frac{r}{\pi} \int_0^{-\pi} W^{-\alpha} \log b_\alpha(re^{i\vartheta}, \zeta) \sin \vartheta d\vartheta, \tag{1.16}$$

$$\frac{1}{2} \lim_{v \to -\infty} |v| W^{-\alpha} \log |b_\alpha(iv, \zeta)| = -\frac{|\eta|^{1+\alpha}}{\Gamma(2+\alpha)}. \tag{1.17}$$

**Proof.** The function $W^{-\alpha} \log b_\alpha(s, \zeta)$ is continuous by $s$ in $\overline{G^-}$, with possible exception of the point $\zeta \in G^-(r)$ (see Theorem 3.2 in Ch. 2). Therefore, using the equality

$$\lim_{v \to -\infty} |v| \left[ \frac{1}{s - iv} - \frac{1}{s + iv} - \left( \frac{iv}{r^2 - isv} + \frac{iv}{r^2 + isv} \right) \right] = 2i \left( \frac{r^2}{s^2} - 1 \right) \tag{1.18}$$

we get

$$I_\alpha(\zeta,r)=\frac{1}{2\pi}\int_{\gamma_r}W^{-\alpha}\log|b_\alpha(s,\zeta)|\left(\frac{r^2}{s^2}-1\right)ds$$
$$=\frac{r}{\pi}\int_0^{-\pi}W^{-\alpha}\log|b_\alpha(re^{i\vartheta},\zeta)|\sin\vartheta d\vartheta.$$

On the other hand, by formula (2.2) of Ch. 2

$$\frac{1}{2}\lim_{v\to-\infty}|v|W^{-\alpha}\log|b_\alpha(iv,\zeta)|=\frac{1}{2}\lim_{v\to-\infty}\frac{|v|}{\Gamma(1+\alpha)}\mathrm{Re}\int_{-|\eta|}^{|\eta|}\frac{(|\eta|-|t|)^\alpha dt}{t-|v|+i\xi}$$
$$=-\frac{|\eta|^{1+\alpha}}{\Gamma(2+\alpha)}.$$

**Lemma 1.5.** *For any* $\alpha>-1$, $r>0$ *and* $\zeta=\xi+i\eta\in G^-(r)$

$$I_\alpha(\zeta,r)=\frac{r^2}{\Gamma(1+\alpha)}\mathrm{Re}\int_0^{|\eta|}\frac{\tau^\alpha d\tau}{(\tau-i\zeta)^2}\quad \textit{if}\quad |\xi|\geq r, \tag{1.19}$$

$$I_\alpha(\zeta,r)=\frac{r^2}{\Gamma(1+\alpha)}\mathrm{Re}\int_0^{|\eta|-\sqrt{r^2-\xi^2}}\frac{\tau^\alpha d\tau}{(\tau-i\zeta)^2}$$
$$-\frac{1}{\Gamma(2+\alpha)}\left[|\eta|^{1+\alpha}-\left(|\eta|-\sqrt{r^2-\xi^2}\right)^{1+\alpha}\right]\quad \textit{if}\ |\xi|<r. \tag{1.19'}$$

**Proof.** Since $W^{-\alpha}\log|b_\alpha(re^{i\vartheta},\zeta)|=\mathrm{Re}\left\{W^{-\alpha}\log b_\alpha(re^{i\vartheta},\zeta)\right\}$, (1.16) can be written in the form

$$I_\alpha(\zeta,r)=\frac{1}{2\pi}\mathrm{Re}\int_{\gamma_r}W^{-\alpha}\log b_\alpha(s,\zeta)\left(\frac{r^2}{s^2}-1\right)ds. \tag{1.20}$$

Substitute the representation of $W^{-\alpha}\log b_\alpha(s,\zeta)$ as a Cauchy type integral (i.e. formula (2.2) of Ch. 2) and change the order of integration (for the validity of this operation see, for instance, [34], Ch. 1, § 7). This gives

$$I_\alpha(\zeta,r)=\frac{1}{2\pi\Gamma(1+\alpha)}\int_{|-\eta|}^{|\eta|}\mathrm{Re}\left\{\int_{\gamma_r}\left(\frac{r^2}{s^2}-1\right)\frac{ds}{t-i(s-\xi)}\right\}(|\eta|-|t|)^\alpha dt.$$

Changing the variables: $|\eta|-t=\tau$ in $\int_0^{|\eta|}$ and $|\eta|+t=\tau$ in $\int_{-|\eta|}^0$, we get

$$I_\alpha(\zeta,r)=\frac{1}{\Gamma(1+\alpha)}\int_0^{|\eta|}\mathrm{Re}\left\{J(\tau,\zeta)\right\}\tau^\alpha d\tau, \tag{1.21}$$

where

$$J(\tau,\zeta)=\frac{1}{2\pi}\int_{|s|=1}\left(\frac{1}{s^2}-1\right)\frac{ds}{s-(\zeta+i\tau)r^{-1}}. \tag{1.21'}$$

Now set

$$J(\tau,\zeta) = \frac{1}{2\pi}\int_{|s|=1}\frac{ds}{s^2[s-(\zeta+i\tau)r^{-1}]} - \frac{1}{2\pi}\int_{|s|=1}\frac{ds}{s-(\zeta+i\tau)r^{-1}}$$
$$\equiv J_1(\tau,\zeta)+J_2(\tau,\zeta), \tag{1.22}$$

and calculate $J_1$ and $J_2$ in the assumption that $\tau \in (0,|\eta|)$. Two cases are to be considered.

**(a)** $|\xi| \equiv |\mathrm{Re}\ \zeta| \geq r$. The integrands in $J_1$ and $J_2$ have first order poles at $s_0 = (\zeta+i\tau)r^{-1}$. Besides, the integrand of $J_2$ has a second order pole at $s=0$. Consequently,

$$J_1(\tau,\zeta) = \operatorname*{res}_{s=0}\frac{1}{s^2[s-(\zeta+i\tau)r^{-1}]} = \frac{r^2}{(\tau-i\zeta)^2}, \quad J_2(\tau,\zeta)=0,$$

since in case $|s_0| > |\mathrm{Re}\ s_0| = |\xi r^{-1}| \geq 1$. Hence (1.19) holds by (1.22) and (1.21)–(1.21′).

**(b)** $0 \leq |\xi| \equiv |\mathrm{Re}\ \zeta| < r$. In this case the pole $s_0 = (\zeta+i\tau)r^{-1}$ of the integrands in $J_1$ and $J_2$ is out of the unit disc for $0<\tau<|\eta|-\sqrt{r^2-\xi^2}$, and it lies inside the unit disc for $|\eta|-\sqrt{r^2-\xi^2}<\tau<|\eta|$. Hence

$$J(\tau,\zeta) = \frac{r^2}{(\tau-i\zeta)^2}, \qquad 0\leq|\xi|<r, \quad 0<\tau<|\eta|-\sqrt{r^2-\xi^2} \tag{1.23}$$

if $|\eta|-\sqrt{r^2-\xi^2}<\tau<|\eta|$. And

$$J_1(\tau,\zeta) = \left(\operatorname*{res}_{s=0}+\operatorname*{res}_{s=s_0}\right)\frac{1}{s^2[s-(\zeta+i\tau)r^{-1}]} = 0,$$

$$J_2(\tau,\zeta) = -\operatorname*{res}_{s=s_0}\frac{1}{s-(\zeta+i\tau)r^{-1}} = -1$$

if $|\eta|-\sqrt{r^2-\xi^2}<\tau<|\eta|$. Thus, by (1.21)

$$J(\tau,\zeta) = -1, \qquad 0\leq|\xi|<r, \quad |\eta|-\sqrt{r^2-\xi^2}<\tau<|\eta|. \tag{1.23′}$$

Formulas (1.22), (1.22′), (1.20) and (1.20′) imply (1.19′).

**Remark 1.3.** For $\alpha=0$ (1.19) and (1.19′) coincide and become

$$I_0(\zeta,r) = -r^2\frac{|\eta|}{\zeta^2}. \tag{1.24}$$

Besides, if $|\xi| \geq r$, then $|i\zeta - \tau| \geq r$ for $\tau \in [0, |\eta|]$. And if $|\xi| < r$, then $|i\zeta - \tau| \geq r$ for $\tau \in [0, |\eta| - \sqrt{r^2 - \xi^2}]$. Hence, one can obtain from (1.19) and (1.19′) that for any $\alpha \in (-1, +\infty)$ and $\xi \in (-\infty, +\infty)$

$$|I_\alpha(\zeta, r)| \leq \frac{|\eta|^{1+\alpha}}{\Gamma(2+\alpha)}, \quad \zeta = \xi + i\eta \in G^-, \quad |\zeta| > r. \tag{1.25}$$

*1.6.* **Theorem 1.3.** *Let a function $f(z)$ be meromorphic in the closed half-plane $\overline{G^+}\backslash\{0\} = \{z : \text{Im } z \geq 0,\ z \neq 0\}$, and let for some $R_0 > 0$ this function have no zeros and poles in the semidisc $\overline{G^+(R_0)}\backslash\{0\} = \{z : \text{Im } z \geq 0,\ |z| \leq R_0,\ z \neq 0\}$. Further, let $\alpha \in (-1, +\infty)$ and $\kappa > 1$ be any numbers, and let*

1°. $\log|f(z)| \in \widetilde{K}_{\alpha+\kappa}\{\pi/2, G^+\}$ *if* $\alpha \in [0, +\infty)$,

2°. $\partial/\partial(\text{Im } 1/z) \log|f(z)| \in \widetilde{K}_{1+\alpha+\kappa}\{\pi/2, G^+\}$, *if* $\alpha \in (-1, 0)$.

*Then for any $R > 0$*

$$\sum_{|a_k|<R} \left\{ \frac{1}{\Gamma(2+\alpha)} \left| \text{Im } \frac{1}{a_k} \right|^{1+\alpha} + I_\alpha\left(\frac{1}{a_k}, \frac{1}{R}\right) \right\} - \sum_{|b_n|<R} \left\{ \frac{1}{\Gamma(2+\alpha)} \left| \text{Im } \frac{1}{b_n} \right|^{1+\alpha} + I_\alpha\left(\frac{1}{b_n}, \frac{1}{R}\right) \right\}$$
$$= \frac{1}{\pi R} \int_0^\pi \widetilde{W}^{-\alpha} \log|f(Re^{i\vartheta})| \sin\vartheta d\vartheta + \frac{1}{2\pi} \int_0^R \left(\frac{1}{t^2} - \frac{1}{R^2}\right) \widetilde{W}^{-\alpha} \log|f(t)f(-t)| dt, \tag{1.26}$$

*where $a_k$ are the zeros and $b_n$ are the poles of $f(z)$ and the terms left-hand side of sums are nonnegative.*

**Proof.** $f(z)$ satisfies the conditions of Theorem 1.1. Hence (1.10) is true. Taking $z = iy$ $(0 < y < R)$ in (1.10), let $y \to +0$ after multiplying both its sides by $(2y)^{-1}$. This gives (1.26) by Lemma 1.4 and (1.17). It remains to see that the terms of sums in (1.26) are nonnegative by (1.25).

**Remark 1.4.** Since $\widetilde{W}^0$ is identical and (1.24) is true, for $\alpha = 0$ formula (1.26) coincides with the well-known Carleman formula for the semidisc (see [10], Sec. 1.2.2 or [77], Ch. V, formula (5.02′)).

**Theorem 1.4.** *Let $F(w)$ be meromorphic in $G^- = \{w : \text{Im } w < 0\}$, and let the sets of zeros and poles of $F(w)$ in the closed half-plane $\overline{G_\rho^-} = \{w : \text{Im } w \leq \rho\}$ be finite for some $\rho < 0$. Further, let $\alpha \in (-1, +\infty)$ be any number, and let*

1°. $\log|F(w)| \in M_\alpha\{G^-\}$ *if* $\alpha \in [0, +\infty)$,

2°. $\partial/\partial(\text{Im } w)\log|F(w)| \in M_{1+\alpha}\{G^-\}$ *if* $\alpha \in (-1,0)$.

*If*

$$\sup_{v<\rho} \int_{-\infty}^{+\infty} \left|W^{-\alpha} \log|F(u+iv)|\right| du < +\infty, \tag{1.27}$$

*then*

$$W^{-\alpha} \log|F(u+i\rho)| \in L^1(-\infty,+\infty) \tag{1.27'}$$

*and for any* $-1 < \alpha < +\infty$

$$\frac{1}{\Gamma(2+\alpha)} \sum_{\text{Im } a_k<\rho} (|\text{Im } a_k| + \rho)^{1+\alpha} - \frac{1}{\Gamma(2+\alpha)} \sum_{\text{Im } b_n<\rho} (|\text{Im } b_n| + \rho)^{1+\alpha}$$

$$= \frac{1}{2\pi} \int_{-\infty}^{+\infty} W^{-\alpha} \log|F(u+i\rho)| du - \frac{1}{2} \lim_{v\to-\infty} W^{-\alpha} \log|F(iv)|, \tag{1.28}$$

*where* $a_k$ *and* $b_n$ *are zeros and poles of* $F(w)$, *and the limit exists and is finite.*

**Proof.** $F(w)$ satisfies the conditions of Theorem 1.2. Hence (1.27 ′) and (1.12) hold. Taking $w = iv$ $(v < \rho)$ in (1.12), multiply both sides by $|v|/2$ and let $v \to -\infty$. In view of (1.17) this passage gives (1.28).

**Remark 1.5.** For $\alpha = 0$ and $w = 1/iz$ (1.28) becomes a version of B.Ja.Levin's formula (see [77], Ch. IV, § 2).

## 2. SOME OTHER FORMULAS OF CARLEMAN AND LEVIN TYPE

Here we prove some other similarities of Carleman and Levin formulas which will be written in a semi-ring and in the domain $\{z : |z - iR/2| < R/2, |z| > R_0\}$ $(0 < R_0 < R < +\infty)$. One has to mention that in some cases these domains turn to be more preferable (see, for instance, [38], Ch. 1, Theorems 3.1 and 3.4). Also some applications of obtained formulas are given.

*2.1.* We start by two preliminary lemmas.

**Lemma 2.1.** *For any* $r, r_0$ $(0 < r < r_0 < +\infty)$, *any* $\zeta = \xi + i\eta \in G^-$ $(r < |\zeta| < r_0)$ *and any* $-1 < \alpha < +\infty$

$$Q^{(1)}(r, r_0, W^{-\alpha} \log|b_\alpha(w,\zeta)|) \equiv \frac{1}{2\pi} \int_{-\pi}^{0} \left\{ W^{-\alpha} \log|b_\alpha(r_0 e^{i\vartheta},\zeta)| \left(r_0 + \frac{r^2}{r_0}\right)\right.$$

$$\left. -(r_0^2 - r^2)\frac{\partial}{\partial r_0} W^{-\alpha} \log|b_\alpha(r_0 e^{i\vartheta},\zeta)| \right\} \sin\vartheta d\vartheta = \frac{|\eta|^{1+\alpha}}{\Gamma(2+\alpha)}. \tag{2.1}$$

**Proof.** Set

$$Q^{(1)}(r, r_0, \alpha, \zeta) = I_\alpha^{(1)} + I_\alpha^{(2)}, \tag{2.2}$$

$$I_\alpha^{(1)} = \frac{1}{2\pi} \int_{-\pi}^{0} W^{-\alpha} \log|b_\alpha(r_0 e^{i\vartheta},\zeta)| \left(r_0 + \frac{r^2}{r_0}\right) \sin\vartheta d\vartheta,$$

$$I_\alpha^{(2)} = -\frac{1}{2\pi} \int_{-\pi}^{0} (r_0^2 - r^2)\frac{\partial}{\partial r_0} W^{-\alpha} \log|b_\alpha(r_0 e^{i\vartheta},\zeta)| \sin\vartheta d\vartheta. \tag{2.2'}$$

For writing $I_\alpha^{(1)}$ and $I_\alpha^{(2)}$ in more suitable forms, observe that $|\zeta| < r_0$, and by formula (2.2) of Ch. 2 $W^{-\alpha} \log b_\alpha(w, \zeta)$ is holomorphic outside $[\zeta, \overline{\zeta}]$. Integrating by parts and using Cauchy–Riemann equalities we get

$$I_\alpha^{(1)} = \frac{1}{2\pi} \int_{-\pi}^{0} \text{Re}\,\{W^{-\alpha} \log b_\alpha(r_0 e^{i\vartheta}, \zeta)\} \text{Im}\, \left( r_0 e^{i\vartheta} - \frac{r^2}{r_0 e^{i\vartheta}} \right) d\vartheta,$$

$$I_\alpha^{(2)} = \frac{1}{2\pi} \int_{-\pi}^{0} \text{Im}\,\{W^{-\alpha} \log b_\alpha(r_0 e^{i\vartheta}, \zeta)\} \text{Re}\, \left( r_0 e^{i\vartheta} - \frac{r^2}{r_0 e^{i\vartheta}} \right) d\vartheta. \qquad (2.3)$$

From (2.2) of Ch. 2 it follows that the function $\text{Re}\,\{W^{-\alpha} \log b_\alpha(r_0 e^{i\vartheta}, \zeta)\}$ is odd and $\text{Im}\,\{W^{-\alpha} \log b_\alpha(r_0 e^{i\vartheta}, \zeta)\}$ is even for $\vartheta \in [-\pi, \pi]$. Other multipliers in the integrands of (2.3) have similar properties. Using these propoerties we come to some new representations for $I_\alpha^{(1)}$ and $I_\alpha^{(2)}$, which differ from (2.3) by the multiplier $1/4\pi$ instead of $1/2\pi$ and by the integration contour that becomes the whole $[-\pi, \pi]$. This results in the representation

$$I_\alpha^{(1)} + I_\alpha^{(2)} = \frac{1}{2} \text{Im}\, \left\{ \frac{1}{2\pi i} \int_{|s|=r_0} \left( 1 - \frac{r^2}{s^2} \right) W^{-\alpha} \log b_\alpha(s, \zeta) ds \right\}.$$

Inserting here the formula (2.2) of Ch. 2, changing the integration order and then calculating residues, we arrive at (2.1).

**Lemma 2.2.** *For any $\rho < 0$, $r_0 > |\rho|$ any $\zeta = \xi + i\eta \in G_\rho^-$ ($|\zeta| < r_0$) and any $-1 < \alpha < +\infty$*

$$Q^{(2)}(\rho, r_0, W^{-\alpha} \log |b_\alpha(w - i\rho, \zeta - i\rho)|)$$
$$\equiv \frac{1}{2\pi} \int_{-\pi + \arcsin \frac{|\rho|}{r_0}}^{-\arcsin \frac{|\rho|}{r_0}} \left\{ W^{-\alpha} \log |b_\alpha(r_0 e^{i\vartheta} - i\rho, \zeta - i\rho)| \sin \vartheta - (r_0 \sin \vartheta - \rho) \right.$$
$$\left. \times \frac{\partial}{\partial r_0} W^{-\alpha} \log |b_\alpha(r_0 e^{i\vartheta} - i\rho, \zeta - i\rho)| \right\} r_0 d\vartheta = \frac{(|\eta| + \rho)^{1+\alpha}}{\Gamma(2+\alpha)}. \qquad (2.4)$$

**Proof.** An argument similar to the proof of the previous lemma gives

$$Q^{(2)}(\rho, r_0, W^{-\alpha} \log |b_\alpha(w - i\rho, \zeta - i\rho)|) = -\frac{1}{4\pi} \text{Re}\, \int_\Gamma W^{-\alpha} \log b_\alpha(s, \zeta - i\rho) ds,$$

where $\Gamma$ is the union of the arcs $\gamma^+(\rho, r_0)$ (i.e. the part of $|s - i\rho| = r_0$ in $\overline{G^+}$) and $\gamma^-(\rho, r_0)$ (i.e. the part of $|s + i\rho| = r_0$ in $\overline{G^-}$). Further, one has to insert into this formula the representation (2.2) (Ch. 2) of $W^{-\alpha} \log b_\alpha(s, \zeta)$, to change the integration order and to count residues.

*2.2.* **Theorem 2.1.** *Let a function $f(z)$ be meromorphic in the closed half-plane $\overline{G^+} \backslash \{0\} = \{z : \text{Im}\, z \geq 0, z \neq 0\}$, and let for some $R_0 > 0$ this function* *have no zeros and poles in the semidisc* $\{z : \operatorname{Im} z > 0, |z| \le R_0\} \cup \{-R_0, R_0\}$. *Further, let* $\alpha \in (-1, +\infty)$ *be any number and let*

$1^\circ$. $\log|f(z)| \in \widetilde{M}_\alpha\{G^+\}$ *if* $\alpha \in [0, +\infty)$,

$2^\circ$. $\partial/\partial(\operatorname{Im} 1/z)\log|f(z)| \in \widetilde{M}_{1+\alpha}\{G^+\}$ *if* $\alpha \in (-1, 0)$.

*Then for any* $R > R_0$ *and any* $-1 < \alpha < +\infty$

$$\sum_{R_0<|a_k|<R}\left\{\frac{1}{\Gamma(2+\alpha)}\left|\operatorname{Im}\frac{1}{a_k}\right|^{1+\alpha}+I_\alpha\left(\frac{1}{a_k},\frac{1}{R}\right)\right\}$$
$$-\sum_{R_0<|b_n|<R}\left\{\frac{1}{\Gamma(2+\alpha)}\left|\operatorname{Im}\frac{1}{b_n}\right|^{1+\alpha}+I_\alpha\left(\frac{1}{b_n},\frac{1}{R}\right)\right\}$$
$$=\frac{1}{\pi R}\int_0^\pi \widetilde{W}^{-\alpha}\log|f(Re^{i\vartheta})|\sin\vartheta d\vartheta$$
$$+\frac{1}{2\pi}\int_{R_0}^{R}\left(\frac{1}{t^2}-\frac{1}{R^2}\right)\widetilde{W}^{-\alpha}\log|f(t)f(-t)|dt+\widetilde{Q}^{(1)}, \tag{2.9}$$

*where* $a_k$ *are the zeros and* $b_n$ *are the poles of* $f(z)$, *the terms of sums are nonnegative and as* $R \to +\infty$

$$\widetilde{Q}^{(1)} \equiv Q^{(1)}(R, R_0, \widetilde{W}^{-\alpha}\log|f|) \equiv -\frac{R_0}{2\pi}\int_0^\pi\left\{\widetilde{W}^{-\alpha}\log|f(R_0e^{i\vartheta})|\left(\frac{1}{R_0^2}+\frac{1}{R^2}\right)\right.$$
$$\left.+\left(\frac{1}{R_0}-\frac{R_0}{R^2}\right)\frac{\partial}{\partial R_0}\widetilde{W}^{-\alpha}\log|f(R_0e^{i\vartheta})|\right\}\sin\vartheta d\vartheta = O(1). \tag{2.10}$$

**Proof.** It is evident that $f_\alpha(z)$ defined by (1.11) is meromorphic in the closed semidisc $\overline{G^+(R)}\backslash\{0\} = \{z : \operatorname{Im} z \ge 0, |z| \le R, z \ne 0\}$, where it can have only a finite set of zeros and poles $\{A_m\}_1^q \subset \partial G^+(R)\backslash\{0\}$. Further, $F_\alpha(w) \equiv f_\alpha(w^{-1})$ satisfies the conditions of Lemma 1.1. Therefore, $u(z) = \widetilde{W}^{-\alpha}\log|f_\alpha(z)|$ is harmonic in $\overline{G^+(R)}\backslash\{0\}$, except some points $\{A_m\}_1^q$, in neighborhoods of which one of representations (1.4) is true, as these are integrable singularities. Repeating the argument which proved Theorem 3.2 in Ch. I of [38], one can verify that $u(z)$ admits the representation (3.2) of that theorem, i.e.

$$\frac{1}{2\pi}\int_{R_0}^{R}\left(\frac{1}{t^2}-\frac{1}{r^2}\right)[u(t)+u(-t)]dt+\frac{1}{\pi R}\int_0^\pi u(Re^{i\vartheta})\sin\vartheta d\vartheta$$
$$= -\widetilde{Q}^{(1)}(R, R_0, u),$$

where

$$\widetilde{Q}^{(1)}(R, R_0, u) = -\frac{R_0}{2\pi}\int_0^\pi\left\{u(R_0e^{i\vartheta})\left(\frac{1}{R_0^2}+\frac{1}{R^2}\right)\right.$$
$$\left.+\left(\frac{1}{R_0}-\frac{R_0}{R^2}\right)\frac{\partial}{\partial R_0}u(R_0e^{i\vartheta})\right\}\sin\vartheta d\vartheta.$$

Taking here $u(z) = \widetilde{W}^{-\alpha} \log|f_\alpha(z)|$ and granting that

$$\widetilde{Q}^{(1)}(R, R_0, \widetilde{W}^{-\alpha} \log|b_\alpha(z,\zeta)|) = Q^{(1)}(R^{-1}, R_0^{-1}, W^{-\alpha} \log|b_\alpha(z^{-1}, \zeta^{-1})|),$$

we come to (2.9). It remains to observe that by Theorem 1.3 the terms of the sums in (2.9) are nonnegative and (2.10) is true.

**Remark 2.1.** As $\widetilde{W}^0$ is identical, from (1.24) it follows that formula (2.9) for $\alpha = 0$ coincides with the Carleman formula in the semi-ring (see, for instance, [38], Ch. I, formula (2.1)).

**Theorem 2.2.** *Let $f(z)$ be meromorphic in the upper half-plane $G^+$ and also in the neighborhoods of points $-R_0$ and $R_0$ for some $R_0 > 0$. Further, let $f(z)$ have no zeros and poles in the semidisc $\{z : \text{Im } z > 0, |z| \leq R_0\}$, let $\alpha \in (-1, +\infty)$ be any number and let*

1°. $\log|f(z)| \in \widetilde{M}_\alpha\{G^+\}$ *if* $\alpha \in [0, +\infty)$,

2°. $\partial/\partial(\text{Im } 1/z) \log|f(z)| \in \widetilde{M}_{1+\alpha}\{G^+\}$ *if* $\alpha \in (-1, 0)$.

*Then for any* $R > R_0$

$$\frac{1}{\Gamma(2+\alpha)} \sum_{|a_k - i\frac{R}{2}| < \frac{R}{2}} \left( \left| \text{Im } \frac{1}{a_k} \right| - \frac{1}{R} \right)^{1+\alpha}$$

$$- \frac{1}{\Gamma(2+\alpha)} \sum_{|b_n - i\frac{R}{2}| < \frac{R}{2}} \left( \left| \text{Im } \frac{1}{b_n} \right| - \frac{1}{R} \right)^{1+\alpha}$$

$$= \frac{1}{2\pi} \int_{\arcsin \frac{R_0}{R}}^{\pi - \arcsin \frac{R_0}{R}} \widetilde{W}^{-\alpha} \log|f(R \sin\vartheta e^{i\vartheta})| \frac{d\vartheta}{R \sin^2 \vartheta}$$

$$+ \widetilde{Q}^{(2)}(R, R_0, \widetilde{W}^{-\alpha} \log|f|), \tag{2.11}$$

*where $\{a_k\}$ and $\{b_n\}$ are the zeros and poles of $f(z)$ and as $R \to +\infty$*

$$\widetilde{Q}^{(2)}(R, R_0, \widetilde{W}^{-\alpha} \log|f|) \equiv \frac{R_0}{2\pi} \int_{\arcsin \frac{R_0}{R}}^{\arcsin \frac{R_0}{R}} \left\{ \widetilde{W}^{-\alpha} \log|f(R_0 e^{i\vartheta})| \left( -\frac{\sin\vartheta}{R_0^2} \right) \right.$$

$$\left. - \left( \frac{\sin\vartheta}{R_0} - \frac{1}{R} \right) \frac{\partial}{\partial R_0} \widetilde{W}^{-\alpha} \log|f(R_0 e^{i\vartheta})| \right\} d\vartheta = O(1). \tag{2.12}$$

**Proof.** The function $F(w) \equiv f(w^{-1})$ is meromorphic in the lower half-plane. Assuming $\rho = -R^{-1}$, define $F_\alpha(w)$ as in (1.13). The latter function is meromorphic in $\overline{G_\rho^-} = \{w : \text{Im } w \leq \rho\}$, where it can have only a finite set of zeros and poles $\{c_m\}_1^q \subset \{w = u + i\rho : |u| < \sqrt{r_0^2 - \rho^2}\}$ $(r_0 = R_0^{-1})$. Further, $F_\alpha(w)$

satisfies the conditions of Lemma 1.1. Therefore, $u(z) = \widetilde{W}^{-\alpha} \log|F_\alpha(z^{-1})|$ is harmonic in $\{z : |z - iR/2| \leq R/2, z \neq 0\}$, except the boundary points $\{A_m\}_1^q \equiv \{c_m^{-1}\}_1^q$ ($|A_m| > R_0$, $1 \leq m \leq q$). Besides, $u(z)$ admits one of the representations of (1.4) in neighborhoods of points $A_m$. Repeating the argument proving Theorem 3.5 in Ch. I of [38], for $u(z)$ we obtain formula (3.4) of that theorem, i.e.

$$0 = \frac{1}{2\pi} \int_{\arcsin \frac{R_0}{R}}^{\pi - \arcsin \frac{R_0}{R}} u(R \sin \vartheta e^{i\vartheta}) \frac{d\vartheta}{R \sin^2 \vartheta} + \widetilde{Q}^{(2)}(R, R_0, u),$$

where $\widetilde{Q}^{(2)}(R, R_0, u)$ is the expression of (2.12) with $u = \widetilde{W}^{-\alpha} \log|f|$. Hence (2.11)–(2.12) holds by the definition of $u(z)$ and (2.4). And it remains to observe that the quantity $\widetilde{Q}^{(2)}(R, R_0, \widetilde{W}^{-\alpha} \log|f|)$ is bounded as $R \to +\infty$.

**Remark 2.2.** For $\alpha = 0$ (2.11)–(2.12) coincides with B.Ja.Levin's formula (see, for instance, [38], Ch. I, Theorem 3.4).

*2.3.* The value distribution of functions which are meromorphic in the closed half-plane is one of the main fields where Carleman and B.Ja.Levin formulas are applied (see [38], Ch. I, § 5, Ch. III, § 3). But the generalizations of these formulas given by Theorems 1.3, 2.1 and 1.4, 2.2 are valid under additional restrictions on the behavior of the considered meromorphic function $f(z)$ near the origin. Namely, when $f(z)$ has no zeros and poles in some semidisc $\{z : \operatorname{Im} z > 0, |z| \leq R_0\}$ ($0 < R_0 < +\infty$) and one of the inclusions 1°, 2° of these theorems is true for a given $\alpha \in (-1, +\infty)$. These assumptions permit to apply $\widetilde{W}^{-\alpha}$ to $\log|f(z)|$. Our next lemma shows that one can get rid of those additional restrictions if $f(z)$ is meromorphic in a neighborhood of the origin.

**Lemma 2.3.** *Let $f(z)$ be meromorphic in $|z| \leq R_0$. Then, multiplying $f(z)$ by definite rational functions $R(z)$ one can attain any of inclusions*

1°. $\log|R(z)f(z)| \in \widetilde{K}_\beta\{\pi/2, G^+\}$ *for a given* $\beta \in [0, +\infty)$,
2°. $\partial/\partial(\operatorname{Im} 1/z) \log|R(z)f(z)| \in \widetilde{K}_\beta\{\pi/2, G^+\}$ *for a given* $\beta \in (0, 1)$.

**Proof.** It suffices to consider the case when $f(z)$ ($f(0) \neq 0$) is holomorphic in $|z| \leq R_0$.
1°. Let $\beta \in [0, +\infty)$, $n > \beta$ be a natural number, and let $f(z) = \sum_{k=0}^{+\infty} a_k z^k$ ($a_0 \neq 0$, $|z| < R_0$). Observe that, multiplying $f(z)$ by the $n$-th partial sum $P_n(z)$ of the Taylor expansion of $1/f(z)$, one can find $f(z)P_n(z) = 1 + b_n z^n + b_{n+1} z^{n+1} + \cdots$ ($|z| < R_0$). Then $\log|P_n(z)f(z)| \sim |b_{n_1}||z|^{n_1}$ as $z \to 0$, where $n_1 \geq n$. Therefore, 1° holds by the definition of $\widetilde{K}_\beta$.

2°. If $f(z) = \sum_{k=0}^{+\infty} a_k z^k$ ($a_0 \neq 0$, $|z| < R_0$), then

$$\frac{\partial \log|f(z)|}{\partial(\operatorname{Im} 1/z)} = -\operatorname{Re} \left\{ iz^2 \frac{a_1 + 2a_2 z + 3a_3 z^2 + \cdots}{a_0 + a_1 z + a_2 z^2 + \cdots} \right\} = O(|z|^{n_1}) \text{ as } z \to 0,$$

where $n_1 \geq 2$. Thus, 2° automatically holds for any $\beta \in (0, 2)$.

The following lemma contains a useful estimate.

**Lemma 2.4.** *Let $F(z)$ be holomorphic in $\overline{G^-}\backslash\{0\} = \{w : \text{Im } w \leq 0, w \neq 0\}$, and let this function have no zeros in $\{w : \text{Im } w < 0, |w| > r\}$ for some $r > 0$. Further, let $\log|F(w) \in M_\alpha\{\overline{G^-}\backslash\{0\}\}$ for an $\alpha \in (0, +\infty)$, and let*

$$|F(w) \leq e^{\sigma|w|^{-(\alpha+\lambda)}}, \quad |w| < R, \quad w \in \overline{G^-}\backslash\{0\}, \tag{2.14}$$

*for some numbers $R$, $\sigma$ and $\lambda$. Then*

$$W^{-\alpha} \log|F(w)| \leq \sigma_1 |w|^{-\lambda}, \quad |w| < R_1, \quad w \in \overline{G^-}\backslash\{0\}, \tag{2.15}$$

*where $\sigma_1$ is a positive number and $R_1 = \min\{R/2, R_0 \tan \delta_0, R_0 + 1\}$ (with $R_0$ and $\delta_0$ from the definition of $M_\alpha$).*

**Proof.** By Lemma 1.1, $W^{-\alpha} \log|F(w)|$ is continuous in $\overline{G^-}\backslash\{0\}$. Assuming $|w| < R_1$, set

$$\begin{aligned} W^{-\alpha} \log|F(w)| =& \frac{1}{\Gamma(\alpha)} \left( \int_{R_0+1}^{+\infty} + \int_{R_1}^{R_0+1} + \int_0^{R_1} \right) t^{\alpha-1} \log|F(w - it)| \\ \equiv& I_1(w) + I_2(w) + I_3(w). \end{aligned}$$

Since the compact $\mathbf{K}(R_0, R_1) = \{\zeta = w - iR_0 : w \in \overline{G^-}, |w| \leq R_1\}$ lies in $\Lambda(\delta_0, R_0) = \{\zeta : |\arg\zeta + \pi/2| \leq \delta_0, |\zeta| \geq R_0\}$,

$$\begin{aligned} |I_1(w)| &\leq \frac{1}{\Gamma(\alpha)} \int_1^{+\infty} (\tau + R_0)^{\alpha_0 - 1} |\log|F(w - iR_0 - i\tau)|| \, d\tau \\ &\leq \frac{\max\{1, (R_0+1)^{\alpha-1}\}}{\Gamma(\alpha)} \sup_{\zeta \in \mathbf{K}(R_0, R_1)} \int_1^{+\infty} \tau^{\alpha-1} |\log|F(\zeta - i\tau)|| \, d\tau = N_1 < +\infty \end{aligned}$$

by the definition of $M_\alpha$. Further, it is obvious that

$$|I_2(w)| \leq \frac{\max\{R_1^{\alpha-1}, (R_0+1)^{\alpha-1}\}}{\Gamma(\alpha)} \int_{R_1}^{R_0+1} |\log|F(w - it)|| \, dt.$$

The last integral is continuous as a function of $w$ in $\overline{G^-}\backslash\{0\} + iR_1 = \{w : \text{Im } w \leq R_1, w \neq iR_1\}$. Hence $|I_2(w)| \leq N_2 < +\infty$. Further,

$$\begin{aligned} I_3(w) \leq \frac{\sigma}{\Gamma(\alpha)} \int_0^{R_1} \frac{t^{\alpha-1}}{|w - it|^{\alpha+\lambda}} dt <& \frac{\sigma C_{\alpha+\lambda}}{\Gamma(\alpha)} \left[ \int_0^{+\infty} \frac{x^{\alpha-1} dx}{1 + x^{\alpha+\lambda}} \right] |w|^{-\lambda} \\ \equiv& N_3 |w|^{-\lambda} \end{aligned}$$

by (2.14), and (2.15) follows from the estimates of $I_1$, $I_2$, $I_3$.

*2.4.* It is natural to consider the Carleman and Levin type formulas of Theorems 1.3, 2.1 and 1.4, 2.2 as *equilibrium relations* between the density of zeros and poles and the growth of the meromorphic function. These formulas can be written also as equilibrium relations for some Nevanlinna and Tsuji type characteristics. Below we give only the relations arising from the Levin type formula.

Let $f(z)$ be meromorphic in the upper half-plane $G^+$ and also in some neighborhoods of the points $-R_0$ and $R_0$ for an $R_0 > 0$. Further, let $f(z)$ have no zeros and poles in the semidisc $\{z : \operatorname{Im} z > 0, |z| \leq R_0\}$, and let $f(z)$ satisfy the remaining conditions of Theorem 2.2 for a given $\alpha \in (-1, +\infty)$. Assuming $\mathbf{n}(t, f)$ the number of poles (according to their multiplicities) of $f(z)$, which lie in $\{z : |z - it/2| = t/2, |z| > R_0\}$, set

$$\mathbf{N}_\alpha(R, f) = \frac{1}{\Gamma(1+\alpha)} \int_{R_0}^{R} \left(\frac{1}{t} - \frac{1}{R}\right)^{\alpha} \mathbf{n}(t, f) \frac{dt}{t^2}, \quad R_0 < R < +\infty.$$

Observe that, being everywhere constant the function $\mathbf{n}(t, f)$ grows stepwise at $t = \rho_n / \sin \psi_n$, where the jumps are equal to the quantity of poles $b_n = \rho_n e^{i\psi_n}$ of $f(z)$, which lie on the arcs $\{z : |z - it/2| = t/2, |z| > R_0\}$. Therefore, integration by parts gives

$$\frac{1}{\Gamma(2+\alpha)} \sum_{|b_n - i\frac{R}{2}| < \frac{R}{2}} \left(\frac{\sin \psi_n}{\rho_n} - \frac{1}{R}\right)^{1+\alpha}$$
$$= \frac{1}{\Gamma(2+\alpha)} \int_{R_0}^{R} \left(\frac{1}{t} - \frac{1}{R}\right)^{1+\alpha} d\mathbf{n}(t, f) = \mathbf{N}_\alpha(R, f).$$

Hence, formula (2.11)–(2.12) can be written in the form

$$\mathbf{N}_\alpha(R, f^{-1}) - \mathbf{N}_\alpha(R, f)$$
$$= \frac{1}{2\pi} \int_{\kappa(R)}^{\pi - \kappa(R)} \widetilde{W}^{-\alpha} \log |f(R \sin \vartheta e^{i\vartheta})| \frac{d\vartheta}{R \sin^2 \vartheta} + O(1), \qquad (2.16)$$

where $\kappa(R) = \arcsin(R_0/R)$. For $a^+ = \max\{a, 0\}$, $a^- = \max\{-a, 0\}$ set

$$\mathbf{m}_\alpha(R, f) = \frac{1}{2\pi} \int_{\kappa(R)}^{\pi - \kappa(R)} \left\{\widetilde{W}^{-\alpha} \log |f(R \sin \vartheta e^{i\vartheta})|\right\}^+ \frac{d\vartheta}{R \sin^2 \vartheta},$$
$$\mathcal{L}_\alpha(R, f) = \mathbf{m}_\alpha(R, f) + \mathbf{N}_\alpha(R, f).$$

Then (2.16) becomes $\mathbf{m}_\alpha(R, f^{-1}) + \mathbf{N}_\alpha(R, f^{-1}) = \mathbf{m}_\alpha(R, f) + \mathbf{N}_\alpha(R, f) + O(1)$ or, which is the same, the following equilibrium relation

$$\mathcal{L}_\alpha(R, f^{-1}) = \mathcal{L}_\alpha(R, f) + O(1).$$

*2.5.* One can use the boundedness of above mentioned Nevanlinna and Tsuji type characteristics for defining some general classes of meromorphic functions and finding their representations by passages in R. and F. Nevanlinna formulas of Sec. 1. Nevertheless, we prefer a different way for establishing our factorizations. As our first step, from Carleman and Levin type formulas we find out some sufficient conditions under which zeros $\{z_k\} \in G^+$ of a holomorphic function satisfy the condition

$$\sum_k \left| \mathrm{Im}\, \frac{1}{z_k} \right|^{1+\alpha} < +\infty \tag{2.17}$$

for a given $\alpha \in (-1, +\infty)$. If zeros $\{z_k\} \in \mathbb{C}$ of an entire function $f(z)$ satisfy (2.17) for $\alpha = 0$, then it is said that $f(z)$ is of the class $A$. Similar to this, we shall say that *a function holomorphic in the upper half-plane $G^+$ is of the class $A_\alpha\{G^+\}$ $(-1 < \alpha < +\infty)$, if its zeros $\{z_k\} \in G^+$ satisfy (2.17).* Note that the Blaschke type products of Ch. 2 are of $A_\alpha\{G^+\}$.

**Theorem 2.3.** *Let $f(z)$ be holomorphic in the closed half-plane $\overline{G^+}\backslash\{0\} = \{z : \mathrm{Im}\ z \geq 0,\ z \neq 0\}$, and let for some $R_0 > 0$ this function have no zeros in the semidisc $\{z : \mathrm{Im}\ z > 0,\ |z| \leq R_0\} \cup \{-R_0, R_0\}$. Further, let $\alpha \in (-1, +\infty)$ be any number and let*

1°. $\log|f(z)| \in \widetilde{M}_\alpha\{G^+\}$ *if* $\alpha \in [0, +\infty)$,
2°. $\partial/\partial(\mathrm{Im}\ 1/z) \log|f(z)| \in \widetilde{M}_\alpha\{G^+\}$ *if* $\alpha \in (-1, 0)$.

*If*

$$\liminf_{R\to+\infty} \frac{1}{\pi R} \int_0^\pi \left\{ \widetilde{W}^{-\alpha} \log|f(Re^{i\vartheta})| \right\}^+ \sin\vartheta d\vartheta < +\infty \tag{2.18}$$

*and*

$$\left( \int_{-\infty}^{-R_0} + \int_{R_0}^{+\infty} \right) \left\{ \widetilde{W}^{-\alpha} \log|f(t)| \right\}^+ \frac{dt}{t^2} < +\infty, \tag{2.19}$$

*then*

1°. $f(z) \in A_\alpha\{G^+\}$,

$$2°.\ \left( \int_{-\infty}^{-R_0} + \int_{R_0}^{+\infty} \right) \left| \widetilde{W}^{-\alpha} \log|f(t)| \right| \frac{dt}{t^2} < +\infty, \tag{2.20}$$

$$\lim_{R\to+\infty} \int_{R_0}^{R} \left( \frac{1}{t^2} - \frac{1}{R^2} \right) \widetilde{W}^{-\alpha} \log|f(t)f(-t)| dt = \int_{R_0}^{+\infty} \widetilde{W}^{-\alpha} \log|f(t)f(-t)| \frac{dt}{t^2}. \tag{2.21}$$

3°. *The following relations hold*

$$\frac{2}{\pi}\lim_{R\to+\infty}\frac{1}{R}\int_0^\pi \widetilde{W}^{-\alpha}\log|f(Re^{i\vartheta})|\sin\vartheta d\vartheta = h \neq \infty, \tag{2.22}$$

$$\lim_{R\to+\infty}\sum_{|z_k|<R}\left\{\frac{1}{\Gamma(2+\alpha)}\left|\operatorname{Im}\frac{1}{z_k}\right|^{1+\alpha}+I_\alpha\left(\frac{1}{z_k},\frac{1}{R}\right)\right\}$$
$$=\frac{1}{\Gamma(2+\alpha)}\sum_k\left|\operatorname{Im}\frac{1}{z_k}\right|^{1+\alpha}<+\infty. \tag{2.23}$$

**Proof.** We start by 2°. Let $\{R_n\}_1^\infty$ be a sequence by which the lower limit (2.18) is attained. Then by formula (2.9) (where the terms of left-hand side sums are nonnegative)

$$\left(\int_{-R_n/2}^{-R_0}+\int_{R_0}^{R_n/2}\right)\left(\frac{1}{t^2}-\frac{1}{R_n^2}\right)\left\{\widetilde{W}^{-\alpha}\log|f(t)|\right\}^- dt$$
$$\leq\left(\int_{-R_n}^{-R_0}+\int_{R_0}^{R_n}\right)\left(\frac{1}{t^2}-\frac{1}{R_n^2}\right)\left\{\widetilde{W}^{-\alpha}\log|f(t)|\right\}^- dt$$
$$\leq\left(\int_{-R_n}^{-R_0}+\int_{R_0}^{R_n}\right)\left(\frac{1}{t^2}-\frac{1}{R_n^2}\right)\left\{\widetilde{W}^{-\alpha}\log|f(t)|\right\}^+ dt+O(1)$$
$$\leq\left(\int_{-\infty}^{-R_0}+\int_{R_0}^{+\infty}\right)\left\{\widetilde{W}^{-\alpha}\log|f(t)|\right\}^+\frac{dt}{t^2}+O(1)=O(1).$$

As $4/3(1/t^2-1/R^2)\geq 1/t^2$ for $R_0\leq|t|\leq R_n/2$, we have

$$\left(\int_{-R_n/2}^{-R_0}+\int_{R_0}^{R_n/2}\right)\left(\frac{1}{t^2}-\frac{1}{R^2}\right)\left\{\widetilde{W}^{-\alpha}\log|f(t)|\right\}^-\frac{dt}{t^2}$$
$$\leq\frac{4}{3}\left(\int_{-R_n/2}^{-R_0}+\int_{R_0}^{R_n/2}\right)\left(\frac{1}{t^2}-\frac{1}{R^2}\right)\left\{\widetilde{W}^{-\alpha}\log|f(t)|\right\}^-\frac{dt}{t^2}=O(1).$$

Hence the convergence of (2.19) for $\{\widetilde{W}^{-\alpha}\log|f(t)|\}^-$ follows. Together with (2.19), this provides the convergence of (2.20).

For establishing (2.21) (where the right-hand side integral is already proved to be absolutely convergent), set

$$\psi(t)=\int_t^{+\infty}\widetilde{W}^{-\alpha}\log|f(x)f(-x)|\frac{dx}{x^2},\quad R_0\leq t<+\infty.$$

Evidently

$$\int_{R_0}^R\left(\frac{1}{t^2}-\frac{1}{R^2}\right)\widetilde{W}^{-\alpha}\log|f(t)f(-t)|dt-\int_{R_0}^R\widetilde{W}^{-\alpha}\log|f(t)f(-t)|\frac{dt}{t^2}$$
$$=\frac{1}{R^2}\int_{R_0}^R t^2 d\psi(t)=\psi(R)-\left(\frac{R_0}{R}\right)^2\psi(R_0)-\frac{2}{R^2}\int_{R_0}^R t\psi(t)dt.$$

Hence (2.21) follows since

$$\psi(R)-\left(\frac{R_0}{R}\right)^2\psi(R_0)=o(1) \text{ and } \frac{1}{R^2}\int_{R_0}^{R}t\psi(t)dt=o(1) \quad \text{as} \quad R\to+\infty.$$

1°. If (2.18) is attained on $\{R_n\}_1^\infty$, then by (2.9)–(2.10) and (2.18), (2.21)

$$\sum_{|z_k|<R_n}\left\{\frac{1}{\Gamma(2+\alpha)}\left|\operatorname{Im}\frac{1}{z_k}\right|^{1+\alpha}+I_\alpha\left(\frac{1}{z_k},\frac{1}{R_n}\right)\right\}=O(1) \text{ as } n\to\infty. \quad (2.24)$$

The terms of this sum are nonnegative and

$$\lim_{R\to+\infty}I_\alpha\left(\frac{1}{z_k},\frac{1}{R}\right)=0. \quad (2.24')$$

Fixed $R>R_0$, observe that $R_n>R$ for enough great $n\geq 1$. Hence

$$\begin{aligned}0&\leq\sum_{|z_k|<R}\left\{\frac{1}{\Gamma(2+\alpha)}\left|\operatorname{Im}\frac{1}{z_k}\right|^{1+\alpha}+I_\alpha\left(\frac{1}{z_k},\frac{1}{R_n}\right)\right\}\\&\leq\sum_{|z_k|<R_n}\left\{\frac{1}{\Gamma(2+\alpha)}\left|\operatorname{Im}\frac{1}{z_k}\right|^{1+\alpha}+I_\alpha\left(\frac{1}{z_k},\frac{1}{R_n}\right)\right\}=O(1).\end{aligned}$$

The passage $n\to\infty$ gives that for any $R>R_0$

$$\frac{1}{\Gamma(2+\alpha)}\sum_{|z_k|<R}\left|\operatorname{Im}\frac{1}{z_k}\right|^{1+\alpha}\leq M<+\infty.$$

Hence (2.17) follows, and $f(z)\in A_\alpha\{G^+\}$.

3°. (2.23) will hold if $\sum_{|z_k|<R}I_\alpha\left(z_k^{-1},R^{-1}\right)=o(1)$ as $R\to+\infty$. To prove the latter relation, for an arbitrary $\varepsilon>0$ choose $R_\varepsilon>R_0$ enough great to provide

$$\frac{1}{\Gamma(2+\alpha)}\sum_{|z_k|>R_\varepsilon}\left|\operatorname{Im}\frac{1}{z_k}\right|^{1+\alpha}<\varepsilon.$$

Then, by (1.25)

$$\left|\sum_{|z_k|<R}I_\alpha\left(\frac{1}{z_k},\frac{1}{R}\right)\right|\leq\sum_{|z_k|<R_\varepsilon}\left|I_\alpha\left(\frac{1}{z_k},\frac{1}{R}\right)\right|+\varepsilon.$$

for any $R>R_\varepsilon$. Consequently $\limsup_{R\to+\infty}\left|\sum_{|z_k|<R}I_\alpha\left(z_k^{-1},R^{-1}\right)\right|\leq\varepsilon$ by (2.24′). Hence (2.23) follows. For proving (2.22), observe that

$$\begin{aligned}&\lim_{R\to+\infty}\widetilde{Q}^{(1)}(R,R_0,\widetilde{W}^{-\alpha}\log|f|)\\&=-\frac{1}{2\pi}\int_0^\pi\left\{\widetilde{W}^{-\alpha}\log|f(R_0e^{i\vartheta})|R_0^{-1}+\frac{\partial}{\partial R_0}\widetilde{W}^{-\alpha}\log|f(R_0e^{i\vartheta})|\right\}\sin\vartheta d\vartheta,\end{aligned}$$

where the right-hand side integral is absolutely convergent. Hence (2.22) follows by (2.9) and the finiteness of the limits (2.21), (2.23).

The proof of the next theorem is quite similar.

**Theorem 2.4.** *Let $F(w)$ be holomorphic in $G^- = \{w : \text{Im } w < 0\}$, and let for any $\rho < 0$ this function have not more than a finite set of zeros in $G_\rho^- = \{w : \text{Im } w < \rho\}$. Further, let $\alpha \in (-1, +\infty)$ and let*

$1°$. $\log|F(w)| \in M_\alpha\{G^-\}$ *if* $\alpha \in [0, +\infty)$,
$2°$. $\partial/\partial(\text{Im } w) \log|F(w)| \in M_{1+\alpha}\{G^-\}$ *if* $\alpha \in (-1, 0)$.

*If*

$$\sup_{\rho<0} \int_{-\infty}^{+\infty} \left|W^{-\alpha} \log|F(u+i\rho)|\right| du < +\infty, \quad -\infty < \rho < 0, \tag{2.25}$$

*then:* $1°$. *Zeros* $\{w_k\} \subset G^-$ *of* $F(w)$ *satisfy*

$$\sum_k |\text{Im } w_k|^{1+\alpha} < +\infty. \tag{2.26}$$

$2°$. *The following relations hold:*

$$\lim_{\rho \to -0} \sum_{\text{Im } w_k < \rho} (|\text{Im } w_k| + \rho)^{1+\alpha} = \sum_k |\text{Im } w_k|^{1+\alpha},$$

$$\lim_{\rho \to -0} \int_{-\infty}^{+\infty} \left|W^{-\alpha} \log|F(u+i\rho)|\right| du = H \neq \infty.$$

## 3. FACTORIZATIONS

Theorems 3.2, 3.5 and 3.6 of this section present some solutions of II-nd and III-rd Problems (see Introduction) for several weighted classes of functions holomorphic in a half-plane.

*3.1.* We start by a representation of harmonic functions, which follows from Nevanlinna factorization in the half-plane (see [83] or [10], Sec. 6.5).

**Theorem 3.1.** *Let $u(z)$ be a real-valued harmonic function in $G^+$, and let this function be continuous in $\overline{G^+} = \{z : \text{Im } z \geq 0\}$, with possible exception of a countable set of real points $\{c_m\}_1^N$ (if $N = \infty$, then $\lim_{m\to\infty} |c_m| = \infty$), in which $u(z)$ admits one of representations (1.4). If*

$$\liminf_{R \to +\infty} \frac{1}{R} \int_0^\pi u^+(Re^{i\vartheta}) \sin\vartheta < +\infty \quad \text{and} \quad \int_{-\infty}^{+\infty} \frac{u^+(t)}{1+t^2} dt < +\infty,$$

*then*

$$\int_{-\infty}^{+\infty} \frac{|u(t)|}{1+t^2} dt < +\infty \tag{3.1}$$

*and*

$$u(z) = hy + \frac{y}{\pi}\int_{-\infty}^{+\infty} \frac{u(t)dt}{(x-t)^2+y^2}, \quad z = x+iy \in G^+, \tag{3.2}$$

*where $h$ is a real number deduced from*

$$h = \frac{2}{\pi}\lim_{R\to+\infty}\frac{1}{R}\int_0^{\pi} u(Re^{i\vartheta}) sin\vartheta. \tag{3.3}$$

**Theorem 3.2.** *Let $f(z)$ be holomorphic in $\overline{G^+}\backslash\{0\} = \{z : \text{Im } z \geq 0, |z| \leq R_0, z \neq 0\}$, and let this function satisfy one of the following conditions:*

$1^\circ$. $\log|f(z)| \in \widetilde{K}_{\alpha+\kappa}\{\pi/2, G^+\}$ *for some* $\alpha \in [0, +\infty)$ *and* $\kappa \geq 1-\alpha+p$,

$2^\circ$. $\partial/\partial(\text{Im } 1/z)\log|f(z)| \in \widetilde{K}_{1+\alpha+\kappa}\{\pi/2, G^+\}$ *for some* $\alpha \in (-1, 0)$ *and* $\kappa \geq 1-\alpha$.

*If*

$$\liminf_{R\to+\infty} R^{-1}\int_0^{\pi}\left\{\widetilde{W}^{-\alpha}\log|f(Re^{i\vartheta})|\right\}^{+}\sin\vartheta d\vartheta < +\infty \tag{3.4}$$

*and*

$$\int_{-\infty}^{+\infty}\frac{\{\widetilde{W}^{-\alpha}\log|f(t)|\}^{+}}{1+t^2}dt < +\infty, \tag{3.5}$$

*then*

$$\int_{-\infty}^{+\infty}\frac{|\widetilde{W}^{-\alpha}\log|f(t)||}{|t|^{\kappa}(1+|t|^{2-\kappa})}dt < +\infty, \tag{3.6}$$

*and the zeros $\{z_k\} \subset G^+$ of $f(z)$ satisfy the condition*

$$\sum_k \left|\text{Im } \frac{1}{z_k}\right|^{1+\alpha} < +\infty. \tag{3.7}$$

*Besides, for $z \in G^+$*

$$f(z) = \widetilde{B}_{\alpha}(z, \{z_k\})\exp\Bigg\{iC + \\ + \Gamma(1+\alpha)e^{-i\frac{\pi}{2}(1+\alpha)}\left[hz^{1+\alpha} + \frac{1}{\pi}\int_{-\infty}^{+\infty}\frac{\widetilde{W}^{-\alpha}\log|f(t)|}{t^2(1/z-1/t)^{1+\alpha}}dt\right] + iC\Bigg\}, \tag{3.8}$$

*where $C$ and $h$ are real numbers, and*

$$h = \frac{2}{\pi}\lim_{R\to+\infty}\frac{1}{R}\int_0^{\pi}\widetilde{W}^{-\alpha}\log|f(Re^{i\vartheta})|\sin\vartheta d\vartheta. \tag{3.9}$$

**Proof.** By Lemma 1.9 of Ch. 1

$$\left|\widetilde{W}^{-\alpha}\log|f(z)|\right|\leq O(|z|^{\kappa})\quad as\quad z\to 0\quad (z\in\overline{G^{+}}).\tag{3.10}$$

In view of (1.27) in Ch. 1, $f(z)$ satisfies the requirements of Theorems 2.1 and 2.3. By Theorem 2.3, $f(z)\in A_{\alpha}\{G^{+}\}$, i.e. (3.7) is true. Hence the product $\widetilde{B}_{\alpha}(z,\{z_k\})$ is convergent and the function

$$f_{\alpha}(z)=\frac{f(z)}{\widetilde{B}_{\alpha}(z,\{z_k\})}\ (\neq 0,\quad z\in G^{+})\tag{3.11}$$

is holomorphic in $\overline{G^{+}}\backslash\{0\}$, except some points $\{(\text{Re } 1/z_k)^{-1}\}$. On the other hand, $\log|f_{\alpha}(z)|\in\widetilde{K}_{\alpha+\gamma}\{\pi/2,G^{+}\}$ if $\alpha\geq 0$, and $\partial/\partial(\text{Im } 1/z)\log|f_{\alpha}(z)|\in\widetilde{K}_{1+\alpha+\gamma}\{\pi/2,G^{+}\}$ ($\gamma=\min\{\kappa,1\}$) if $-1<\alpha<0$ by (1.2), (1.2′) and our requirements 1°, 2°. Therefore, by Lemma 1.1 and Theorem 3.2 of Ch. 2, the function

$$u(z)=\widetilde{W}^{-\alpha}\log|f_{\alpha}(z)|=\widetilde{W}^{-\alpha}\log|f(z)|-\widetilde{W}^{-\alpha}\log|\widetilde{B}_{\alpha}(z,\{z_k\})|\tag{3.12}$$

is harmonic in $\overline{G^{+}}\backslash\{0\}$, except the points $\{c_m\}_1^N\equiv\{(\text{Re } 1/z_m)^{-1}\}_1^N$ ($N\leq\infty$, $\{z_m\}_1^N\subset\overline{G^{+}}\backslash\{0\}$, $\text{Re } z_m\neq 0$) which are the zeros of $f(z)$ in $(-\infty,+\infty)$. Besides, $u(z)$ has the representations (1.4) in the neighborhoods of these points, and by Lemma 1.9 of Ch. 1 $u(z)$ is continuous at the origin, besides $u(0)=0$.

By (1.16), (1.19)–(1.19′) and Remark 3.2 in Ch. 2

$$\lim_{R\to+\infty}\frac{1}{R}\int_0^{\pi}\widetilde{W}^{-\alpha}\log|\widetilde{B}_{\alpha}(Re^{i\vartheta},\{z_k\})|\sin\vartheta d\vartheta=0.\tag{3.13}$$

Hence by (2.11′) of Ch. 2 $u(z)$ satisfies the conditions of Theorem 3.1 and therefore is representable in the form (3.1)–(3.3). Besides, by (3.15) the number $h$ in (3.1)–(3.3) can be deduced by (3.9). Further, by (3.10) the integral (3.6) is convergent since the last term in (3.12) vanishes on the real axis. Thus,

$$\widetilde{W}^{-\alpha}\log|f_{\alpha}(z)|=hy+\frac{y}{\pi}\int_{-\infty}^{+\infty}\frac{\widetilde{W}^{-\alpha}\log|f(t)|}{(x-t)^2+y^2}dt\quad z=x+iy\in G^{+}.\tag{3.14}$$

For proving the remaining formula (3.8), introduce the function

$$U(w)\equiv\log|f_{\alpha}(w^{-1})|\tag{3.15}$$

which is harmonic in $G^{-}$. As $W^{-\alpha}U(w)\equiv\widetilde{W}^{-\alpha}\log|f_{\alpha}(w^{-1})|$, (3.14) becomes

$$W^{-\alpha}U(w)=\text{Re}\left\{\frac{h}{iw}+\frac{1}{\pi i}\int_{-\infty}^{+\infty}\frac{W^{-\alpha}\log|f(t^{-1})|}{w-t}dt\right\},\quad w\in G^{-}.$$

On the other hand, (3.6) implies

$$\int_{-\infty}^{+\infty} \frac{|W^{-\alpha} \log |f(t^{-1})||}{1+|t|^{2-\kappa}} dt < +\infty,$$

where $2-\kappa \leq 1+\alpha-p$ if $\alpha \geq 0$ and $2-\kappa \leq 1+\alpha$ if $-1<\alpha<0$. Therefore, using Lemmas 1.10 and 3.1 of Ch. 1 and the obvious relation

$$\lim_{w\to\infty,\, w\in\overline{G^-}} U(w) = \lim_{z\to 0,\, w\in\overline{G^+}} \log|f(z)| = 0,$$

for $w \in G^-$ we come to the representation

$$U(w) = \Gamma(1+\alpha)\mathrm{Re}\left\{\frac{h}{(iw)^{1+\alpha}} + \frac{1}{\pi}\int_{-\infty}^{+\infty} \frac{\widetilde{W}^{-\alpha}\log|f(t^{-1})|}{[i(w-t)]^{1+\alpha}} dt\right\}, \quad w \in G^-.$$

Hence (3.8) follows by (3.15) and (3.11).

**Remark 3.1.** Similar to the case $\alpha = 0$, it is natural to call the number $h$ in the factorization (3.18) *$\alpha$-mean-type of $f(z)$*. Along with (3.9), there is another relation for $h$:

$$h = \limsup_{y\to+\infty} y^{-1}\widetilde{W}^{-\alpha}\log|f(iy)|. \tag{3.9'}$$

**Remark 3.2.** Let $f(z)$ be holomorphic in $\overline{G^+}$, where this function is of finite order $\rho \geq 0$, i.e.

$$\rho = \limsup_{R\to+\infty} \frac{\log\log M(R,f)}{\log R}, \quad M(R,f) = \max_{0\leq\vartheta\leq\pi} |f(Re^{i\vartheta})|.$$

Then for any $\alpha > \max\{\rho-1, 0\}$ by the coefficients of Taylor's expansion of $f(z)$ in $z=0$ one can construct a polynomial $P_n(z)$ of order $n = [\alpha]+3$, such that $f(z)P_n(z)$ admits a factorization of the form (3.6)–(3.9).

**Proof.** Denote $\rho' = \max\{\rho, 1\}$. If $\alpha > (\rho-1)^+$, then $\alpha = \rho'+\varepsilon-1$, where $\varepsilon > 0$. By the assertion 1° of Lemma 2.3, multiplying $f(z)$ by a definite polynomial $P_n(z)$ of order $n = [\alpha]+3$ one can attain the validity of the requirement 1° of Theorem 3.1 for $f(z)P_n(z)$. On the other hand, it is obvious that for any number $\gamma \in ((\varepsilon-1)^+, \varepsilon)$

$$|f(Re^{i\vartheta})P_n(Re^{i\vartheta})| < \exp\left\{R^{\rho'+\gamma}\right\}, \quad R > R_0,\ 0 \leq \vartheta \leq \pi,$$

for enough great $R_0 > 0$. Hence

$$\widetilde{W}^{-\alpha}\log|f(Re^{i\vartheta})P_n(Re^{i\vartheta})| < \sigma_1 R^{\gamma-\varepsilon+1}, \quad R > R_0,\ 0 \leq \vartheta \leq \pi,$$

where $\sigma_1$ is a positive constant and $\gamma - \varepsilon + 1 \in (0, 1)$. By the properties of $\widetilde{W}^{-\alpha} \log|f(z)P_n(z)|$, which follow from Lemma 1.9 of Ch. 1 and Lemma 1.1, the function $f(z)P_n(z)$ satisfies all conditions of Theorem 3.2. Hence the desired factorization holds.

*3.2.* Consider two classes of functions harmonic in the lower half-plane.

**Definition 3.1.** *The classes $\mathfrak{U}^+$ and $\mathfrak{U}^m$ are the sets of functions $U(w)$ which are harmonic in $G^-$ and satisfy the following conditions:*

$$\mathfrak{U}^+ : \qquad \sup_{v<0} \int_{-\infty}^{+\infty} U^+(u + iv)du < +\infty, \tag{3.16}$$

$$\mathfrak{U}^m : \qquad \sup_{v<0} \int_{-\infty}^{+\infty} |U(u + iv)|du < +\infty. \tag{3.17}$$

Obviously $\mathfrak{U}^m \subset \mathfrak{U}^+$. Considering the function $f(z) \equiv \exp\{U(-z) + V(z)\}$ (where $U(w) \in \mathfrak{U}^+$ or $U(w) \in \mathfrak{U}^m$ and $V(z)$ is the conjugate harmonic of $\varphi(z) \equiv U(-z)$ in $G^+$), which is holomorphic in $G^+$, one can state the below theorems as direct consequences of Theorems XVII and XX of [76].

**Theorem 3.3.** *The class $\mathfrak{U}^+$ coincides with the set of functions representable in the form*

$$U(w) = a|v| + \frac{|v|}{\pi} \int_{-\infty}^{+\infty} \frac{d\mu(t)}{(u-t)^2 + v^2}, \qquad w = u + iv \in G^-, \tag{3.18}$$

*where $a$ is a nonpositive number and $\mu(t)$ can be written as the difference $\mu(t) = \mu_1(t) - \mu_2(t)$ of two nondecreasing functions for which*

$$\int_{-\infty}^{+\infty} d\mu_1(t) < +\infty \quad \textit{and} \quad \int_{-\infty}^{+\infty} \frac{d\mu_2(t)}{1+t^2} < +\infty. \tag{3.18$'$}$$

**Theorem 3.4.** *The class $\mathfrak{U}^m$ coincides with the set of functions representable in the form (3.18), where $a = 0$, and $\mu(t)$ is of bounded variation in $(-\infty, +\infty)$.*

We also use the following subclass of $\mathfrak{U}^+$.

**Definition 3.2.** $\mathfrak{U}^+_\kappa$ $(-1 < \kappa < 1)$ *is the set of those $U(w) \in \mathfrak{U}^+$ for which*

$$\int_1^{+\infty} U^-(-it)\frac{dt}{t^{1+\kappa}} < +\infty. \tag{3.19}$$

Using some results from [72], for $U(w) \in \mathfrak{U}^m$ one can show that $|U(-it)| \in L(t^{-(1+\kappa)}dt,\ (1, +\infty))$ for any $\kappa \in (-1, 1)$. Hence, $\mathfrak{U}^m \subset \bigcap_{-1<\kappa<1} \mathfrak{U}^+_\kappa$. Besides, the below statement is true.

**Lemma 3.1.** *For any $\kappa \in (-1,1)$ the class $\mathfrak{U}_\kappa^+$ coincides with the set of functions representable in the form (3.18), where $a=0$ and $\mu(t)$ can be written as the difference $\mu(t)=\mu_1(t)-\mu_2(t)$ of two nondecreasing functions such that*

$$\int_{-\infty}^{+\infty} d\mu_1(t) < +\infty \quad \text{and} \quad \int_{-\infty}^{+\infty} \frac{d\mu_2(t)}{1+|t|^{1+\kappa}} < +\infty. \tag{3.18''}$$

**Proof.** Let $U(w) \in \mathfrak{U}_\kappa^+$ for some $\kappa \in (-1,1)$. Then $U(w)$ is representable in the form (3.18)–(3.18′) since $\mathfrak{U}_\kappa^+ \subset \mathfrak{U}^+$. For proving that in this representation $a=0$ and $\mu_2(t)$ satisfies the second requirement in (3.18″), consider the harmonic functions

$$U_1(w) = \frac{|v|}{\pi}\int_{-\infty}^{+\infty} \frac{d\mu_1(t)}{(u-t)^2+v^2}, \qquad w=u+iv \in G^-,$$

$$U_2(w) = -a|v| + \frac{|v|}{\pi}\int_{-\infty}^{+\infty} \frac{d\mu_2(t)}{(u-t)^2+v^2}, \qquad w=u+iv \in G^-.$$

Obviously $U_j(w) \geq 0$ $(j=1,2)$ and $U(w)=U_1(w)-U_2(w)$. Besides, $U(w) = U^+(w)-U^-(w)$, where $U_1(w) \geq U^+(w)$, $U_2(w) \geq U^-(w)$. Therefore, setting $\Phi(w)=U_1(w)-U^+(w)$ we conclude

$$U^-(w)+\Phi(w) = U_2(w), \qquad w \in G^-. \tag{3.20}$$

But $0 \leq \Phi(w) \leq U_1(w)$, and for $U_1$ the integral of (3.19) is convergent in view of Theorem 1 in [72]. Therefore, the same integral is convergent also for $\Phi(w)$. Further, by (3.20) the integral (3.19) is convergent also for $U_2$. Hence, by same theorem $a=0$ and $\mu_2(t)$ satisfies the second requirement in (3.18″). The converse statement is obvious since

$$0 \leq \int_1^{+\infty} U^-(-it)\frac{dt}{t^{1+\kappa}} \leq \int_1^{+\infty} U_2(-it)\frac{dt}{t^{1+\kappa}} < +\infty.$$

*3.3.* **Definition 3.3.** $\mathfrak{N}_\alpha$ $(-1<\alpha<+\infty)$ *is the set of those functions $F(w)$ holomorphic in $G^-$, which satisfy the following conditions:*

(i) *The zeros of $F(w)$ lie in a semidisc $\{w \in G^- : |w| < R_0\}$ $(R_0 > 0)$.*

(ii) *If $0 \leq \alpha < +\infty$, then $\log|F(w)| \in M_p\{G^-\}$, where $p$ is the integer deduced from $p-1<\alpha\leq p$. If $-1<\alpha<0$, then $\partial/\partial(\operatorname{Im} w)\log|F(w)| \in M_1\{G^-\}$.*

(iii)
$$\sup_{v<0}\int_{-\infty}^{+\infty}\left[W^{-\alpha}\log|F(u+iv)|\right]^+ du < +\infty. \tag{3.21}$$

(iv) *If $\alpha = p \geq 1$ is a natural number, then additionally*

$$\int_1^{+\infty}\left[W^{-\alpha}\log|F(-it)|\right]^- \frac{dt}{t} < +\infty. \tag{3.22}$$

**Definition 3.4.** *The class* $\mathfrak{N}_\alpha^m$ $(-1 < \alpha < +\infty)$ *is the set of those functions* $F(w)$ *holomorphic in* $G^-$, *which satisfy the following conditions:*

(i) $F(w)$ *has not more than finite number of zeros in any half-plane* $G_\rho^- \equiv \{\text{Im } w < \rho\}$ $(\rho < 0)$.

(ii) *If* $0 \le \alpha < +\infty$, *then* $\log|F(w)| \in M_p\{G^-\}$ $(p \ge 0,\ p-1 < \alpha \le p)$. *If* $-1 < \alpha < 0$, *then* $\partial/\partial(\text{Im } w)\log|F(w)| \in M_1\{G^-\}$.

(iii)
$$\sup_{v<0}\int_{-\infty}^{+\infty}\left|W^{-\alpha}\log|F(u+iv)|\right|du < +\infty. \tag{3.23}$$

**Lemma 3.2.** *Let* $F(w)$ *be holomorphic in* $G^-$, *and let* $\{w_k\} \equiv \{u_k + iv_k\} \subset G^-$ *be the zeros of this function. If* $F(w) \in \mathfrak{N}_\alpha$ *or* $F(w) \in \mathfrak{N}_\alpha^m$ *for some* $\alpha \in (-1, +\infty)$, *then*

$$\sum_k |\text{Im } w_k|^{1+\alpha} \equiv \sum_k |v_k|^{1+\alpha} < +\infty. \tag{3.24}$$

**Proof.** One can be convinced that for $F(w) \in \mathfrak{N}_\alpha^m$ (3.24) follows from Theorem 2.4. Let $F(w) \in \mathfrak{N}_\alpha$. Similar to Proof of Theorem 1.2, introduce the function

$$F_\alpha(w) \equiv \frac{F(w)}{\prod_{v_k<\rho} b_\alpha(w - i\rho, w_k - i\rho)}, \quad -\infty < \rho < 0,$$

which is holomorphic in $G_\rho^-$. By Lemma 1.1, $W^{-\alpha}\log|F_\alpha(w)| \equiv U(w)$ is harmonic in $\overline{G_\rho^-}$, with possible exception of a finite set of boundary points, where one of representations (1.4′) is true. By (3.14) of Ch. 2

$$\begin{aligned}\sup_{v<\rho}\int_{-\infty}^{+\infty} U^+(u+iv)du &\le \sup_{v<\rho}\int_{-\infty}^{+\infty}\left[W^{-\alpha}\log|F_\alpha(u+iv)|\right]^+ du \\ &+ \sum_{v_k<\rho}\sup_{v<0}\int_{-\infty}^{+\infty}\left|W^{-\alpha}\log|b_\alpha(u+iv, w_k-\rho)|\right|du < +\infty.\end{aligned}$$

Therefore $V(w) \equiv U(w+i\rho) \in \mathfrak{U}^+$. Hence $V(w)$ is representable in the form (3.18)–(3.18′), where $d\mu(t) = W^{-\alpha}\log|F_\alpha(t+i\rho)|dt$. Consequently,

$$\begin{aligned}&-\sum_{v_k<\rho} W^{-\alpha}\log|b_\alpha(w - i\rho, w_k - \rho)| \\ &\le \frac{|v-\rho|}{\pi}\int_{-\infty}^{+\infty}\frac{\left[W^{-\alpha}\log|F_\alpha(t+i\rho)|\right]^+}{(u-t)^2+(v-\rho)^2}dt - W^{-\alpha}\log|F_\alpha(w)| \\ &\le \frac{1}{\pi|v-\rho|}\int_{-\infty}^{+\infty}\left[W^{-\alpha}\log|F_\alpha(t+i\rho)|\right]^+ dt - W^{-\alpha}\log|F_\alpha(w)|\end{aligned}$$

for any $w \in G_\rho^-$. Now let $w = -i(R_0+1)$ (see. Definition 3.3(i) of $\mathfrak{N}_\alpha$), and let $\rho \in (-1,0)$. Then, by (2.5) of Ch. 2 and (3.21)

$$\frac{2(R_0+1+\rho)}{\Gamma(1+\alpha)} \sum_{v_k<\rho} A_k^{(\alpha)}(\rho) \le C < +\infty,$$

where $C$ is a constant independent of $\rho$ and

$$A_k^{(\alpha)}(\rho) = \int_0^{|v_k|+\rho} \frac{(R_0+1+\rho)^2 + u_k^2 - t^2}{|t^2 + [i(R_0+1+\rho)+u_k]^2|^2}(|v_k|+\rho-t)^\alpha dt \ge 0.$$

Obviously

$$\lim_{\rho\to -0} A_k^{(\alpha)}(\rho) = \int_0^{|v_k|} \frac{(R_0+1)^2 + u_k^2 - t^2}{|t^2 + [i(R_0+1)+u_k]^2|^2}(|v_k|-t)^\alpha dt$$
$$> \frac{2R_0+1}{13(R_0+1)^4} \frac{|v_k|^{1+\alpha}}{1+\alpha} > 0.$$

Consequently, for any $\rho_0 < 0$

$$\frac{2(2R_0+1)}{13(R_0+1)^3} \frac{1}{\Gamma(1+\alpha)} \sum_{v_k<\rho_0} |v_k|^{1+\alpha} < \frac{2(R_0+1)}{\Gamma(1+\alpha)} \sum_{v_k<\rho_0} \lim_{\rho\to -0} A_k^{(\alpha)}(\rho) \le C.$$

Hence our assertion follows by the arbitrariness of $\rho_0 < 0$.

*3.4.* The representations of the below theorems are solutions of III-rd Problem of Introduction for the weighted classes $\mathfrak{N}_\alpha$ and $\mathfrak{N}_\alpha^m$.

**Theorem 3.5.** *$\mathfrak{N}_\alpha$ $(-1 < \alpha < +\infty)$ coincides with the set of functions admitting one of the following factorizations in $G^-$: for $0 \le \alpha < +\infty$*

$$F(w) = B_\alpha(w) \exp\left\{\frac{\Gamma(1+\alpha)}{\pi} e^{-i\frac{\pi}{2}(1+\alpha)} \int_{-\infty}^{+\infty} \frac{d\mu(t)}{(w-t)^{1+\alpha}} + iC\right\}, \tag{3.25}$$

*and for $-1 < \alpha < 0$*

$$F(w) = B_\alpha(w) \exp\left\{a_1 w + a_0 + \frac{\Gamma(1+\alpha)}{\pi} e^{-i\frac{\pi}{2}(1+\alpha)} \int_{-\infty}^{+\infty} \frac{d\mu(t)}{(w-t)^{1+\alpha}} + iC\right\}. \tag{3.26}$$

*In these factorizations $B_\alpha(w)$ are convergent Blaschke type products with a bounded sets of zeros satisfying (3.24), $a_0$, $a_1$, and $C$ are real numbers and $\mu(t) = \mu_1(t) - \mu_2(t)$, where $\mu_1(t)$ are $\mu_2(t)$ nondecreasing functions satisfying*

$$\int_{-\infty}^{+\infty} d\mu_1(t) < +\infty, \quad \int_{-\infty}^{+\infty} \frac{d\mu_2(t)}{1+|t|^{1+\alpha-p}} < +\infty \quad if \quad 0 \le \alpha < +\infty \tag{3.25'}$$

*$(p \geq 0,\, p-1 < \alpha \leq p)$ or*

$$\int_{-\infty}^{+\infty} d\mu_1(t) < +\infty, \quad \int_{-\infty}^{+\infty} \frac{d\mu_2(t)}{1+|t|^{1+\alpha}} < +\infty \quad if \quad -1 < \alpha < 0. \tag{3.26'}$$

*If (3.26)-(3.26') is valid, then*

$$\lim_{t\to+\infty} \log|F(u-it)| = a_1 u + a_0, \quad -\infty < u < +\infty. \tag{3.26''}$$

**Proof.** If $F(w) \in \mathfrak{N}_\alpha$ $(-1 < \alpha < +\infty)$, then the zeros $\{w_k\} \subset G^-$ of $F(w)$ satisfy (3.24) by Lemma 3.2. Hence, by the product $B_\alpha(w)$ is convergent and the function

$$F_\alpha(w) = \frac{F(w)}{B_\alpha(w)} \; (\neq 0, \quad w \in G^-), \quad -1 < \alpha < +\infty,$$

is holomorphic in $G^-$.

First assume $\alpha \geq 0$. Then $\log|F_\alpha(w)|$ (which is harmonic in $G^-$) belongs to $M_p\{G^-\}$ since $\log|F(w)|$ and $\log|B_\alpha(w)|$ are of $M_p\{G^-\}$. Consequently, by Lemma 1.10 of Ch. 1 also the function

$$U(w) = W^{-\alpha}\log|F_\alpha(w)| = W^{-\alpha}\log|F(w)| - W^{-\alpha}\log|B_\alpha(w)| \tag{3.27}$$

is harmonic in $G^-$. By (3.21) and formula (3.14) of Ch. 2

$$\sup_{v<0} \int_{-\infty}^{+\infty} [U(u+iv)]^+ du \leq \sup_{v<0} \int_{-\infty}^{+\infty} \left[W^{-\alpha}\log|F(u+iv)|\right]^+ du$$
$$+ \sup_{v<0} \int_{-\infty}^{+\infty} \left|W^{-\alpha}\log|B_\alpha(u+iv)|\right| du < +\infty.$$

Besides, if $\alpha = p \geq 1$ is a natural number, then by (3.22) and nonpositivity of $W^{-\alpha}\log|B_\alpha(w)|$ in $G^-$ (see Theorem 3.2 in Ch. 2)

$$\int_1^{+\infty} [U(-it)]^- \frac{dt}{t} \leq \int_1^{+\infty} \left[W^{-\alpha}\log|F_\alpha(-it)|\right]^- \frac{dt}{t} < +\infty.$$

If $\alpha \geq 0$ is not a natural number, then by Lemma 3.2 of Ch. 1

$$\int_1^{+\infty} [U(-it)]^- \frac{dt}{t^{1+\alpha-p}} \leq \int_1^{+\infty} W^{-\alpha}\left|\log|F_\alpha(-it)|\right| \frac{dt}{t^{1+\alpha-p}} < +\infty.$$

Thus, $U(w) \in \mathfrak{U}^+_{\alpha-p}$ since all requirements of Definition 3.2 are satisfied. Hence, $U(w)$ is representable in the form (3.18), (3.18''), where $\alpha = 0$, i.e.

$$W^{-\alpha}\log|F_\alpha(w)| = \operatorname{Re}\left\{\frac{1}{\pi i}\int_{-\infty}^{+\infty} \frac{d\mu(t)}{w-t}\right\}, \quad w \in G^-,$$

where $\mu(t)$ is as required in (3.25)–(3.25″). In view of Lemmas 1.10 and 3.1 of Ch. 1, applying $(\partial^p/\partial v^p)W^{-(p-\alpha)}$ to both sides of the last formula we come to (3.25).

Now let $-1 < \alpha < 0$. Then the proof of formulas (3.26)–(3.26′) and (3.26″) is quite similar to the previous case. The only difference is that instead of nonpositivity of $W^{-\alpha}\log|B_\alpha(w)|$ one has to use its continuity in $G^-\backslash\{w_k\}$ and the fact that the points $\{w_k\}$ (i.e. zeros of $F(w)$) are integrable singularities. Also the inclusion $W^{-\alpha}\log|B_\alpha(w)| \in K_1\{\pi/2, G^-\}$ has to be used, which is a consequence of (1.2′) and Lemma 1.9 of Ch. 1.

For proving the converse statements, first assume that $\alpha \geq 0$ and (3.25)–(3.25′) is valid. Then $\log|F(w)| \in M_p\{G^-\}$ by Lemma 3.3 of Ch. 1. The function $U(w)$ defined by (3.27) is harmonic in $G^-$ and representable in the form (3.18)–(3.18″), and $U(w) \in \mathfrak{U}^+_{\alpha-p}$ by Lemma 3.1. Therefore, by the estimate (3.14) of Ch. 2

$$\sup_{v<0}\int_{-\infty}^{+\infty}\left[W^{-\alpha}\log|F(u+iv)|\right]^+ du$$
$$\leq \sup_{v<0}\int_{-\infty}^{+\infty} U^+(u+iv)du + \sup_{v<0}\int_{-\infty}^{+\infty}\left|W^{-\alpha}\log|B_\alpha(u+iv)|\right| du < +\infty.$$

On the other hand, by (1.2) and Lemma 1.9 of Ch. 1

$$\int_1^{+\infty}\left[W^{-\alpha}\log|F(-it)|\right]^- \frac{dt}{t^{1+\alpha-p}}$$
$$\leq \int_1^{+\infty} U^-(-it)\frac{dt}{t^{1+\alpha-p}} + \int_1^{+\infty}\left|W^{-\alpha}\log|B_\alpha(-it)|\right|\frac{dt}{t^{1+\alpha-p}} < +\infty.$$

Thus, $F(w) \in \mathfrak{N}_\alpha$ $(0 \leq \alpha < +\alpha)$, as all requirements of Definition 3.3 are satisfied. For $-1 < \alpha < 0$ the proof is similar.

**Remark 3.3.** Let $F(w)$ be holomorphic in $G^-$ and also in a neighborhood of $\infty$. If

$$|F(w)| \leq \exp\left\{\frac{c_0}{|\mathrm{Im}\ w|^\rho}\right\}, \quad w \in G^-, \quad |\mathrm{Im}\ w| < r_0, \tag{3.28}$$

for some $\rho$, $r_0$, $c_0 > 0$, then for any $\alpha > \rho$ there exists a definite rational function $R(w)$ such that $F_1(w) = R(w)F(w) \in \mathfrak{N}_\alpha$, and hence $F_1(w)$ admits the factorization (3.25)–(3.25′).

**Proof.** By Lemma 2.3, $|\log|F_1(w)|| \leq c_1|w|^{3-[\alpha]}$ $(|w| > R_0)$ for a rational function $R(w)$ defined by the coefficients of the expansion of $F(w)$ at $\infty$. Therefore, by Lemma 1.9 of Ch. 1

$$\left|W^{-\alpha}\log|F_1(w)|\right| \leq \frac{c_2}{|w|^2}, \quad w \in \overline{G^-}, \quad |w| > R_0. \tag{3.29}$$

Besides, the estimate (3.28) remains true for $F_1(w)$ (only $c_0$ changes). Assuming $v<0$ any number, set

$$\int_{-\infty}^{+\infty}\left[W^{-\alpha}\log|F_1(u+iv)|\right]^+du$$
$$=\left(\int_{|u|>R_0}+\int_{-R_0}^{R_0}\right)\left[W^{-\alpha}\log|F_1(u+iv)|\right]^+du\equiv I_1(v)+I_2(v). \qquad (3.30)$$

From (3.29) it follows that $I_1(v)\le c_3\equiv 2c_2/R_0$ $(-\infty<v<0)$ and also that $I_2(v)$ is a continuous function of $v$ $(-\infty\le v<0)$, such that $I_2(-\infty)=0$. For evaluating this function as $v\to -0$, observe that for $-r_0/2<v<0$

$$I_2(v)\le\frac{2R_0}{\Gamma(\alpha)}\left\{\frac{c_1R_0^{-(3+[\alpha]-\alpha)}}{3+[\alpha]-\alpha}+\frac{c_0}{\alpha-\rho}\left(\frac{2}{r_0}\right)^{\alpha-p}\right\}+\varphi(v),$$

where

$$\varphi(v)\equiv\frac{1}{\Gamma(\alpha)}\int_{-R_0}^{R_0}du\int_{r_0/2}^{R_0}|\log|F_1(u+i(v-\sigma))||\,d\sigma$$

is continuous by $v$ on the whole interval $(-\infty,r_0/2)$. Therefore, $I_2(v)\le c_4$ $(-r_0<v<0)$ and $I_1(v)+I_2(v)\le c_3+c_4\equiv c_5$ for $-\infty<v<0$. Hence by (3.30) the relation (3.21) holds for $F_1(w)$, and $F_1(w)\in\mathfrak{N}_\alpha$ since this function satisfies all the remaining requirements of Definition 3.3 of $\mathfrak{N}_\alpha$.

**Theorem 3.6.** $\mathfrak{N}_\alpha^m$ $(-1<\alpha<+\infty)$ *coincides with the set of functions admitting in $G^-$ the factorization (3.25)–(3.26), where $B_\alpha(w)$ are convergent Blaschke type products with zeros satisfying (3.24), $a_0$, $a_1$, and $C$ are real numbers, and $\mu(t)$ is a function of bounded variation on the whole axis $-\infty<t<+\infty$. Besides, the relation (3.26$''$) is true.*

**Proof** does not essentially differ from that of Theorem 3.5. Therefore, we shall just outline the main stages. As in the proof of Theorem 3.5, consider the functions $F_\alpha(w)$ and $U(w)=W^{-\alpha}\log|F_\alpha(w)|$. One can be convinced that $U(w)\in\mathfrak{U}^m$ and hence is representable in the form (3.18), where $\alpha=0$ and the function $\mu(t)$ is of bounded variation on the whole axis $-\infty<t<+\infty$. In view of Lemmas 1.10 and 3.1 of Ch. 1, application of $\partial^p/\partial v^p W^{-(p-\alpha)}$ (if $\alpha\ge 0$) or $W^\alpha$ (if $-1<\alpha<0$) to both sides of this representation leads to the desired representations. For proving the converse statements, one has to verify that the function $U(w)$ defined by (3.33) belongs to $\mathfrak{U}^m$. Hence the inclusion $F(w)\in\mathfrak{N}_\alpha^m$ $(-1<\alpha<+\infty)$ holds in view of the properties of the Blaschke type product.

NOTES. One can find the general scheme of the proof of Theorem 2.3 in Nevanlinna's hard accessible paper [80] and in the monograph of Boas [10], where some misprints occur.

Qualitatively different from Theorem 3.2 factorizations for functions of any finite order, meromorphic in the closed half-plane, first were found by R.Nevanlinna [83]. Later N.V.Govorov [39] extent them to the case when the function is meromorphic in the open half-plane.

In view of Lemma 3.4 the classes $\mathfrak{N}_\alpha$ and $\mathfrak{N}_\alpha^m$ considered in Theorems 3.5 and 3.6 for $\alpha$=0 coincide, in essence, with the classes $\mathfrak{N}$ and $\mathfrak{N}_m$ of V.I.Krilov. This is precise for $\mathfrak{N}_0^m = \mathfrak{N}^m$ while $\mathfrak{N}_0$ coincides with the subset of functions from $\mathfrak{N}$, having bounded sets of zeros.

# CHAPTER 4

# MEROMORPHIC FUNCTIONS WITH SUMMABLE TSUJI CHARACTERISTICS

## 1. LEVIN TYPE FORMULAS AND TSUJI TYPE CHARACTERISTICS

All over this section, we assume that $F(w)$ is a function meromorphic in the lower half-plane $G^- = \{w : \text{Im } w < 0\}$, having finite sets of zeros $\{a_\mu\}$ and poles $\{b_\nu\}$ in any half-plane $G_\rho^- = \{w : \text{Im } w < \rho\}$ $(\rho < 0)$.

*1.1.* Let $\rho \in (-\infty, 0)$ and $R > \max_{\mu,\nu} \{|\text{Im } a_\mu|, |\text{Im } b_\nu|\}$ be fixed numbers. Then the mapping $z = [w + i(R-\rho)]/R = (w - i\rho)/R + i$ (a parallel shift with a contraction) transforms the disc $|w + i(R-\rho)| \le R$ into $|z| < 1$. On the other hand, there is a finite set of zeros and poles of $F$ in the initial disc, and they lie upper its center by our choice of $R$. Hence, the function

$$f(z) \equiv F(w), \qquad z = \frac{w - i\rho}{R} + i, \tag{1.1}$$

is meromorphic in the closed unit disc $|z| \le 1$, where $f$ has a finite set of zeros $a_\mu^* = (a_\mu - i\rho)/R + i$ and poles $b_\nu^* = (b_\nu - i\rho)/R + i$ lying in the upper half-plane. It is obvious that $f$ admits the factorization (7) of Introduction for any $\alpha > -1$ (where $\lambda = 0$, $k_\alpha = 1$ and $C_\lambda = C_0 = f(0)$). Replacing $\alpha$ by $\alpha - 1$ in that formula, we get

$$\begin{aligned}\log f(z) \equiv & \frac{2\alpha}{\pi} \iint_{|\zeta|<1} (1 - |\zeta|^2)^{\alpha-1} \frac{\log|f(\zeta)|}{(1 - z\overline{\zeta})^{1+\alpha}} d\sigma(\zeta) \\ & + \log \pi_{\alpha-1}\left(z, \{a_\mu^*\}\right) - \log \pi_{\alpha-1}\left(z, \{b_\nu^*\}\right) - \log f(0)\end{aligned} \tag{1.2}$$

$(\alpha > 0,\ |z| < 1)$, where $d\sigma(\zeta)$ is Lebesgue's area measure and

$$\log \pi_{\alpha-1}\left(z, \{z_k\}\right) \equiv -\sum_{|z_k|<1} \int_{|z_k|^2}^1 \frac{(1-t)^\alpha}{\left(1 - \frac{z}{z_k} t\right)^{1+\alpha}} \frac{dt}{t} \equiv -\sum_{|z_k|<1} \Phi_\alpha(z, z_k), \tag{1.3}$$

where $\{z_k\} = \{a_\mu^*\}$ or $\{z_k\} = \{b_\nu^*\}$. Using (1.1), one can write (1.2)–(1.3) as a representation for $F$. As for $\zeta \in O(R, \rho) = \{w : |w + i(R - \rho)| < R\}$

$$1 - \left|\frac{\zeta - i\rho}{R} + i\right|^2 = 2\frac{|\text{Im } \zeta - \rho|}{R} - \frac{|\zeta - i\rho|^2}{R^2} \in (0, 1], \tag{1.4}$$

$$1-\left(\frac{w-i\rho}{R}+i\right)\left(\frac{\overline{\zeta}+i\rho}{R}-i\right)$$
$$=\frac{1}{R}\left[i(w-\overline{\zeta}-2i\rho)-\frac{(w-i\rho)(\overline{\zeta}+i\rho)}{R}\right], \qquad (1.4')$$

for $0<\alpha<+\infty$ and $w\in O(R,\rho)$

$$\log F(w)\equiv\frac{\alpha 2^{\alpha}}{\pi}\iint_{O(R,\rho)}\left(1-\frac{|\zeta-i\rho|^2}{2R|\mathrm{Im}\ \zeta-\rho|}\right)^{\alpha-1}$$
$$\times\frac{|\mathrm{Im}\ \zeta-\rho|^{\alpha-1}\log|F(\zeta)|}{\left[i(w-\overline{\zeta}-2i\rho)-\frac{(w-i\rho)(\overline{\zeta}+i\rho)}{R}\right]^{1+\alpha}}d\sigma(\zeta)$$
$$-\sum_{a_\mu\in O(R,\rho)}\Phi_\alpha\left(\frac{w-i\rho}{R}+i\frac{a_\mu-i\rho}{R}+i\right)$$
$$+\sum_{b_\nu\in O(R,\rho)}\Phi_\alpha\left(\frac{w-i\rho}{R}+i\frac{b_\nu-i\rho}{R}+i\right)-\log F(-i(R-\rho)). \qquad (1.5)$$

In view of the definition (1.3) of $\Phi_\alpha$, taking $w=-i(R-\rho)$ we get

$$\Phi_\alpha\left(0,\frac{s-i\rho}{R}+i\right)=\left(\frac{2}{R}\right)^{1+\alpha}|\mathrm{Im}\ s-\rho|^{1+\alpha}\int_0^{\frac{1-|s-i\rho|^2}{2R|\mathrm{Im}\ s-\rho|}}\frac{\tau^\alpha d\tau}{1-\frac{2}{R}|\mathrm{Im}\ s-\rho|\tau}$$
$$\equiv\left(\frac{2}{R}\right)^{1+\alpha}|\mathrm{Im}\ s-\rho|^{1+\alpha}I_\alpha(s,\rho,R),\quad 0<\alpha<+\infty,\quad s\in O(R,\rho). \qquad (1.6)$$

Consequently, the same substitution in (1.5) gives

$$0=\frac{\alpha}{2\pi}\iint_{O(R,\rho)}\left(1-\frac{|\zeta-i\rho|^2}{2R|\mathrm{Im}\ \zeta-\rho|}\right)^{\alpha-1}|\mathrm{Im}\ \zeta-\rho|^{\alpha-1}\log|F(\zeta)|d\sigma(\zeta)$$
$$-\sum_{a_\mu\in O(R,\rho)}|\mathrm{Im}\ a_\mu-\rho|^{1+\alpha}I_\alpha(a_\mu,\rho,R)+\sum_{b_\nu\in O(R,\rho)}|\mathrm{Im}\ b_\nu-\rho|^{1+\alpha}I_\alpha(a_\nu,\rho,R)$$
$$-\frac{R^{1+\alpha}}{2^\alpha}\log|F(-i(R-\rho))|,\qquad 0<\alpha<+\infty. \qquad (1.7)$$

*1.2.* The following two lemmas are necessary for the passage $R\to+\infty$ in (1.7), which will lead to a family of Levin type formulas.

**Lemma 1.1.** *Let $\varphi(t)\geq 0$ be a measurable function on $(0,+\infty)$. Then for any $\alpha\in[1,+\infty)$*

$$\int_0^{+\infty}t^{\alpha-1}\varphi(t)dt=\sup_{\varepsilon>0}\int_\varepsilon^{+\infty}(t-\varepsilon)^{\alpha-1}\varphi(t)dt.$$

**Proof.** It is obvious that the right-hand side integral monotonely increases and tends to left-hand side integral for any sequence $\varepsilon \equiv \varepsilon_n \to +0$.

**Lemma 1.2.** *Let $\Psi(\zeta)$ be a measurable function in the half-plane $G_\rho^-$ ($-\infty < \rho < 0$), and let*

$$\iint_{G_\rho^-} |\text{Im } \zeta - \rho|^{\alpha-1} |\Psi(\zeta)| d\sigma(\zeta) < +\infty$$

*for some $\alpha \in [1, +\infty)$. Then*

$$\lim_{R\to+\infty} \iint_{O(R,\rho)} \left(1 - \frac{|\zeta - i\rho|^2}{2R|\text{Im } \zeta - \rho|}\right)^{\alpha-1} |\text{Im } \zeta - \rho|^{\alpha-1} \Psi(\zeta) d\sigma(\zeta)$$
$$= \iint_{G_\rho^-} |\text{Im } \zeta - \rho|^{\alpha-1} |\Psi(\zeta)| d\sigma(\zeta).$$

**Proof.** If

$$P_{\alpha,\rho}(\zeta, R) = \begin{cases} \left(1 - \dfrac{|\zeta - i\rho|^2}{2R|\text{Im } \zeta - \rho|}\right)^{\alpha-1} - 1 & \text{for} \quad \zeta \in O(R,\rho) \\ -1 & \text{for} \quad \zeta \notin O(R,\rho), \end{cases}$$

then our assertion is equivalent to

$$\lim_{R\to+\infty} \iint_{G_\rho^-} P_{\alpha,\rho}(\zeta, R) |\text{Im } \zeta - \rho|^{\alpha-1} \Psi(\zeta) d\sigma(\zeta) = 0. \tag{1.8}$$

For proving the latter relation, observe that $|P_{\alpha,\rho}(\zeta, R)| \leq 1$ for $\zeta \in G_\rho^-$. But

$$\lim_{R\to+\infty} P_{\alpha,\rho}(\zeta, R) = 0$$

for any fixed $\zeta \in G_\rho^-$. Therefore, (1.8) follows by Lebesgue's theorem.

**Theorem 1.1.** *Let $F$ be a meromorphic function in $G^-$, having a finite set of zeros $\{a_\mu\}$ and poles $\{b\nu\}$ in $G_\rho^- \subset G^-$ for any $\rho < 0$. If*

$$\iint_{G^-} |\text{Im } \zeta|^{\alpha-1} |\log |F(\zeta)|| \, d\sigma(\zeta) < +\infty \tag{1.9}$$

*for a given $\alpha \in [1, +\infty)$, then for any $\rho < 0$*

$$0 = \frac{\alpha}{2\pi} \iint_{G^-} |\text{Im } \zeta|^{\alpha-1} \log |F(\zeta)| d\sigma(\zeta) - \frac{1}{1+\alpha} \sum_{a_\mu \in G_\rho^-} |\text{Im } a_\mu - \rho|^{1+\alpha}$$
$$+ \frac{1}{1+\alpha} \sum_{b_\nu \in G_\rho^-} |\text{Im } b_\mu - \rho|^{1+\alpha} - 2^{-\alpha} \lim_{t\to+\infty} t^{1+\alpha} \log |F(-it)|, \tag{1.10}$$

*where the limit exists and is finite.*

**Proof.** By Lemma 1.1, (1.9) implies the requirements of Lemma 1.2 for $\Psi = \log|F|$. Consequently, the integral in (1.7) tends to that in (1.10) as $R \to +\infty$. It is obvious from (1.6), that $\lim_{R\to+\infty} I_\alpha(s,\rho,R) = (1+\alpha)^{-1}$ for any fixed $\rho < 0$ and $s \in G_\rho^-$. Therefore, the passage $R \to +\infty$ transforms the (finite) sums of (1.7) into those in (1.10). The existence and the finiteness of the limit in (1.10) is provided by the existence and the finiteness of all other limits.

*1.3.* The below theorem, which easily follows from Theorem XX of [76] by a passage, contains a special case of Levin's formula which gives rise to Tsuji characteristics.

**Theorem 1.2.** *Let $F(w)$ be meromorphic in $G^-$, and let $\{a_\mu\}$ and $\{b_\nu\}$ be zeros and poles of $F(w)$. If for any $\rho < 0$ in $G_\rho^-$ lies a finite set of points $b_\nu$ and for $\Phi(w) \equiv \log|F(w+i\rho)|$ the second condition of (4) in Introduction is fulfilled, then for any $\rho < 0$*

$$\sum_{a_\mu \in G_\rho^-} |a_\mu - \rho| - \sum_{b_\nu \in G_\rho^-} |b_\nu - \rho|$$

$$= \frac{1}{2\pi}\int_{-\infty}^{+\infty} \log|F(u+i\rho)|du - \frac{1}{2}\lim_{t\to+\infty} t\log|F(-it)|, \quad (1.11)$$

*where the limit exists and is finite.*

Assuming that $\mathbf{n}(t,F)$ is the number of poles $b_\nu$ of $F(w)$ (counted according to their multiplicities) lying in the half-plane $G_t^-$, consider the functions

$$\begin{aligned} \mathfrak{N}(\rho,F) &= \sum_{b_\nu \in G_\rho^-} |\mathrm{Im}\, b_\nu - \rho| = \int_{-\infty}^{\rho} (\rho - t)d\mathbf{n}(t,F), \\ \mathfrak{N}(\rho,F^{-1}) &= \sum_{a_\mu \in G_\rho^-} |\mathrm{Im}\, a_\nu - \rho| = \int_{-\infty}^{\rho} (\rho - t)d\mathbf{n}(t,F^{-1}), \end{aligned} \quad (1.12)$$

$$\begin{aligned} \mathbf{m}(\rho,F) &= \frac{1}{2\pi}\int_{-\infty}^{+\infty} \log^+|F(u+i\rho)|du, \\ \mathbf{m}(\rho,F^{-1}) &= \frac{1}{2\pi}\int_{-\infty}^{+\infty} \log^+\frac{1}{|F(u+i\rho)|}du, \end{aligned} \quad (1.13)$$

$$\mathcal{L}(\rho,F) = \mathbf{m}(\rho,F) + \mathfrak{N}(\rho,F), \;\; \mathcal{L}(\rho,F^{-1}) = \mathbf{m}(\rho,F^{-1}) + \mathfrak{N}(\rho,F^{-1}). \quad (1.14)$$

Formula (1.11) can be written as the equilibrium relation

$$\mathcal{L}(\rho,F) = \mathcal{L}(\rho,F^{-1}) + C_F, \quad -\infty < \rho < 0. \quad (1.15)$$

*1.4.* We will reduce a family of Tsuji type characteristics depending on $\alpha \in [1, +\infty)$ from the Levin *type* formula (1.10). For this, observe that

$$\frac{\alpha}{2\pi} \iint_{G_\rho^-} |\mathrm{Im}\ \zeta - \rho|^{\alpha-1} \log |F(\zeta)| d\sigma(\zeta) = \alpha \left[\mathbf{m}_\alpha(\rho, F) - \mathbf{m}_\alpha(\rho, F^{-1})\right],$$

where the functions $\mathbf{m}_\alpha$ are some integrals of Tsuji characteristics (1.13):

$$\mathbf{m}_\alpha(\rho, F) = \int_{-\infty}^{\rho} (\rho - t)^{\alpha-1} \mathbf{m}(t, F) dt,$$
$$\mathbf{m}_\alpha(\rho, F^{-1}) = \int_{-\infty}^{\rho} (\rho - t)^{\alpha-1} \mathbf{m}(t, F^{-1}) dt.$$

It is obvious that

$$\sum_{b_\nu \in G_\rho^-} |\mathrm{Im}\ b_\nu - \rho|^{1+\alpha} = \int_{-\infty}^{\rho} (\rho - t)^{\alpha+1} d\mathbf{n}(t, F) = \int_{-\infty}^{\rho} (\rho - t)^{\alpha} d\mathfrak{N}(t, F),$$

where $\mathfrak{N}(t, F)$ is the Tsuji characteristic (1.12) for the poles of $F$. An integration by parts gives

$$\frac{1}{1+\alpha} \sum_{b_\nu \in G_\rho^-} |\mathrm{Im}\ b_\nu - \rho|^{1+\alpha} = \alpha \mathfrak{N}_\alpha(\rho, F),$$

where

$$\mathfrak{N}_\alpha(\rho, F) = \frac{1}{1+\alpha} \int_{-\infty}^{\rho} (\rho - t)^{\alpha-1} \mathfrak{N}(t, F) dt.$$

Similarly,

$$\frac{1}{1+\alpha} \sum_{a_\mu \in G_\rho^-} |\mathrm{Im}\ a_\mu - \rho|^{1+\alpha} = \alpha \mathfrak{N}_\alpha(\rho, F^{-1})$$

for zeros of $F$. Additionally, denoting $(\alpha 2^\alpha)^{-1} \lim_{t \to +\infty} t^{1+\alpha} \log |F(-it)| = C_\alpha(F)$, one can write (1.10) in the form

$$\mathbf{m}(\rho, F) + \mathfrak{N}(\rho, F) = \mathbf{m}(\rho, F^{-1}) + \mathfrak{N}(\rho, F^{-1}) + C_\alpha(F), \quad -\infty < \rho < 0,$$

or, which is the same, in the form of an equilibrium relation:

$$\mathcal{L}_\alpha(\rho, F) = \mathcal{L}_\alpha(\rho, F^{-1}) + C_\alpha(F), \quad -\infty < \rho < 0, \tag{1.16}$$

where $\mathcal{L}_\alpha$ are the Tsuji type characteristics

$$\mathcal{L}_\alpha(\rho, F) = \int_{-\infty}^{\rho} (\rho - t)^{\alpha-1} \left[\mathbf{m}(t, F) + \frac{1}{1+\alpha} \mathfrak{N}(t, F)\right] dt,$$
$$\mathcal{L}_\alpha(\rho, F^{-1}) = \int_{-\infty}^{\rho} (\rho - t)^{\alpha-1} \left[\mathbf{m}(t, F^{-1}) + \frac{1}{1+\alpha} \mathfrak{N}(t, F^{-1})\right] dt. \tag{1.17}$$

*1.5.* The equilibrium relation

$$T(r,f) = T(r,f^{-1}) + O(1), \quad 0 < r < 1,$$

between Nevanlinna characteristics of a function $f$ meromorphic in the unit disc $\mathbb{D} = \{z : |z| < 1\}$ is well known. In view of this relation, Nevanlinna's condition (5) (Introduction) defining his weighted class is equivalent to

$$\int_0^1 (1-r)^\alpha \left[T(r,f) + T(r,f^{-1})\right] dr < +\infty, \quad -1 < \alpha < +\infty,$$

and for $0 \le \alpha < +\infty$ it is equivalent to

$$\sup_{0<r<1} \left[T_\alpha(r,f) + T_\alpha(r,f^{-1})\right] < +\infty,$$

where $T_\alpha$ is a Nevanlinna type characteristic of the form

$$T_\alpha(r,f) = \int_0^r (r-t)^\alpha T(t,f) dt.$$

But one can be convinced that the equilibrium relation (1.15) between Tsuji characteristics is true *not for all* functions meromorphic in the half-plane $G^-$. Therefore, the similarity of Nevanlinna's weighted class is to be defined as the set of meromorphic functions $F$ in $G^-$, for which

$$\int_{-\infty}^0 |t|^{\alpha-1} \left[\mathcal{L}(t,F) + \mathcal{L}(t,F^{-1})\right] dt < +\infty, \quad 0 < \alpha < +\infty. \tag{1.18}$$

By Lemma 1.1, for $1 \le \alpha < +\infty$ the last condition is equivalent to

$$\sup_{-\infty<\rho<0} \left[\mathcal{L}_\alpha(t,F) + \mathcal{L}_\alpha(t,F^{-1})\right] < +\infty.$$

One can verify that (1.18) is equivalent also to requirements of Theorem 1.1 which provide the validity of the Levin type formula (1.10) and the equilibrium relations (1.16) for the above Tsuji type characteristics.

## 2. CANONICAL FACTORIZATIONS

In this section, successive passages $R \to +\infty$ and $\rho \to -0$ in (1.5) will lead to canonical factorizations of the below classes of meromorphic functions.

*2.1.* **Definition 2.1.** *A function $F$ meromorphic in the half-plane $G^-$ is said to be of the class $\mathfrak{N}^m_{\alpha,\beta}$ $(0 < \alpha < +\infty, 0 \le \beta \le 1+\alpha)$, if*

$$\int_{-\infty}^0 \frac{|t|^{\alpha-1}}{1+|t|^\beta} \left[\mathcal{L}(t,F) + \mathcal{L}(t,F^{-1})\right] dt < +\infty. \tag{2.1}$$

The following inclusions are obvious:

$$\mathfrak{N}^m_{\alpha,\beta_1} \subseteq \mathfrak{N}^m_{\alpha,\beta_2}, \qquad 0 \le \beta_1 \le \beta_2 \le 1+\alpha, \tag{2.2}$$

$$\mathfrak{N}^m_{\alpha,\beta} \subseteq \mathfrak{N}^m_{\alpha+\delta,\beta+\delta}, \qquad 0 \le \beta \le 1+\alpha, \quad 0 < \delta < +\infty. \tag{2.2'}$$

**Lemma 2.1.** *If $F \in \mathfrak{N}^m_{\alpha,\beta}$ $(\alpha > 0,\ 0 \le \beta \le 1+\alpha)$, then zeros $\{a_\mu\} \subset G^-$ and poles $\{b_\nu\} \subset G^-$ of this function satisfy the density conditions*

$$\sum_\mu |\mathrm{Im}\ a_\mu|^{1+\alpha} < +\infty \quad \textit{and} \quad \sum_\nu |\mathrm{Im}\ b_\nu|^{1+\alpha} < +\infty. \tag{2.3}$$

*Besides,*

$$\iint_{G^-} \frac{|\mathrm{Im}\ \zeta|^{\alpha-1}}{1+|\mathrm{Im}\ \zeta|^\beta} |\log|F(\zeta)||\, d\sigma(\zeta) < +\infty. \tag{2.4}$$

**Proof.** (2.4) follows from (2.1). From (2.1) also follows that

$$\int_{-\infty}^0 \frac{|t|^{\alpha-1}}{1+|t|^\beta} \mathfrak{N}\left(t, F^{\pm 1}\right) dt < +\infty. \tag{2.5}$$

Hence (2.3) holds. Indeed, for proving, for instance, the boundedness of the first sum in (2.3) suppose that for some $\rho_0 < 0$ there is an infinite set of zeros $a_\mu$ of $F$ in $G^-_{\rho_0}$. Then for any $t \in (\rho_0/2, 0)$

$$\begin{aligned}\mathfrak{N}(t,F^{-1}) &= \sum_{a_\mu \in G^-_t} (|\mathrm{Im}\ a_\mu| - |t|) \ge \sum_{a_\mu \in G^-_t} (|\mathrm{Im}\ a_\mu| - |\rho_0|/2) \\ &\ge \sum_{a_\mu \in G^-_{\rho_0}} (|\mathrm{Im}\ a_\mu| - |\rho_0|/2) \ge \sum_{a_\mu \in G^-_{\rho_0}} (|\rho_0| - |\rho_0|/2) = +\infty\end{aligned}$$

which contradicts (2.5). Thus, $\max_\mu |\mathrm{Im}\ a_\mu| = m_0 < +\infty$, and

$$\begin{aligned}\int_{-\infty}^0 \frac{|t|^{\alpha-1}}{1+|t|^\beta} \mathfrak{N}(t,F^{-1})dt &= \sum_\mu \int_0^{|\mathrm{Im}\ a_\mu|} \frac{|t|^{\alpha-1}}{1+|t|^\beta} (|\mathrm{Im}\ a_\mu| - t)\, dt \\ &\ge (1+m_0^\beta)^{-1} \sum_\mu \int_0^{|\mathrm{Im}\ a_\mu|} |t|^{\alpha-1} (|\mathrm{Im}\ a_\mu| - t)\, dt \\ &= \left\{\alpha(1+\alpha)(1+m_0^\beta)\right\}^{-1} \sum_\mu |\mathrm{Im}\ a_\mu|^{1+\alpha}.\end{aligned}$$

*2.2.* Now the passage $R \to +\infty$ in (1.5) will lead to some similarities of the Jensen–Nevanlinna formula for functions from $\mathfrak{N}^m_{\alpha,1+\alpha}$. These formulas, as well as following them canonical representations of $\mathfrak{N}^m_{\alpha,\beta}$ $(\alpha > 0,\ \beta \in [0, 1+$

$\alpha$]) contain some Blaschke type products which are considered in [69]. These products are similar to those considered in Ch. 1 and consist of factors of the form

$$a_\alpha(w,\zeta) = \exp\left\{-\int_0^{2|\mathrm{Im}\,\zeta|} \frac{\tau^\alpha d\tau}{[\tau + i(w-\zeta)]^{1+\alpha}}\right\}. \tag{2.6}$$

As the factor of Ch. 1, for any fixed $\zeta \in G^-$ and $\alpha > -1$ the function $a_\alpha(w,\zeta)$ is holomorphic in $G^-$ and vanishes only at $w = \zeta$ which is a first order zero. Besides, the product

$$B_\alpha(w,\{w_k\}) = \prod_k a_\alpha(w,w_k), \quad \{w_k\} \subset G^-, \quad -1 < \alpha < +\infty, \tag{2.7}$$

is convergent when

$$\sum_k |\mathrm{Im}\ w_k|^{1+\alpha} < +\infty. \tag{2.8}$$

We start by some preliminary lemmas.

**Lemma 2.2.** *Let $w$ and $s$ are fixed points in $G_\rho^-$ $(\rho \le 0)$. Then for any $\alpha > -1$*

$$\lim_{R\to+\infty} \Phi_\alpha\left(\frac{w-i\rho}{R} + i, \frac{s-i\rho}{R} + i\right) = -\log a_\alpha(w-i\rho, s-i\rho). \tag{2.9}$$

**Proof.** Let $z, \zeta \in G^-$ be any fixed points and $R > 0$ be chosen enough great to provide $|z + iR| < R$ and $|\zeta + iR| < R$. Then by (1.3)

$$\Phi_\alpha\left(\frac{z}{R} + i, \frac{\zeta}{R} + i\right) = \int\limits_{\left|\frac{\zeta}{R}+i\right|^2}^{1} \frac{(1-t)^\alpha}{\left(1 - \frac{z+iR}{\zeta+iR}t\right)^{1+\alpha}} \frac{dt}{t}.$$

Hence

$$\Phi_\alpha\left(\frac{z}{R} + i, \frac{\zeta}{R} + i\right) = \int_0^{2|\mathrm{Im}\,\zeta| - \frac{|\zeta|^2}{R}} \frac{\tau^\alpha}{\left[\frac{\tau + i(z-\zeta) - i\tau z R^{-1}}{1 - i\zeta R^{-1}}\right]^{1+\alpha}} \frac{d\tau}{1 - \frac{\tau}{R}}$$

by (1.4) and $\tau = R(1-t)$. Thus,

$$\lim_{R\to+\infty} \Phi_\alpha\left(\frac{z}{R} + i, \frac{s}{R} + i\right) = \int_0^{2|\mathrm{Im}\,\zeta|} \frac{\tau^\alpha d\tau}{[\tau + i(z-\zeta)]^{1+\alpha}}.$$

In view of (2.6), taking $z = w - i\rho$, $\zeta = s - i\rho$ we arrive at (2.9).

**Lemma 2.3.** *Let $F \in \mathfrak{N}^m_{\alpha,1+\alpha}$ $(\alpha \geq 1)$. Then for any $\rho < 0$*

$$\lim_{R\to+\infty} \iint\limits_{O(R,\rho)} \left(1 - \frac{|\zeta - i\rho|^2}{2R|\mathrm{Im}\ \zeta - \rho|}\right)^{\alpha-1} \frac{|\mathrm{Im}\ \zeta - \rho|^{\alpha-1} \log|F(\zeta)|}{\left[i(w - \overline{\zeta} - 2i\rho) - \frac{(w-i\rho)(\overline{\zeta}+i\rho)}{R}\right]^{1+\alpha}} d\sigma(\zeta)$$
$$= \iint_{G^-} |\mathrm{Im}\ \zeta - \rho|^{\alpha-1} \frac{\log|F(\zeta)|}{\left[i(w - \overline{\zeta} - 2i\rho)\right]^{1+\alpha}} d\sigma(\zeta), \quad w \in G^-_\rho. \tag{2.10}$$

**Proof** is similar to that of Lemma 1.2. Namely, introducing the function

$$\Psi_{\alpha,\rho}(w,\zeta,R) + \frac{1}{\left[i(w - \overline{\zeta} - 2i\rho)\right]^{1+\alpha}} = \begin{cases} \dfrac{\left(1 - \frac{|\zeta - i\rho|^2}{2R|\mathrm{Im}\ \zeta - \rho|}\right)^{\alpha-1}}{\left[i(w - \overline{\zeta} - 2i\rho) - \frac{(w-i\rho)(\overline{\zeta}+i\rho)}{R}\right]^{1+\alpha}} & \text{if} \quad \zeta \in O(R,\rho) \\ 0 & \text{if} \quad \zeta \notin O(R,\rho) \end{cases}$$

write (2.10) in the following equivalent form:

$$\lim_{R\to+\infty} \iint_{G^-_\rho} \Psi_{\alpha,\rho}(w,\zeta,R) |\mathrm{Im}\ \zeta - \rho|^{\alpha-1} \log|F(\zeta)| d\sigma(\zeta) = 0. \tag{2.11}$$

For proving this relation, evaluate $\Psi_{\alpha,\rho}$ in assumption that $w \in G^-_\rho$ is a fixed point and $R > 2|w - i\rho|^2 / |\mathrm{Im}\ w - \rho|$. For $\zeta \in O(R,\rho)$ obviously

$$|\Psi_{\alpha\rho}(w,\zeta,R)| \leq \left|(w - i\rho) - \overline{(\zeta - i\rho)}\right|^{-(1+\alpha)}$$
$$\times \left\{1 + \left(1 - \frac{|\zeta - i\rho|^2}{2R|\mathrm{Im}\ \zeta - \rho|}\right)^{\alpha-1} \left[1 - \frac{1}{R}\left|\frac{(w - i\rho)\overline{(\zeta - i\rho)}}{(w - i\rho) - \overline{(\zeta - i\rho)}}\right|\right]^{-(1+\alpha)}\right\}.$$

Besides,

$$\frac{1}{R}\left|\frac{(w - i\rho)\overline{(\zeta - i\rho)}}{(w - i\rho) - \overline{(\zeta - i\rho)}}\right| = \frac{1}{R}\left|\frac{1}{w - i\rho} - \frac{1}{\overline{(\zeta - i\rho)}}\right|^{-1}$$
$$< \frac{1}{R}\left(\mathrm{Im}\ \frac{1}{w - i\rho}\right)^{-1} = \frac{|w - i\rho|^2}{R|\mathrm{Im}\ w - \rho|} < \frac{1}{2}$$

since $\mathrm{Im}\ (w - i\rho) < 0$ and $\mathrm{Im}\ (\zeta - i\rho) < 0$. But

$$|(w - i\rho) - \overline{(\zeta - i\rho)}| \geq |\mathrm{Im}\ w - \rho| + |\mathrm{Im}\ \zeta - \rho|.$$

Therefore by (2.4)

$$|\Psi_{\alpha,\rho}(w,\zeta,R)| < \frac{1+2^{1+\alpha}}{\left(|\mathrm{Im}\ w-\rho|+|\mathrm{Im}\ \zeta-\rho|\right)^{1+\alpha}}$$

for any fixed $\rho<0$, $w\in G_\rho^-$ and any $R>2|w-i\rho|^2/\,|\mathrm{Im}\ w-\rho|$, $\zeta\in O(R,\rho)$. In view of the definition of $\Psi_{\alpha,\rho}$, this estimate remains true for $\zeta\in G_\rho^-\backslash O(R,\rho)$. Further, the use of the inequality

$$\frac{1}{(a+b)^{1+\alpha}}\le\frac{1}{a^{1+\alpha}+b^{1+\alpha}}\le\frac{\max\left\{1,a^{-(1+\alpha)}\right\}}{1+b^{1+\alpha}}\quad(0<a,b<+\infty,\ \alpha\ge 1)$$

gives

$$|\Psi_{\alpha,\rho}(w,\zeta,R)| < \frac{C_{\alpha,\rho}(w)}{1+|\mathrm{Im}\ \zeta-\rho|^{1+\alpha}}, \tag{2.12}$$

where $C_{\alpha,\rho}(w)=(1+2^{1+\alpha})\max\left\{1,|\mathrm{Im}\ w-\rho|^{-(1+\alpha)}\right\}$. Now set

$$g(x)=\frac{1+x^{1+\alpha}}{1+(x-a)^{1+\alpha}},\quad 0<a\le x<+\infty.$$

Evidently $g(x)<\left(1+x^{1+\alpha}\right)(x-a)^{-(1+\alpha)}<1+(1+a)^{1+\alpha}$ for $x>1+a$, and this is true also for $a\le x\le 1+a$. Consequently

$$\frac{1}{1+(x-a)^{1+\alpha}}<\frac{1+(1+a)^{1+\alpha}}{1+x^{1+\alpha}},\quad 0<a\le x<+\infty,\quad 1\le\alpha<+\infty.$$

Therefore, by (2.12)

$$|\Psi_{\alpha,\rho}(w,\zeta,R)| < \frac{C_{\alpha,\rho}(w)\left[1+(1+|\rho|)^{1+\alpha}\right]}{1+|\mathrm{Im}\ \zeta|^{1+\alpha}}.$$

From this estimate, from the convergence of (2.4) (with $\beta=1+\alpha$) and from Lemma 1.1 it follows that the integrand in (2.11) has a summable majorant independent of $R>2|w-i\rho|^2|\mathrm{Im}\ w-\rho|^{-1}$. On the other hand, it is obvious that $\lim_{R\to+\infty}\Psi_{\alpha,\rho}(w,\zeta,R)=0$ for any fixed $\rho<0$ and $w,\zeta\in G_\rho^-$. Therefore, (2.11) follows by Lebesgue's theorem.

**Theorem 2.1.** *Let $F\in\mathfrak{N}^m_{\alpha,1+\alpha}$ $(\alpha\ge1)$. Then for any $\rho<0$*

$$\begin{aligned}\log F(w)\equiv&\frac{\alpha 2^\alpha}{\pi}\iint_{G_\rho^-}|\mathrm{Im}\ \zeta-\rho|^{\alpha-1}\frac{\log|F(\zeta)|}{\left[i(w-\overline{\zeta}-2i\rho)\right]^{1+\alpha}}d\sigma(\zeta)\\&+\sum_{a_\mu\in G_\rho^-}\log a_\alpha(w-i\rho,a_\mu-i\rho)-\sum_{b_\nu\in G_\rho^-}\log a_\alpha(w-i\rho,b_\nu-i\rho)\\&-\log\left\{\lim_{t\to+\infty}F(-it)\right\},\quad w\in G_\rho^-,\end{aligned} \tag{2.13}$$

*where $\{a_\mu\}$ and $\{b_\nu\}$ are the zeros and poles of $F$ and the limit exists, is finite and not equal to zero.*

**Proof** follows from formulas (1.5), (2.9) and (2.10).

**Lemma 2.4.** *Let $\{w_k\} \subset G^-$ satisfy (2.8) for a given $\alpha \geq -1$. Then*

$$\lim_{\rho \to -0} \sum_{w_k \in G_\rho^-} \log a_\alpha(w - i\rho, w_k - i\rho) = \sum_k \log a_\alpha(w, w_k) \tag{2.14}$$

*for any $w \in G^-$ which does not lie in an intercept $l_k = \{w : \text{Re } w = \text{Re } w_k, \text{Im } w_k \leq \text{Im } w < 0\}$ $(k \geq 1)$.*

**Proof.** Assume $\text{Im } w \leq -4m_0$, where $m_0 = \max_k |\text{Im } w_k|$. Then

$$\left| \sum_{w_K \in G_\rho^-} \log a_\alpha(w - i\rho, w_k - i\rho) - \sum_k \log a_\alpha(w, w_k) \right|$$
$$\leq \sum_{\text{Im } w_k < \rho} |\log a_\alpha(w - i\rho, w_k - i\rho) - \log a_\alpha(w, w_k)|$$
$$+ \sum_{\rho \leq \text{Im } w_k < 0} |\log a_\alpha(w, w_k)| \equiv S_1 + S_2.$$

One can verify that the series $S_2$ is convergent. Hence, for any $\varepsilon > 0$ there exists $\rho_1 < 0$ such that

$$S_2 < \varepsilon/2 \quad \text{for} \quad \rho_1 < \rho < 0. \tag{2.15}$$

Now observe that by (2.6)

$$S_1 \leq \sum_{\text{Im } w_k < \rho} \int_{2(|\text{Im } w_k| - |\rho|)}^{2|\text{Im } w_k|} \frac{\tau^\alpha d\tau}{|\tau + i(w - w_k)|^{1+\alpha}},$$

where $|w - w_k| - \tau \geq m_0$ since $|w - w_k| \geq 3m_0$ and $0 \leq \tau \leq 2m_0$. Therefore

$$S_1 \leq \frac{2^{1+\alpha}}{(1+\alpha)m_0^{1+\alpha}} \left\{ \sum_{\text{Im } w_k < \rho} |\text{Im } w_k|^{1+\alpha} - \sum_{\text{Im } w_k < \rho} (|\text{Im } w_k| - |\rho|)^{1+\alpha} \right\}.$$

One can verify that under the condition (2.8)

$$\lim_{\rho \to -0} \sum_{\text{Im } w_k < \rho} (|\text{Im } w_k| - |\rho|)^{1+\alpha} = \sum_k |\text{Im } w_k|^{1+\alpha},$$

and the first summand in the above figure brackets tends to the same limit. Consequently, there exists $\rho_2 < 0$ such that $S_1 < \varepsilon/2$ for $\rho_2 < \rho < 0$. Hence by (2.15) the relation (2.14) holds for any $w$ such that $\text{Im } w \leq -4m_0$. Besides, if $\{w_k\}$ is a finite sequence then (2.14) is true for all $w \in G^- \backslash \cup_k \{l_k\}$. Therefore, (2.14) is valid out of the set $\cup_k \{l_k\}$ also in the case of infinite sequence $\{w_k\}$.

**Lemma 2.5.** *If* $F \in \mathfrak{N}^m_{\alpha,1+\alpha}$ $(\alpha \geq 1)$, *then*

$$\lim_{\rho \to -0} \iint_{G_\rho^-} |\text{Im } \zeta - \rho|^{\alpha-1} \frac{\log|F(\zeta)|}{[i(w - \overline{\zeta} - 2i\rho)]^{1+\alpha}} d\sigma(\zeta)$$

$$= \iint_{G^-} |\text{Im } \zeta|^{\alpha-1} \frac{\log|F(\zeta)|}{[i(w - \overline{\zeta})]^{1+\alpha}} d\sigma(\zeta), \quad w \in G^-. \tag{2.16}$$

**Proof** is similar to that of Lemma 2.2.

**Lemma 2.6.** *Let* $\{w_k\} \subset G^-$ *satisfy (2.8) for a given* $\alpha \geq -1$ *and let* $w$ *be fixed and such that* $\text{Im } w \leq -4m_0$ $(m_0 = \max\{1, \max_k |\text{Im } w_k|\})$. *Then*

$$\log B_\gamma(w, \{w_k\}) \equiv -\sum_k \int_0^{2|\text{Im } w_k|} \frac{\tau^\alpha d\tau}{[\tau + i(w - w_k)]^{1+\gamma}} \tag{2.17}$$

*is a continuous function of* $\gamma \in [\alpha, +\infty)$.

**Proof.** All summands in (2.17) obviously have the required property. Therefore, it suffices to prove the absolute and uniform convergence of this series in any segment $[\alpha, \beta] \subset [\alpha, +\infty)$. Assuming that $\gamma \in [\alpha, \beta]$, we have $|\tau + i(w - w_k)|^{1+\gamma} \geq m_0^{1+\alpha}$. Hence for $0 < |\text{Im } w_k| < 1/2$

$$\left| \int_0^{2|\text{Im } w_k|} \frac{\tau^\alpha d\tau}{[\tau + i(w - w_k)]^{1+\gamma}} \right| \leq \frac{2^{1+\alpha}}{(1+\alpha) m_0^{1+\alpha}} |\text{Im } w_k|^{1+\alpha}.$$

*2.4.* **Theorem 2.2.** *Let* $F \in \mathfrak{N}^m_{\alpha,1+\alpha}$ $(\alpha > 0)$. *Then*

$$F(w) \equiv C_F \frac{B_\alpha(w, \{a_\mu\})}{B_\alpha(w, \{b_\nu\})} \times \exp\left\{ \frac{\alpha 2^\alpha}{\pi} e^{-i\frac{\pi}{2}(1+\alpha)} \iint_{G^-} |\text{Im } \zeta|^{\alpha-1} \frac{\log|F(\zeta)|}{(w - \overline{\zeta})^{1+\alpha}} d\sigma(\zeta) \right\}, \tag{2.18}$$

*where* $B_\alpha$ *are convergent products constructed by zeros and poles of* $F$, *and*

$$C_F = \lim_{t \to +\infty} [F(-it)]^{-1} \neq 0$$

*is a finite number.*

**Proof.** Let $\alpha \geq 1$. Then (2.18) follows by the passage $\rho \to -0$ in (2.13), in view of (2.16), (2.14) and the uniqueness of a holomorphic function. Now let $0 < \alpha < 1$. Then, by (2.2$'$) $F \in \mathfrak{N}^m_{\gamma,1+\gamma}$ for any $\gamma \geq \alpha$, and hence $F$ admits the representation (2.18) for any $\gamma \geq 1$, i.e.

$$F(w) \equiv C_F \frac{B_\gamma(w,\{a_\mu\})}{B_\gamma(w,\{b_\nu\})} \exp\left\{\frac{\gamma 2^\gamma}{\pi}\iint_{G^-} |\text{Im}\ \zeta|^{\gamma-1}\frac{\log|F(\zeta)|}{\left[i(w-\overline{\zeta})\right]^{1+\alpha}}d\sigma(\zeta)\right\}. \quad (2.19)$$

Setting $m_1 = \max_{\mu,\nu}\{|\text{Im}\ a_\mu|, |\text{Im}\ b_\nu|\}$, fix any $w \in G^-$ such that $\text{Im}\ w \leq -4m_1$. Then the right-hand side of (2.19) becomes identically constant. On the other hand, by Lemma 2.6 the factors in (2.19) are continuous functions of $\gamma \in [\alpha, +\infty)$. Hence formula (2.18) holds.

**Remark 2.1.** In view of inclusion (2.2), the factorization (2.18) is true for any function from $\mathfrak{N}^m_{\alpha,\beta}$ ($\alpha > 0$, $0 \leq \beta \leq 1+\alpha$).

**Remark 2.2.** The argument used to prove the case $0 < \alpha < 1$ in Theorem 2.2 permits to extend the assertion of Theorem 2.1 (i.e. formula (2.13)) to the case $0 < \alpha < 1$.

*2.5.* Below we prove a useful estimate for the exponential factor of (2.18).

**Lemma 2.7.** *If $F \in \mathfrak{N}^m_{\alpha,\beta}$ ($\alpha > 0$, $\beta \in [0, 1+\alpha]$), then for any $w \in G^-$*

$$\left|\frac{\alpha 2^\alpha}{\pi}\iint_{G^-} |\text{Im}\ \zeta|^{\alpha-1}\frac{\log|F(\zeta)|}{\left[i(w-\overline{\zeta})\right]^{1+\alpha}}d\sigma(\zeta)\right| \leq \frac{\alpha 2^{\alpha+|\beta-1|}}{\pi}\frac{1+|\text{Im}\ w|^\beta}{|\text{Im}\ w|^{1+\alpha}}\iint_{G^-}\frac{|\text{Im}\ \zeta|^{\alpha-1}}{1+|\text{Im}\ \zeta|^\beta}\left|\log|F(\zeta)|\right| d\sigma(\zeta). \quad (2.20)$$

**Proof.** Obviously $|w-\overline{\zeta}| \geq |\text{Im}\ w| + |\text{Im}\ \zeta|$ for any $w, \zeta \in G^-$. Therefore, using the inequality $a+b > a(1+b)/(1+a)$ $(a,b>0)$ we get

$$\frac{1}{|w-\overline{\zeta}|^\beta} \leq \frac{1}{|\text{Im}\ w|^\beta}\frac{(1+|\text{Im}\ w|)^\beta}{(1+|\text{Im}\ \zeta|)^\beta}.$$

Further, $(1+x)^\beta/(1+x^\beta) \in [1, 2^{\beta-1}]$ $(x > 0,\ \beta \geq 0)$. Thus,

$$\frac{1}{|w-\overline{\zeta}|^\beta} \leq 2^{|\beta-1|}\frac{1+|\text{Im}\ w|^\beta}{|\text{Im}\ w|^\beta}\frac{1}{1+|\text{Im}\ \zeta|^\beta}$$

and

$$\frac{1}{|w-\overline{\zeta}|^{1+\alpha}} \leq 2^{|\beta-1|}\frac{1+|\text{Im}\ w|^\beta}{|\text{Im}\ w|^{1+\alpha}}\frac{1}{1+|\text{Im}\ \zeta|^\beta}.$$

Using the last inequality we come to (2.20).

**Remark 2.1.** If $\{w_k\} \subset G^-$ satisfies (2.8) for an $\alpha > -1$, then one can show that

$$|\log B_\alpha(w, w_k\})| \leq \frac{8^{1+\alpha}}{1+\alpha}\left\{\sum_k |\text{Im } w_k|^{1+\alpha}\right\} |\text{Im } w|^{-(1+\alpha)} \qquad (2.21)$$

for $\text{Im } w < -5\max_k |\text{Im } w_k|$. By (2.20) and (2.21), $(\lim_{t\to+\infty} F(-it))^2 = 1$ in the factorization (2.18), i.e.

$$C_F = \lim_{t\to+\infty} [F(-it)]^{-1} = \pm 1 \qquad (2.22)$$

for any $F \in \mathfrak{N}^m_{\alpha,\beta}$ $(\alpha > 0,\ \beta \in [0, 1+\alpha))$.

*2.6.* **Definition 2.2.** *A function $F$ holomorphic in $G^-$ is of the class $\mathfrak{N}^+_\alpha$ $(0 < \alpha < +\infty)$ if for any $\rho < 0$*

$$\sup_{v<\rho} \int_{-\infty}^{+\infty} |\log |F(u+iv)|| \, du < +\infty \qquad (2.23)$$

*and*

$$\int_{-\infty}^{0} \frac{|t|^{\alpha-1}}{1+|t|^{1+\alpha}} \mathcal{L}(t, F) dt \equiv \int_{-\infty}^{0} \frac{|t|^{\alpha-1}}{1+|t|^{1+\alpha}} dt \int_{-\infty}^{+\infty} \log^+ |F(u+it)| du < +\infty. \qquad (2.24)$$

By Theorem 1.2, Levin formula (1.11) is true for any function $F \in \mathfrak{N}^+_\alpha$ $(0 < \alpha < +\infty)$. This implies the equilibrium relation (1.15). Consequently

$$\mathfrak{N}^+_\alpha \subset \mathfrak{N}^m_{\alpha,1+\alpha}, \quad 0 < \alpha < +\infty. \qquad (2.25)$$

Hence, all assertions of Lemma 2.1, as well as the factorization (2.18) of Theorem 2.2 are true for these new classes $\mathfrak{N}^+_\alpha$.

## 3. DESCRIPTIVE REPRESENTATIONS

We start by some assertions related to inclusion of multipliers of the factorization (2.18) in $\mathfrak{N}^m_{\alpha,\beta}$ $(\alpha > 0,\ \beta \in [0, 1+\alpha])$.

*3.1.* We start by a Tsuji type estimate which is similar to that found by F.A.Shamoian [94] in $|z| < 1$.

**Lemma 3.1.** *Let a sequence* $\{w_k\} \subset G^-$ *satisfy the condition*

$$\sum_k |\text{Im } w_k|^{1+\alpha} < +\infty \tag{3.1}$$

*for a given* $\alpha > -1$. *Then*

$$\log |B_\alpha(w, \{w_k\})| \le C_\alpha \sum_k \frac{|\text{Im } w_k|^{1+\alpha}}{|w - \overline{w_k}|^{1+\alpha}}, \quad w \in G^-, \tag{3.2}$$

*where* $C_\alpha > 0$ *is a constant depending solely on* $\alpha$.

**Proof.** It suffices to prove (3.2) for a single factor of $B_\alpha$. To this end, we consider two cases in assumption that $\zeta = \xi + i\eta \in G^-$ is fixed.

**(a)** Let $2|\eta|/|w - \overline{\zeta}| > 1/5$, then we use the recurrent formula

$$a_\alpha(w, \zeta) = a_{\alpha-p}(w, \zeta) \exp\left\{ \sum_{n=1}^{p} \frac{1}{\alpha - p + n} \left( \frac{2i\eta}{w - \overline{\zeta}} \right) \right\}, \tag{3.3}$$

where $p \ge 0$ is the integer from $p - 1 < \alpha \le p$ (this formula is easily derived from (2.6) by integration by parts). By (3.3)

$$\log |a_p(w, \zeta)| \le \sum_{n=1}^{p} \frac{1}{n} \left( \frac{2|\eta|}{|w - \overline{\zeta}|} \right)^n,$$

if $\alpha = p \ge 0$ is an integer. But in our case

$$5^{1+p} \left( \frac{2|\eta|}{|w - \overline{\zeta}|} \right)^{1+p} \ge 5^n \left( \frac{2|\eta|}{|w - \overline{\zeta}|} \right)^n > \left( \frac{2|\eta|}{|w - \overline{\zeta}|} \right)^n, \quad 1 \le n \le p.$$

Therefore

$$\log |a_p(w, \zeta)| \le p 10^{1+p} \left( \frac{|\eta|}{|w - \overline{\zeta}|} \right)^{1+p}, \quad p = 0, 1, 2, \dots \tag{3.4}$$

If $\alpha > -1$ is not an integer, then from (3.3) one can derive

$$\log |a_p(w, \zeta)| \le \log |a_\delta(w, \zeta)| + C'_\alpha \left( \frac{|\eta|}{|w - \overline{\zeta}|} \right)^{1+\alpha}, \quad \delta = \alpha - p \in (-1, 0). \tag{3.5}$$

For evaluation of $\log |a_\delta(w, \zeta)|$ we change $t = \tau / i(w - \zeta)$ in (2.6). Since $i(w - \zeta) = -\text{Im } w + \eta > 0$ for $\text{Im } w < \eta$ and $\text{Re } w = \xi$, by uniqueness of holomorphic function we get

$$\log |a_\delta(w, \zeta)| = -\text{Re} \int_0^{\frac{2|\eta|}{i(w-\zeta)}} \frac{t^\delta dt}{(1+t)^{1+\delta}}.$$

If $2|\eta|/|w-\zeta| \leq 3$, then

$$|\log |a_\delta(w,\zeta)|| \leq \int_0^3 \frac{x^\delta \, dx}{|1-x|^{\delta+1}} \equiv C_\delta < +\infty. \tag{3.6}$$

And if $2|\eta|/|w-\zeta| > 3$, then

$$\begin{aligned}\log |a_\delta(w,\zeta)| \leq & C_\delta - \mathrm{Re} \int_{3\frac{|w-\zeta|}{i(w-\zeta)}}^{\frac{2|\eta|}{i(w-\zeta)}} \frac{t^\delta \, dt}{(1+t)^{1+\delta}} \\ &= C_\delta - \mathrm{Re} \int_{3\frac{|w-\zeta|}{i(w-\zeta)}}^{\frac{2|\eta|}{i(w-\zeta)}} \left(1 - \frac{1}{1+t}\right)^{1+\delta} \frac{d|t|}{|t|}\end{aligned}$$

again by uniqueness of holomorphic function. As $3 \leq |t| \leq 2|\eta|/|w-\zeta|$, we conclude that $|1+t| \geq 2$ and

$$\mathrm{Re}\left(1-\frac{1}{1+t}\right)^{1+\delta} = \left|1-\frac{1}{1+t}\right|^{1+\delta} \cos\left[(1+\delta)\arg\left(1-\frac{1}{1+t}\right)\right] > \frac{\sqrt{3}}{4}.$$

Consequently,

$$\log |a_\delta(w,\zeta)| \leq C_\delta - \frac{\sqrt{3}}{4} \int_3^{\frac{2|\eta|}{|w-\zeta|}} \frac{dx}{x} < C_\delta.$$

Hence $\log |a_\delta(w,\zeta)| \leq C_\delta$ by (3.6). Therefore, by (3.5) and (3.4)

$$\log |a_\alpha(w,\zeta)| \leq C''_\alpha \left(\frac{|\eta|}{|w-\bar{\zeta}|}\right)^{1+\alpha}, \qquad \frac{2|\eta|}{|w-\bar{\zeta}|} > \frac{1}{5}, \quad \alpha > -1. \tag{3.7}$$

**(b)** Now let $2|\eta|/|w-\bar{\zeta}| \leq 1/5$. Then one can verify that

$$\left|\frac{w-\zeta}{w-\bar{\zeta}}\right|^2 = 1 - \frac{4\eta \mathrm{Im}\, w}{|w-\bar{\zeta}|^2} \quad \text{and} \quad \frac{2|\mathrm{Im}\, w|}{|w-\bar{\zeta}|} < \frac{2|\eta|}{|w-\bar{\zeta}|} + 2.$$

Therefore

$$\left|\frac{w-\zeta}{w-\bar{\zeta}}\right|^2 \geq 1 - \frac{1}{5}\frac{2|\mathrm{Im}\, w|}{|w-\bar{\zeta}|} > 1 - \frac{1}{5}\left(\frac{2|\eta|}{|w-\bar{\zeta}|} + 2\right) \geq \frac{14}{25}$$

and consequently

$$\left|\frac{w-\zeta}{w-\bar{\zeta}} - \frac{2i|\eta|x}{w-\bar{\zeta}}\right| \geq \left|\frac{w-\zeta}{w-\bar{\zeta}}\right| - \frac{2|\eta|}{|w-\bar{\zeta}|} > \frac{1}{2}, \quad x \in [0,1].$$

Thus, for $\alpha > -1$

$$|\log|a_\alpha(w,\zeta)|| \leq \left| \int_0^{2|\eta|} \frac{\tau^\alpha d\tau}{\left[i(w-\overline{\zeta})\left(\frac{w-\zeta}{w-\overline{\zeta}} - \frac{i\tau}{w-\overline{\zeta}}\right)\right]^{1+\alpha}} \right|$$

$$\leq \frac{(2|\eta|)^{1+\alpha}}{|w-\overline{\zeta}|^{1+\alpha}} \int_0^1 \frac{x^\alpha dx}{\left|\frac{w-\zeta}{w-\overline{\zeta}} - \frac{2i|\eta|x}{w-\overline{\zeta}}\right|^{1+\alpha}} < \frac{4^{1+\alpha}}{1+\alpha}\left(\frac{|\eta|}{|w-\overline{\zeta}|}\right)^{1+\alpha}.$$

Hence by (3.7) we come to the desired estimate:

$$\log|a_\alpha(w,\zeta)| \leq C_\alpha \left(\frac{|\eta|}{|w-\overline{\zeta}|}\right)^{1+\alpha}, \qquad w,\zeta \in G^-, \quad \alpha > -1.$$

*3.2.* **Lemma 3.2.** *Let a sequence $\{w_k\} \subset G^-$ satisfy (3.1) for a given $\alpha > 0$. Then the following equilibrium relation is true for $B_\alpha(w,\{w_k\})$:*

$$\mathcal{L}(\rho, B_\alpha) = \mathcal{L}(\rho, B_\alpha^{-1}), \quad -\infty < \rho < 0. \tag{3.8}$$

**Proof.** First we show that for any $\rho < 0$

$$\sup_{v<\rho} \int_{-\infty}^{+\infty} |\log|B_\alpha(u+iv,\{w_k\})||\, du < +\infty. \tag{3.9}$$

To this end, for a fixed $\rho < 0$ we choose a natural $N_\rho$ enough great to provide $|\mathrm{Im}\ w_k| < |\rho|/2$ for $k \geq N_\rho + 1$. Then we write

$$|\log|B_\alpha(w,\{w_k\})|| \leq \left(\sum_{k=1}^{N_\rho} + \sum_{k=N_\rho+1}^{\infty}\right) |\log|a_\alpha(w,w_k)||\,. \tag{3.10}$$

If $k \geq N_\rho + 1$, then $|u+iv-\overline{w_k}| \geq |v| + |\mathrm{Im}\ w_k| > 2|\mathrm{Im}\ w_k| + |\rho|/2$ for $v < \rho$. Consequently, by (2.6)

$$\int_{-\infty}^{+\infty} |\log|a_\alpha(u+iv,w_k)||du$$

$$\leq \frac{2^{1+\alpha}}{1+\alpha}|\mathrm{Im}\ w_k|^{1+\alpha} \int_{-\infty}^{+\infty} \frac{du}{[|u+i(|v|+|\mathrm{Im}\ w_k|)| - 2|\mathrm{Im}\ w_k|]^{1+\alpha}}.$$

Observe that

$$\int_{-\infty}^{+\infty} \frac{du}{[|u+i(|v|+|\mathrm{Im}\ w_k|)| - 2|\mathrm{Im}\ w_k|]^{1+\alpha}}$$

$$< \left(\frac{2}{|\rho|}\right)^{1+\alpha} \int_{|u|<2|\rho|} du + \int_{|u|>2|\rho|} \frac{du}{(|u|-|\rho|)^{1+\alpha}} < \frac{2^{3+\alpha}}{|\rho|^\alpha} + \frac{4}{\alpha|\rho|^\alpha}.$$

Therefore

$$\sup_{k=N_\rho+1}\int_{-\infty}^{+\infty} |\log|a_\alpha(u+iv,w_k)||\,du$$
$$\le \frac{2^{3+\alpha}}{1+\alpha}|\rho|^{-\alpha}\left(\frac{1}{\alpha}+2^{1+\alpha}\right)\sum_{k=N_\rho+1}^{\infty} |\text{Im } w_k|^{1+\alpha} < +\infty. \quad (3.11)$$

Let $v<\rho$, let $k$ $(1\le k\le N_\rho)$ be fixed, and let $m_k = \max_k |\text{Im } w_k|$. Then

$$\int_{-\infty}^{+\infty} |\log|a_\alpha(u+iv,w_k)||du$$
$$\le \int_{|u|>4m_0} du \int_0^{2|\text{Im } w_k|} \frac{\tau^\alpha d\tau}{|\tau+|v|+|\text{Im } w_k|+iu|^{1+\alpha}}$$
$$+\int_{|u|<4m_0} |\log|a_\alpha(u+iv, i\text{Im } w_k)||\,du \equiv I_1+I_2. \quad (3.12)$$

Obviously

$$I_1 \le \int_{|u|>4m_0} du \int_0^{2m_0} \frac{\tau^\alpha d\tau}{[|u|+i(|v|+|\text{Im } w_k|)|-\tau]^{1+\alpha}}$$
$$< \int_{|u|>4m_0} du \int_0^{2m_0} \frac{\tau^\alpha d\tau}{(|u|-2m_0)^{1+\alpha}} = \frac{2^{2+\alpha}m_0^{1+\alpha}}{\alpha(1+\alpha)}, \quad (3.12')$$

and $I_2 \equiv \Psi_{\alpha k}(v)$ is a continuous function of $v(-\infty<v<0)$. Besides, $\lim_{v\to\infty}\Psi_{\alpha,k}(v)=0$ in view of (2.21). Consequently

$$I_2 \le \max_{1\le k\le N_\rho}\ \max_{-\infty<v\le\rho} \Psi_{\alpha,k}(v) < +\infty.$$

Hence (3.9) follows by (3.12)–(3.12′), (3.11), and (3.10). It remains to observe that by (3.9) $B_\alpha(w,\{w_k\})$ satisfies the requirements of Theorem 1.2, and hence (1.11) and (1.15) are true for this function. In view of (2.21) the relation (1.15) can be written in the form (3.9).

**Lemma 3.3.** *Let $\{w_k\}\subset G^-$ be an arbitrary sequence. Then:*

1°. *If (3.1) is true for an $\alpha>0$, then $B_{\alpha_0}(w,\{w_k\})\in\mathfrak{N}^m_{\alpha,0}$ for any $\alpha_0>\alpha$.*
2°. *If*

$$\sum_k |\text{Im } w_k|^{1+\alpha}\log\frac{1}{|\text{Im } w_k|} < +\infty$$

*for an $\alpha>0$, then $B_\alpha(w,\{w_k\})\in\mathfrak{N}^m_{\alpha,\beta}$ for any $\beta>0$.*

**Proof.** 1°. Let $\alpha_0 > \alpha$ be arbitrary. By (3.2),

$$\int_0^{+\infty} t^{\alpha-1}dt \int_{-\infty}^{+\infty} \log^+ |B_{\alpha_0}(u-it,\{w_k\})|du \leq C_\alpha \sum_k |\mathrm{Im}\ w_k|^{1+\alpha_0} \int_0^{+\infty} t^{\alpha-1} I_{\alpha_0,k}(t)dt, \tag{3.13}$$

where

$$I_{\alpha_0,k}(t) = \int_{-\infty}^{+\infty} \frac{dx}{[(t+|\mathrm{Im}\ w_k|)^2+x^2]^{\frac{1+\alpha_0}{2}}} = (t+|\mathrm{Im}\ w_k|)^{-\alpha_0} \int_{-\infty}^{+\infty} \frac{dy}{(1+y^2)^{\frac{1+\alpha_0}{2}}} \equiv C'_{\alpha_0}(t+|\mathrm{Im}\ w_k|)^{-\alpha_0}. \tag{3.13'}$$

Hence

$$\int_0^{+\infty} t^{\alpha-1} I_{\alpha_0,k}(t)dt = C'_{\alpha_0} \int_0^{+\infty} \frac{t^{\alpha-1}dt}{(t+|\mathrm{Im}\ w_k|)^{\alpha_0}} = C'_{\alpha_0}|\mathrm{Im}\ w_k|^{\alpha-\alpha_0} \int_0^{+\infty} \frac{x^{\alpha-1}dx}{(1+x)^{\alpha_0}} \equiv C(\alpha,\alpha_0)|\mathrm{Im}\ w_k|^{\alpha-\alpha_0}.$$

Consequently

$$\int_0^{+\infty} t^{\alpha-1}\mathcal{L}(-t,B_{\alpha_0})dt \equiv \int_0^{+\infty} t^{\alpha-1}dt \int_{-\infty}^{+\infty} \log^+|B_{\alpha_0}(u-it,\{w_k\})|\,du \leq C'_\alpha C(\alpha,\alpha_0)\sum_k |\mathrm{Im}\ w_k|^{1+\alpha} < +\infty.$$

Hence $B_{\alpha_0}(w,\{w_k\}) \in \mathfrak{N}^m_{\alpha,0}$ by (3.8).

2°. Let $\beta > 0$ be arbitrary. Then by (3.2) and (3.13)–(3.13′)

$$\int_0^{+\infty} \frac{t^{\alpha-1}}{1+t^\beta}\mathcal{L}(-t,B_\alpha)dt \equiv \int_0^{+\infty} \frac{t^{\alpha-1}}{1+t^\beta}dt \int_{-\infty}^{+\infty} \log^+|B_{\alpha_0}(u-it,\{w_k\})|\,du \leq C_\alpha C'_\alpha \sum_k |\mathrm{Im}\ w_k|^{1+\alpha} \int_0^{+\infty} \frac{t^{\alpha-1}}{1+t^\beta}\frac{dt}{(t+|\mathrm{Im}\ w_k|)^\alpha}.$$

One can verify that

$$\int_1^{+\infty} \frac{t^{\alpha-1}}{1+t^\beta}\frac{dt}{(t+|\mathrm{Im}\ w_k|)^\alpha} < \int_1^{+\infty}\frac{dt}{t^{1+\beta}} = \frac{1}{\beta},$$

$$\int_0^1 \frac{t^{\alpha-1}}{1+t^\beta}\frac{dt}{(t+|\mathrm{Im}\ w_k|)^\alpha} < \int_0^{|\mathrm{Im}\ w|^{-1}} \frac{x^{\alpha-1}dx}{(1+x)^\alpha} < \int_0^1 x^{\alpha-1}dx + \left|\int_{|\mathrm{Im}\ w_k|}^1 \frac{dy}{y(1+y)^\alpha}\right| < \frac{1}{\alpha} + \left|\log\frac{1}{|\mathrm{Im}\ w_k|}\right|.$$

Consequently

$$\int_0^{+\infty} \frac{t^{\alpha-1}}{1+t^\beta} \mathcal{L}(t, B_\alpha) dt$$
$$< C_\alpha C'_\alpha \sum_k |\text{Im } w_k|^{1+\alpha} \left\{ \frac{1}{\alpha} + \frac{1}{\beta} + \left| \log \frac{1}{|\text{Im } w_k|} \right| \right\} < +\infty.$$

Hence $B_\alpha(w, \{w_k\}) \in \mathfrak{N}^m_{\alpha,\beta}$ by (3.8).

**Remark 3.1.** The assertion 1° of the pervious lemma and Lemma 2.1 prove that the condition (2.3) presents a complete characterization of the density of zeros $\{a_\mu\}$ and poles $\{b_\nu\}$ of functions from $\mathfrak{N}^m_{\alpha,\beta}$ $(0 < \alpha < +\infty, 0 \le \beta \le 1+\alpha)$.

**Lemma 3.4.** *For any $\alpha > 0$ there exists a sequence $\{w_k\}_1^\infty \subset G^-$ satisfying (3.1) but such that $B_\alpha(w, \{w_k\}_1^\infty) \notin \mathfrak{N}^m_{\alpha,\alpha+1}$.*

**Proof.** Let the sequence $\{w_k\}_1^\infty \equiv \{-ir_k\}_1^\infty$ $(0 < r_k < 1)$ be such that

$$\sum_{k=1}^\infty r_k^{1+\alpha} < +\infty \quad \text{but} \quad \sum_{k=1}^\infty r_k^{1+\alpha} \log \frac{1}{r_k} = +\infty.$$

Consider the function

$$G_\alpha(w) \equiv \frac{B_{\alpha+1}(w, \{w_k\})}{B_\alpha(w, \{w_k\})} \qquad (w = re^{i\vartheta}, \quad -\pi < \vartheta < 0).$$

By (3.3)

$$g_\alpha(w, w_k) \equiv \frac{a_{\alpha+1}(w, w_k)}{a_\alpha(w, w_k)} = \exp\left\{ \frac{1}{1+\alpha} \left( \frac{2r_k}{iw + r_k} \right)^{1+\alpha} \right\}.$$

Consequently

$$\log |g_\alpha(re^{i\vartheta}, -ir_k)| = \frac{2^{1+\alpha}}{1+\alpha} \frac{r_k^{1+\alpha} \cos\left[(1+\alpha) \arctan\left(\frac{r\cos\vartheta}{r|\sin\vartheta| + r_k}\right)\right]}{\left| re^{i(\vartheta + \frac{\pi}{2})} + r_k \right|^{1+\alpha}}.$$

Hence

$$\log |G_\alpha(re^{i\vartheta})| = \frac{2^{1+\alpha}}{1+\alpha} \int_{-1}^0 \frac{|t|^{1+\alpha} \cos\left[(1+\alpha) \arctan\left(\frac{r\cos\vartheta}{r|\sin\vartheta| + |t|}\right)\right]}{\left| re^{i(\vartheta + \frac{\pi}{2})} - t \right|^{1+\alpha}} d\mathbf{n}(t),$$

where $\mathbf{n}(t)$ is the number of those $r_k$ for which $-r_k < t \in (-1, 0)$. Observe that

$$\cos\left[(1+\alpha) \arctan\left(\frac{r\cos\vartheta}{r|\sin\vartheta| + |t|}\right)\right] > \frac{1}{\sqrt{2}} \quad \text{for} \quad \left|\vartheta + \frac{\pi}{2}\right| < \frac{\pi}{4(1+\alpha)}.$$

Therefore

$$
\begin{aligned}
I \equiv & \iint_{G^-} \frac{|\mathrm{Im}\ \zeta|^{\alpha-1}}{1+|\mathrm{Im}\ \zeta|^{\alpha+1}} \log^+ |G_\alpha(\zeta)| d\sigma(\zeta) \\
\geq & \int_{-\frac{\pi}{2}-\frac{\pi}{4(1+\alpha)}}^{\frac{\pi}{2}+\frac{\pi}{4(1+\alpha)}} \int_0^1 |r \sin \vartheta|^{\alpha-1} \log^+ |G_\alpha(re^{i\vartheta})| r dr d\vartheta \\
\geq & \frac{2^\alpha}{1+\alpha} C_\alpha \int_{-\frac{\pi}{4(1+\alpha)}}^{\frac{\pi}{4(1+\alpha)}} \int_0^1 r^\alpha \left\{ \int_{-1}^0 \frac{|t|^{1+\alpha} d\mathbf{n}(t)}{|re^{i\vartheta}-t|^{1+\alpha}} \right\} dr d\vartheta,
\end{aligned}
$$

where $C_\alpha = \min\left\{ \left[\cos \frac{\pi}{4(1+\alpha)}\right]^{\alpha-1}, 1\right\}$. And it is obvious that

$$
\begin{aligned}
I \geq & \frac{2^{\alpha-1}}{(1+\alpha)^2} \pi C_\alpha \int_{-1}^0 |t|^{1+\alpha} d\mathbf{n}(t) \int_0^1 \frac{r^\alpha dr}{(r+|t|)^{1+\alpha}} \\
\geq & \frac{2^{\alpha-1}}{(1+\alpha)^2} \pi C_\alpha \int_{-1}^0 |t|^{\alpha+1} d\mathbf{n}(t) \int_1^{|t|^{-1}} \frac{x^\alpha dx}{(1+x)^{1+\alpha}} \\
\geq & \frac{\pi C_\alpha}{4(1+\alpha)^2} \int_{-1}^0 |t|^{1+\alpha} \log \frac{1}{|t|} d\mathbf{n}(t) = +\infty.
\end{aligned}
$$

Hence $G_\alpha(w) \notin \mathfrak{N}^m_{\alpha,1+\alpha}$, and $B_\alpha(w, \{w_k\}_1^\infty)$ does not belong to $\mathfrak{N}^m_{\alpha,1+\alpha}$ since $B_{\alpha+1}(w, \{w_k\}_1^\infty) \in \mathfrak{N}^m_{\alpha,1+\alpha}$ by the assertion 1° of Lemma 3.3.

*3.3.* The below theorem is the main result of this section.

**Theorem 3.1.** $\mathfrak{N}^m_{\alpha,\beta}$ $(0<\alpha<+\infty,\ 0\leq\beta<\alpha)$ *coincides with the set of those functions which for some* $\alpha_0>\alpha$ *are representable in the form*

$$
F(w) = \frac{B_{\alpha_0}(w,\{a_\mu\})}{B_{\alpha_0}(w,\{b_\nu\})} \times \exp\left\{ \frac{\alpha_0 2^{\alpha_0}}{\pi} e^{-i\frac{\pi}{2}(1+\alpha_0)} \iint_{G^-} \frac{|\mathrm{Im}\ \zeta|^{\alpha_0-1}}{(w-\bar{\zeta})^{1+\alpha_0}} d\mu(\zeta) \right\}, \quad w \in G^-, \tag{3.14}
$$

*where* $B_{\alpha_0}(w,\{a_\mu\})$ *and* $B_{\alpha_0}(w,\{b_\nu\})$ *are convergent Blaschke type products with zeros* $\{a_\mu\}$, $\{b_\nu\} \subset G^-$ *satisfying*

$$
\sum_\mu |\mathrm{Im}\ a_\mu|^{1+\alpha} < +\infty \quad \textit{and} \quad \sum_\nu |\mathrm{Im}\ b_\nu|^{1+\alpha} < +\infty, \tag{3.15}
$$

*and* $\mu(\zeta) \equiv \mu_{\alpha_0}(\zeta)$ *is a function of bounded variation in any compact from* $G^-$, *such that*

$$
\iint_{G^-} \frac{|\mathrm{Im}\ \zeta|^{\alpha-1}}{1+|\mathrm{Im}\ \zeta|^\beta} |d\mu(\zeta)| < +\infty. \tag{3.16}
$$

**Proof.** Let $F(w) \in \mathfrak{N}^m_{\alpha,\beta}$ ($\alpha > 0$, $0 \le \beta < \alpha$). Then the sequences $\{a_\mu\}$ and $\{b_\nu\}$ satisfy (3.15) by Lemma 2.1. Besides, the integral (2.4) is convergent. Further, $\mathfrak{N}^m_{\alpha,\beta} \subseteq \mathfrak{N}^m_{\alpha,1+\alpha} \subseteq \mathfrak{N}^m_{\alpha_0,1+\alpha_0}$ ($\alpha_0 > \alpha$) by (2.2)–(2.2′). Therefore, by Theorem 2.2 and (2.22) we come to the conclusion that $F(w)$ is representable in the form (3.14)–(3.16), whatever be $\alpha_0 > \alpha$, and $d\mu(\zeta) \equiv \log|F(\zeta)|d\sigma(\zeta)$. For proving the converse statement, note that

$$B_{\alpha_0}(w, \{a_\mu\}),\ B_{\alpha_0}(w, \{b_\nu\}) \in \mathfrak{N}^m_{\alpha,\beta} \tag{3.17}$$

by the assertion 1° of Lemma 3.3. Thus, it suffices to prove that also

$$\varphi(w) \equiv \exp\left\{\frac{\alpha_0 2^{\alpha_0}}{\pi} \iint_{G^-} \frac{|\mathrm{Im}\ \zeta|^{\alpha_0 - 1}}{[i(w-\bar{\zeta})]^{1+\alpha_0}} d\mu(\zeta)\right\} \in \mathfrak{N}^m_{\alpha,\beta}. \tag{3.17′}$$

To this end, observe that

$$\iint_{G^-} \frac{|\mathrm{Im}\ w|^{\alpha-1}}{1+|\mathrm{Im}\ w|^\beta} \log^+|\varphi(w)|d\sigma(w) \le \frac{\alpha_0 2^{\alpha_0}}{\pi} \iint_{G^-} |\mathrm{Im}\ \zeta|^{\alpha_0-1} I(\alpha,\alpha_0,\beta,\zeta)|d\mu(\zeta)|, \tag{3.18}$$

where

$$I(\alpha,\alpha_0,\beta,\zeta) \equiv \iint_{G^-} \frac{|\mathrm{Im}\ w|^{\alpha-1}}{1+|\mathrm{Im}\ w|^\beta} \frac{d\sigma(w)}{|w-\bar{\zeta}|^{1+\alpha_0}}. \tag{3.18′}$$

For $a \in (0, 1]$ a suitable estimate of the last integral follows from the inequality

$$\mathcal{J} \equiv \int_0^{+\infty} \frac{t^{\alpha-1}dt}{(1+t^\beta)(t+a)^{\alpha_0}} \le a^{\alpha-\alpha_0} \int_0^{+\infty} \frac{x^{\alpha-1}dx}{1+x^{\alpha_0}} \equiv C_1 a^{\alpha-\alpha_0} \tag{3.19}$$

(which is true for any $a > 0$). If $a > 1$, then one can be convinced that

$$\int_0^a \frac{t^{\alpha-1}dt}{(1+t^\beta)(t+a)^{\alpha_0}} \le a^{\alpha-\alpha_0-\beta} \int_0^1 \frac{x^{\alpha-1}dx}{a^{-\beta}+x^\beta} \equiv a^{-\alpha_0}, \int_{a^{-1}}^{+\infty} \frac{t^{\beta-\alpha-1}dt}{1+t^\beta}$$
$$\le a^{-\alpha_0}\left(\int_1^{+\infty} \frac{dt}{t^{1+\alpha}} + \int_{a^{-1}}^1 t^{\beta-\alpha-1}dt\right) < \frac{a^{\alpha-\alpha_0-\beta}}{\alpha-\beta}.$$

Besides,

$$\int_a^{+\infty} \frac{t^{\alpha-1}dt}{(1+t^\beta)(t+a)^{\alpha_0}} \le \int_0^{a^{-1}} \frac{x^{\alpha_0-\alpha+\beta-1}dx}{(1+ax)^{\alpha_0}} \le \frac{a^{\alpha-\alpha_0-\beta}}{\alpha_0-\alpha+\beta}.$$

Hence, for $a > 1$

$$\mathcal{J}(\alpha,\alpha_0,\beta,a) < \frac{\alpha_0}{(\alpha-\beta)(\alpha_0-\alpha+\beta)} a^{\alpha-\alpha_0-\beta} \equiv C_2 a^{\alpha-\alpha_0-\beta}.$$

By the continuity of $g(a) \equiv \mathcal{J}(\alpha, \alpha_0, \beta, a)$ on $(0, +\infty)$, the last estimate and (3.19) imply

$$\mathcal{J}(\alpha, \alpha_0, \beta, a) \leq C_3 \frac{a^{\alpha-\alpha_0}}{1+a^\beta}, \quad 0 < a < +\infty.$$

Returning to the boundedness of the right-hand side of (3.18), we conclude

$$\begin{aligned} I(\alpha, \alpha_0, \beta, \zeta) &= \int_0^{+\infty} \frac{t^{\alpha-1} dt}{1+t^\beta} \int_{-\infty}^{+\infty} \frac{dx}{|x - it - \xi - i|\eta||^{1+\alpha_0}} \\ &= \int_0^{+\infty} \frac{t^{\alpha-1} dt}{(1+t^\beta)(t+|\eta|)^{\alpha_0}} \int_{-\infty}^{+\infty} \frac{dx}{|x-i|^{1+\alpha_0}} \\ &\equiv C_4 \mathcal{J}(\alpha, \alpha_0, \beta, |\eta|) \leq C_5 \frac{|\eta|^{\alpha-\alpha_0}}{1+|\eta|^\beta} \end{aligned}$$

for $\zeta = \xi + i\eta$ $(\eta < 0)$. Thus

$$\begin{aligned} \frac{\alpha_0 2^{\alpha_0}}{\pi} \iint_{G^-} &|\mathrm{Im}\ \zeta|^{\alpha_0 - 1} I(\alpha, \alpha_0, \beta, \zeta) |d\mu(\zeta)| \\ &\leq C_6 \iint_{G^-} \frac{|\mathrm{Im}\ \zeta|^{\alpha-1}}{1 + |\mathrm{Im}\ \zeta|^\beta} |d\mu(\zeta)| < +\infty. \end{aligned}$$

Hence the inclusion (3.17′) follows by (3.18)–(3.18′). Thus, any function representable in the form (3.14)–(3.16), where $\alpha$, $\beta$ $(\alpha > 0, 0 \leq \beta < \alpha)$ and $\alpha_0 > \alpha$ are any numbers, belongs to $\mathfrak{N}^m_{\alpha,\beta}$.

*3.4.* Now we shall show that for $\alpha < \beta \leq 1 + \alpha$ (3.14)–(3.16) is not a descriptive representation of the classes $\mathfrak{N}^m_{\alpha,\beta}$ $(\alpha > 0)$. To this end, for any $\alpha, \beta$ and $\alpha_0 > \alpha$ consider the measure

$$d\mu_{\alpha_0}(\zeta) = C_{\alpha_0} |\eta|^{\frac{\beta-\alpha}{2}} (1+\xi^2)^{-1} d\xi d\eta \quad (\zeta = \xi + i\eta),$$

where $C_{\alpha_0} = \Gamma(\alpha_0) \left[ 2^{\alpha_0} \Gamma\left(\alpha_0 + \frac{\beta-\alpha}{2}\right) \Gamma\left(1 - \frac{\beta-\alpha}{2}\right) \right]^{-1}$. This measure obviously satisfies (3.16). We shall prove that nevertheless

$$F(w) \equiv \exp\left\{ \frac{\alpha_0 2^{\alpha_0}}{\pi} \iint_{G^-} \frac{|\mathrm{Im}\ \zeta|^{\alpha_0-1}}{[i(w-\overline{\zeta})]^{1+\alpha_0}} d\mu_{\alpha_0}(\zeta) \right\} \notin \mathfrak{N}^m_{\alpha,\beta}. \tag{3.20}$$

Previously we note that the integral in the exponent is absolutely and uniformly convergent in $G^-$, and hence the function

$$G(w) \equiv \frac{\alpha_0 2^{\alpha_0}}{\pi} \iint_{G^-} \frac{|\mathrm{Im}\ \zeta|^{\alpha_0-1}}{[i(w-\overline{\zeta})]^{1+\alpha_0}} d\mu_{\alpha_0}(\zeta)$$

is holomorphic in $G^-$. One can show that

$$G(w) = \frac{\alpha_0 2^{\alpha_0}}{\pi} C_{\alpha_0} \int_0^{+\infty} t^{\alpha_0 + \frac{\beta-\alpha}{2} - 1} I_{\alpha_0}(w - it) dt, \tag{3.21}$$

where

$$I_{\alpha_0}(w - it) \equiv \int_{-\infty}^{+\infty} \frac{d\xi}{\{i[(w - it) - \xi]\}^{1+\alpha_0} (1 + \xi^2)}.$$

The last integral can be calculated by residues. Namely, taking $w - it = s(\in G^-)$ we get

$$I_{\alpha_0}(s) = \int_{-\infty}^{+\infty} \frac{dx}{[i(s - x)]^{1+\alpha_0}(1 + x^2)} \equiv \int_{-\infty}^{+\infty} \Psi_s(x) dx,$$

where $\Psi_s(z) = \left\{[i(s - z)]^{1+\alpha_0}(1 + z^2)\right\}^{-1}$ is holomorphic in the $z$-plane, except the cut $\{z : \text{Re } z = \text{Re } s, \text{ Im } z \leq \text{Im } s\} \subset G^-$, and $\Psi_s(z)$ has simple poles at $z = \pm i$. Calculating the integral of this function along $\Gamma_R = \{Re^{i\vartheta} : 0 < \vartheta < \pi\} \cup \{[-R, R]\}$ and letting $R \to +\infty$, we obtain $I_{\alpha_0}(s) = \pi(1 + is)^{-(1+\alpha_0)}$. Inserting this expression into (3.21) and using the well-known formula for Euler's $\beta$–function we find $G(w) = (1 + iw)^{\frac{\beta-\alpha}{2} - 1}$. Consequently, $F(w) = \exp\{(1 + iw)^{\frac{\beta-\alpha}{2} - 1}\}$ for $w = u + iv \in G^-$ and

$$\log|F(w)| = |1 + iw|^{\frac{\beta-\alpha}{2} - 1} \cos\left[\left(1 - \frac{\beta - \alpha}{2}\right) \arctan\left(\frac{u}{1 + |v|}\right)\right].$$

Besides, $\log|F(w)| > 0$ since $1/2 \leq 1 - (\beta - \alpha)/2 < 1$. Therefore,

$$\begin{aligned} \mathcal{L}(t, F) &= \frac{1}{2\pi} \int_{-\infty}^{+\infty} \log^+ |F(u + iv)| du \\ &\leq \frac{\cos\left[\left(1 - \frac{\beta-\alpha}{2}\right) \frac{\pi}{2}\right]}{2\pi} \int_{-\infty}^{+\infty} \frac{du}{[(1 + |t|)^2 + u^2]^{\frac{1}{2} - \frac{\beta-\alpha}{4}}} = +\infty \end{aligned}$$

for any $t < 0$, and hence (3.20) is true.

The above counterexample shows that for $\alpha < \beta \leq 1 + \alpha$ (3.14)–(3.16) is not a descriptive representation for $\mathfrak{N}^m_{\alpha,\beta}$ $(0 < \alpha < +\infty)$, in contrast to the case $0 \leq \beta < \alpha$ considered in Theorem 3.1. The case $\beta = \alpha$ remains open.

NOTES. A similar method of mapping and successive passage has been used earlier by M.M. and A.E.Djrbashian [24] for proving integral representations of holomorphic functions in $G^+$ from $L^p(G^+, y^\alpha dxdy)$. One can observe that the functions (1.3) are the inversions of a special, infinite case of Tsuji characteristics considered in [38] (Ch. I, Sec. 5). As F.A.Shamoian noticed, the passage of this chapter can be used for proving the representation (2.18) for holomorphic functions from somewhat larger classes (the suitable density condition for zeros must be presupposed). But there were some difficulties in finding descriptive representations. Later a version of the descriptive representation (3.14) was proved [5] for subharmonic extensions of the particular classes $\mathfrak{N}^m_{\alpha,0}$ $(\alpha > 0)$. Namely the approach of F.A.Shamoian [94] was used to prove some descriptive representations over the real axis, where $d\mu$ is replaced by $\varphi(t)dt$ and $\varphi(t)$ is from a definite O.V.Besov space.

# CHAPTER 5

# BOUNDARY VALUES

## 1. MAIN RESULTS

*1.1.* In view of results of Chapters 1 and 3 the below definition is natural.

**Definition 1.1.** A function $F(w)$ meromorphic in the lower half-plane $G^-$ is of $\alpha$-bounded type in $G^-$ or, which is the same, of the class $N_\alpha\{G^-\}$ $(-1 < \alpha < +\infty)$, if it can be represented in $G^-$ in the form

$$F(w) = \frac{B_\alpha(w, \{a_n\})}{B_\alpha(w, \{b_m\})} \exp\left\{c_0 + c_1 w + \int_{-\infty}^{+\infty} \frac{d\sigma(t)}{[i(w-t)]^{1+\alpha}} + iC\right\}, \qquad (1.1)$$

where $c_0$, $c_1$ ($c_1 = 0$ for $\alpha \geq 0$) and $C$ are any real numbers, $\sigma(t)$ is representable as the difference $\sigma = \sigma_1 - \sigma_2$ of two nondecreasing functions satisfying

$$\int_{-\infty}^{+\infty} \frac{d\sigma_{1,2}(t)}{(1+|t|)^{1+\alpha-p}} < +\infty, \qquad (1.2)$$

where $p \geq 0$ is the integer deduced by $p - 1 < \alpha \leq p$ and $B_\alpha$ are convergent Blaschke type products (from Ch. 1) constructed by means of the sequences $\{a_n\}$ $\{b_m\} \subset G^-$ of zeros and poles of $F$, which are assumed to satisfy

$$\sum_n |\mathrm{Im}\, a_n|^{1+\alpha} < +\infty \qquad \text{and} \qquad \sum_m |\mathrm{Im}\, b_m|^{1+\alpha} < +\infty.$$

Note that for $\alpha \in (-1, 0)$ the classes $N_\alpha\{G^-\}$ are subsets of Nevanlinna's class $N$ of functions of bounded type in $G^-$ (i.e. representable in $G^-$ as fractions of two bounded holomorphic functions).

**Definition 1.2.** Let $E \subseteq (-\infty, +\infty)$ be a Borel measurable set ($B$-set) and let $0 < \gamma < 1$. Then we shall say that $E$ is of positive $\gamma$-capacity ($C_\gamma(E) > 0$) if there exists a Borel measure ($B$-measure) $\mu$ supported in $E$ ($\mu \prec E$) and such that

$$\int_{-\infty}^{+\infty} \frac{d\mu(t)}{(1+|t|)^\gamma} = 1 \qquad (1.4)$$

and

$$S_1 \equiv \sup_{w \in G^-} \int_{-\infty}^{+\infty} \frac{d\mu(t)}{|w-t|^\gamma} < +\infty. \qquad (1.5)$$

If there is no such a measure, i.e. $S_1 = +\infty$ for any nonnegative $B$-measure $\mu \prec E$ satisfying (1.4), then we say that $E$ is of zero $\gamma$-capacity ($C_\gamma(E) = 0$).

**Definition 1.3.** Let $E \subseteq [-\pi, \pi]$ be a $B$-set, and let $0 < \gamma < 1$ be any number. It is said that $E$ is of positive $\gamma$-capacity if there exists a nonnegative $B$-measure $\mu \prec E$ such that

$$\int_{-\pi}^{\pi} d\mu(\vartheta) = 1 \tag{1.9}$$

and

$$S_2 \equiv \sup_{z \in \mathbb{D}} \int_{-\pi}^{\pi} \frac{d\mu(\vartheta)}{|z - e^{i\vartheta}|^\gamma} < +\infty. \tag{1.10}$$

If there is no such a measure, i.e. if $S_2 = +\infty$ for any nonnegative $B$-measure $\mu \prec E$ satisfying (1.9), then $E$ is said to be of zero $\gamma$-capacity.

**Remark 1.1.** The above definition of $\gamma$-capacity on the unit circle belongs to O.Frostman [33] and is well-known. One can verify that a conformal mapping of the disc onto the half-plane transforms it to a form which is equivalent to the Definition 1.2 for bounded sets of real axis.

*1.2.* The below theorems are the main results of this chapter.

**Theorem 1.1.** *Any function $F \in N_\alpha\{G^-\}$ $(-1 < \alpha < 0)$ has non-zero, finite nontangential boundary values in all points $u \in (-\infty, +\infty)$, with possible exception of a set of zero $1 + \alpha$-capacity.*

The next theorem gives a more complete result for the Blaschke type products in $G^-$ (Ch. 1), which are special representatives of classes $N_\alpha\{G^-\}$.

**Theorem 1.2.** 1°. *Let $\gamma \in (0,1)$ and $\alpha \in [\gamma - 1, \gamma + 1]$ be any numbers, and let $\{w_k\} \subset G^-$ be a sequence satisfying*

$$\sum_k |\text{Im } w_k|^\gamma < +\infty. \tag{1.11}$$

*Then $B_\alpha(w, \{w_k\})$ has non-zero, finite nontangential boundary values in all points $u \in (-\infty, +\infty)$, with possible exception of a set of zero $\gamma$-capacity.*

2°. *Let $\alpha \in [0, 2]$ be any number, and let $\{w_k\} \subset G^-$ be a sequence satisfying*

$$\sum_k |\text{Im } w_k| < +\infty. \tag{1.12}$$

*Then $B_\alpha(w, \{w_k\})$ has non-zero, finite nontangential boundary values in almost all points of the real axis.*

Also we shall prove a similar theorem in the disc.

**Theorem 1.3.** 1°. *Let $\gamma \in (0,1)$ and $\alpha \in [\gamma - 1, \gamma + 1]$ be any numbers, and let $\{z_k\} \subset \mathbb{D}$ be a sequence satisfying*

$$\sum_k (1 - |z_k|)^\gamma < +\infty. \tag{1.13}$$

*Then the Blaschke-M.M.Djrbashian product $B_\alpha(z, \{z_k\})$ ((10) in Introduction) has non-zero, finite nontangential boundary values in all points of the unit circle, with possible exception of a set of zero $\gamma$-capacity.*

2°. *Let $\alpha \in [0, 2]$ be any number, and let $\{z_k\} \subset \mathbb{D}$ be a sequence satisfying*

$$\sum_k (1 - |z_k|) < +\infty. \tag{1.14}$$

*Then $B_\alpha(z, \{z_k\})$ has non-zero, finite nontangential boundary values in almost all points of the unit circle.*

We shall prove also the below inequality which is of independent interest: *if $-1 < \alpha_1 < \alpha_2 \leq 0$ are any numbers and $\{w_k\} \subset G^-$ is a sequence satisfying (1.11) with $\gamma = 1 + \alpha_1$, then*

$$|B_{\alpha_1}(w, \{w_k\})| \leq |B_{\alpha_2}(w, \{w_k\})|, \quad w \in G^-. \tag{1.15}$$

Note that for $\alpha_2 = 0$ this inequality was proved in the end of Ch. 1.

## 2. PROOFS OF THEOREMS 1.1 AND 1.2

*2.1.* We start by some lemmas necessary for the proof of Theorem 1.1.

**Lemma 2.1.** *Let $E_1, E_2 \subset (-\infty, +\infty)$ be any $B$-sets such that $C_\gamma(E_1) = C_\gamma(E_2) = 0$ for some $\gamma \in (0,1)$. Then $C_\gamma(E_1 \cup E_2) = 0$.*

**Proof.** Let $\mu \prec E_1 \cup E_2$ be a nonnegative $B$-measure satisfying (1.4). It is obvious that, if taken over $E_1$ or $E_2$, the integral (1.4) equals to a number $M \in (0, 1]$. For instance, let it be taken over $E_1$. Then also the measure $\mu_1 = M^{-1}\mu|_{E_1}$ supported in $E_1$ will satisfy (1.4). But $C_\gamma(E_1) = 0$. Therefore

$$\sup_{w \in G^-} \int_{-\infty}^{+\infty} \frac{d\mu(t)}{|w - t|^\gamma} \geq M \sup_{w \in G^-} \int_{-\infty}^{+\infty} \frac{d\mu(t)}{|w - t|^\gamma} = +\infty.$$

The next assertion easily follows by Fatou's lemma and a passage to a limit.

**Lemma 2.2.** *Let $\gamma \in (0,1)$, let $E \subseteq (-\infty, +\infty)$ be a $B$-set such that $C_\gamma(E) > 0$, and let $\mu \prec E$ be a measure from Definition 1.2. Then*

$$\int_{-\infty}^{+\infty} \frac{d\mu(t)}{|u - t|^\gamma} \leq S_1 \equiv \sup_{w \in G^-} \int_{-\infty}^{+\infty} \frac{d\mu(t)}{|w - t|^\gamma} < +\infty, \quad u \in (-\infty, +\infty). \tag{2.1}$$

**Lemma 2.3.** *Let $\beta \in (-1,0)$ be any number, and let $\sigma(t)$ be a nondecreasing function such that*

$$\int_{-\infty}^{+\infty} \frac{d\sigma(t)}{(1+|t|)^{1+\beta}} < +\infty. \tag{2.2}$$

*Further, let $f(w)$ be a holomorphic function given by the formula*

$$f(w) = \int_{-\infty}^{+\infty} \frac{d\sigma(t)}{[i(w-t)]^{1+\beta}}, \quad w \in G^-. \tag{2.3}$$

*Then the set of those $u \in (-\infty,+\infty)$, where the nontangential boundary value $f(u)$ exists as a finite limit, is of zero $1+\beta$-capacity.*

**Proof.** From (2.2) and (2.3) it follows that the function $\exp\{-f(w)\}$ is holomorphic and bounded in $G^-$. Therefore, by Lindelöf's theorem the existence of a finite angular boundary value $f(u)$ in a point $u \in (-\infty,+\infty)$ is equivalent to the existence of the finite limit

$$\lim_{v\to -0} f(u+iv) \ (= f(u)) \tag{2.4}$$

in the same point. For proving the existence of this limit, write

$$f(u+iv) = f(u+iv_0) + i\int_{v_0}^{v} f'(u+iy)dy \tag{2.5}$$

for any $u \in (-\infty,+\infty)$ and $v$, $v_0$ $(-\infty < v_0 < v < 0)$. One can verify that for any $w = u+iv \in G^-$, $t \in (-\infty,+\infty)$ and $\kappa > 0$

$$\frac{1}{|w-t|^{\kappa}} \le \frac{2^{\kappa/2}}{(|u-t|+|v|)^{\kappa}} \le \frac{C_\kappa(w)}{(1+|t|)^{\kappa}}, \tag{2.6}$$

where $C_\kappa(w)$ is a constant depending only on $\kappa$ and $w$. Therefore, by (2.2) and (2.3)

$$\begin{aligned}\int_{-\infty}^{v} |f'(u+iy)|dy &\le (1+\beta)2^{1+\beta/2}\int_{-\infty}^{+\infty} d\sigma(t)\int_{|v|}^{+\infty} \frac{dx}{(|u-t|+x)^{2+\beta}} \\ &= 2^{1+\beta/2}\int_{-\infty}^{+\infty} \frac{d\sigma(t)}{(|u-t|+|v|)^{1+\beta}} \\ &\le \sqrt{2}C_{1+\beta}(u+iv)\int_{-\infty}^{+\infty} \frac{d\sigma(t)}{(1+|t|)^{1+\beta}} < +\infty.\end{aligned}$$

Hence, the passage $v_0 \to -\infty$ in (2.5) gives

$$f(u+iv) = f(u-i\infty) + i\int_{-\infty}^{v} f'(u+iy)dy, \quad v<0, \ -\infty < u < +\infty, \tag{2.7}$$

where the integral is absolutely convergent and $f(u-i\infty)$ is a finite limit.

Now let $E_0$ be the set of those $u \in (-\infty, +\infty)$ for which

$$\int_{-\infty}^{0} |f'(u+iv)|dv = +\infty$$

and let $E$ be the set of those $u \in (-\infty, +\infty)$, where (2.4) is not a finite limit. Then $E \subseteq E_0$ by (2.7). Hence, $C_{1+\beta}(E_0) = 0$ implies $C_{1+\beta}(E) = 0$. Indeed, if we had $C_{1+\beta}(E) > 0$, then there would exist a nonnegative $B$-measure $\mu \prec E$ for which (1.4)–(1.5) is true with $\gamma = 1+\beta$. But $E \subseteq E_0$. Consequently, $\mu \prec E_0$, and $C_{1+\beta}(E_0) > 0$ by Definition 1.2.

Thus, it suffices to prove that $C_{1+\beta}(E_0) = 0$. We shall show that the assumption $C_{1+\beta}(E_0) > 0$ leads to a contradiction. Indeed, if $C_{1+\beta}(E_0) > 0$, then there must exist a nonnegative $B$-measure $\mu_0 \prec E_0$ satisfying (1.4)–(1.5). By (1.4), one can fix an enough great $A > 0$ such that

$$\int_{-A}^{A} d\mu_0(t) = \int_{[-A,A]\cap E_0} d\mu_0(t) > 0.$$

Then

$$\int_{-A}^{A} d\mu_0(u) \int_{-\infty}^{0} |f'(u+iv)|\, dv = +\infty. \tag{2.8}$$

If $|u| > 2A$ and $|t| < A$, then

$$\frac{1}{|u-t|} < \frac{C_1(a)}{(1+|u|)(1+|t|)}, \quad C_1(a) = (1+A^{-1})(1+2A).$$

Hence for $|u| > 2A$

$$\int_{-A}^{A} \frac{d\mu_0(t)}{|u-t|^{1+\beta}} \le \frac{C_1^{1+\beta}(A)}{(1+|u|)^{1+\beta}} \int_{-A}^{A} \frac{d\mu_0(t)}{(1+|t|)^{1+\beta}}$$
$$< \frac{(1+A^{-1})^{1+\beta}(1+2A)^{1+\beta}}{(1+|u|)^{1+\beta}}.$$

On the other hand, if $|u| \le 2A$, then by (2.1)

$$\int_{-A}^{A} \frac{d\mu_0(t)}{|u-t|^{1+\beta}} \le S_1 \le \frac{S_1(1+2A)^{1+\beta}}{(1+|u|)^{1+\beta}}.$$

Hence, particularly for any $u \in (-\infty, +\infty)$

$$\int_{-A}^{A} \frac{d\mu_0(t)}{|u-t|^{1+\beta}} < \frac{\left[(1+A^{-1})^{1+\beta} + S_1\right](1+2A)^{1+\beta}}{(1+|u|)^{1+\beta}} = \frac{C_2(A)}{(1+|u|)^{1+\beta}}. \tag{2.9}$$

But for any $u \in (-\infty, +\infty)$

$$\int_{-\infty}^{0} |f'(u+iv)|\,dv \leq 2^{1+\beta/2} \int_{-\infty}^{+\infty} \frac{d\sigma(t)}{|v-t|^{1+\beta}},$$

where the integrals can be finite or infinite. Therefore, by (2.9)

$$\int_{-A}^{A} d\mu_0(u) \int_{-\infty}^{0} |f'(u+iv)|dv \leq 2^{1+\beta/2} C_2 \int_{-\infty}^{+\infty} \frac{d\sigma(t)}{(1+|t|)^{1+\beta}} < +\infty$$

which contradicts (2.8).

**Lemma 2.4.** *Let $\beta \in (-1, 0)$ be any number, and let the sequence $\{w_k\} \equiv \{u_k + iv_k\} \subset G^-$ be such that*

$$\sum_k |v_k|^{1+\beta} < +\infty. \tag{2.10}$$

*Then $B_\beta(w, \{w_k\})$ has non-zero, finite nontangential boundary values in all points $u \in (-\infty, +\infty)$, with possible exception of a set of zero $1+\beta$-capacity.*

**Proof.** It was proved in Sec. 4 of Ch. 1, that $B_\beta = B_0 \exp\{-f\}$, where $f$ is of the form (2.2)–(2.3). By Lemma 2.3, the exponential factor in (2.2)–(2.3) has the required boundary properties. Therefore, it suffices to prove that $B_0$ has non-zero, finite angular boundary values at all real points, with possible exception of a set of zero $1+\beta$-capacity, and to use Lemma 2.1. To this end, observe that $B_0$ differs from the ordinary Blaschke product in the half-plane only by a constant multiplier. Consequently, by the Frostman's [33] result we can (after mapping conformally the disc onto the half-plane) only state that

$$\sup_{w \in G^-} \left\{ (1+|w|)^{1+\beta} \int_{-\infty}^{+\infty} \frac{d\mu(t)}{|w-t|^{1+\beta}} \right\} = +\infty \tag{2.11}$$

for any nonnegative $B$-measure $\mu$ satisfying (1.4) (where $\gamma = 1+\beta$) and supported on the set of those $u \in (-\infty, +\infty)$, at which $B_0(u) \neq 0$ does not exist. After we choose $A > 0$ enough great to provide that the integral (1.4) (with $\gamma = 1+\beta$) taken over $[-A, A]$ equals a number $M > 0$, we consider the measure $M^{-1}\mu|_{[-A,A]} \equiv \widetilde{\mu} \prec E$. It is evident, that this measure satisfies (1.4) (with $\gamma = 1+\beta$). On the other hand, it is obvious that $\widetilde{\mu}$ satisfies also (2.11). But

$$\frac{1}{|w-t|^{1+\beta}} \leq \frac{(2^{1+\beta} + A^{-1-\beta})(1+A)^{1+\beta}}{(1+|w|)^{1+\beta}(1+|t|)^{1+\beta}}$$

for $|w| \geq 2A$ and $|t| \leq A$. Consequently

$$+\infty = M \sup_{w \in G^-} \left\{ (1+|w|)^{1+\beta} \int_{-\infty}^{+\infty} \frac{d\widetilde{\mu}(t)}{|w-t|^{1+\beta}} \right\}$$

$$= \sup_{w \in G^-} \left\{ (1+|w|)^{1+\beta} \int_{-A}^{A} \frac{d\mu(t)}{|w-t|^{1+\beta}} \right\}$$

$$\leq (2^{1+\beta} + A^{-1-\beta})(1+A)^{1+\beta} + (1+2A)^{1+\beta} \sup_{|w|<2A} \int_{-A}^{A} \frac{d\mu(t)}{|w-t|^{1+\beta}}.$$

Hence $C_{1+\beta}(E) = 0$ in the sense of Definition 1.2, which proves our lemma.

**Proof of Theorem 1.1** is now obvious. By Lemmas 2.3 and 2.4 both the exponential factor and the Blaschke type product in the representation (1.1) of $F \in N_\alpha\{G^-\}$ $(-1 < \alpha < 0)$ have non-zero, finite nontangential boundary values out of some sets of zero $1+\alpha$-capacity. Hence, $F$ has the same property by Lemma 2.1.

*2.2.* Now we turn to lemmas which are necessary for proving Theorem 1.2.

**Lemma 2.5.** 1°. *Let $\alpha \in (-1, +\infty)$ be any number, and let $\{w_k\} \equiv \{u_k + iv_k\} \subset G^-$ be a sequence satisfying (2.10) with $\beta = \alpha$. Then*

$$B_\alpha(w, \{w_k\}) = G_\alpha(w, \{w_k\}) H_\alpha(w, \{w_k\}), \tag{2.12}$$

*where*

$$G_\alpha(w, \{w_k\}) = \prod_k g_\alpha(w, \{w_k\}) \equiv \prod_k \exp\left\{ - \int_0^{|v_k|} \frac{(|v_k| - t)^\alpha dt}{[i(w-u_k) - t]^{1+\alpha}} \right\} \tag{2.13}$$

$$H_\alpha(w, \{w_k\}) = \prod_k h_\alpha(w, \{w_k\}) \equiv \prod_k \exp\left\{ - \int_0^{|v_k|} \frac{(|v_k| - t)^\alpha dt}{[i(w-u_k) + t]^{1+\alpha}} \right\} \tag{2.13'}$$

*both are holomorphic functions in $G^-$, and $H_\alpha(w, \{w_k\}) \neq 0$ $(w \in G^-)$.*

2°. *If $-1 < \beta \leq 0$ and $\beta \leq \alpha < 1 + \beta$ and the sequence $\{w_k\} \equiv \{u_k + iv_k\} \subset G^-$ satisfies (2.10), then the functions $B_\alpha(w, \{w_k\})$, $G_\alpha(w, \{w_k\})$ and $H_\alpha(w, \{w_k\})$ belong to $N_\beta\{G^-\}$.*

**Proof.** 1°. One can easily prove that $B_\alpha$, also the product (2.13), as well as the product (2.13′) in (2.12) are absolutely and uniformly convergent inside $G^-$. Consequently, (2.12) follows form formulas (1.1′) and (3.4) of Ch. 1. Thus, it remains to observe that the integral in the exponent of the representation of each factor of $H_\alpha$ is holomorphic in $G^-$.

2°. Assuming $\zeta = \xi + i\eta \in G^-$ an arbitrary point, we use the well-known formula

$$\frac{1}{\Gamma(\delta)} \int_0^{+\infty} \frac{\sigma^{\delta-1} d\sigma}{(iw+\sigma)^\lambda} = \frac{\Gamma(\lambda-\delta)}{\Gamma(\lambda)} \frac{1}{(iw)^{\lambda-\delta}}, \quad 0 < \delta < \lambda < +\infty. \tag{2.14}$$

Then for $\beta < 0$ we obtain

$$W^{-\beta} \log g_\alpha(w,\zeta) \equiv \frac{i}{\Gamma(1+\beta)} \int_0^{+\infty} \sigma^\beta \left[\log g_\alpha(w - i\sigma, \zeta)\right]' d\sigma$$
$$= -\frac{\Gamma(1+\alpha-\beta)}{\Gamma(1+\alpha)} \int_0^{|\eta|} \frac{(|\eta|-t)^\alpha dt}{[i(w-\xi)-t]^{1+\alpha-\beta}}.$$

Similarly

$$W^{-\beta} \log h_\alpha(w,\zeta) = -\frac{\Gamma(1+\alpha-\beta)}{\Gamma(1+\alpha)} \int_0^{|\eta|} \frac{(|\eta|-t)^\alpha dt}{[i(w-\xi)+t]^{1+\alpha-\beta}}.$$

Besides, the same representations are true also for $\beta = 0$, since $W^0$ is identical. Consider the case $\alpha = \beta$. Clearly

$$\sup_{v<0} \int_{-\infty}^{+\infty} |W^{-\alpha} \log |H_\alpha(u+iv, \{w_k\})||du$$
$$\le \frac{1}{\Gamma(1+\alpha)} \sum_k \sup_{v<0} \int_0^{|v_k|} (|v_k|-t)^\alpha dt \int_{-\infty}^{+\infty} \frac{(t-v)dt}{(u-u_k)^2+(t-v)^2}$$
$$= \frac{\pi}{\Gamma(2+\alpha)} \sum_k |v_k|^{1+\alpha} < +\infty.$$

One can verify that also the remaining conditions of the inclusion $H_\alpha \in \mathfrak{N}_\alpha^m$ are satisfied ($\mathfrak{N}_\alpha^m$ are investigated in Ch. 3). $\mathfrak{N}_\alpha^m \subset N_\alpha\{G^-\}$ by Theorem 3.6 of Ch. 3. Hence $G_\alpha \in N_\alpha\{G^-\}$, since $B_\alpha \in N_\alpha\{G^-\}$. Now let $\beta < \alpha < 1+\beta$. Then, by (2.6)

$$\int_{-\infty}^{+\infty} |W^{-\alpha} \log |g_\alpha(u+iv,\zeta)||du$$
$$\le \frac{\Gamma(1+\alpha-\beta)}{\Gamma(1+\alpha)} 2^{(3+\alpha-\beta)/2} \int_0^{|\eta|} (|\eta|-t)^\alpha dt \int_{-\infty}^{+\infty} \frac{dt}{(x+||v|-t|)^{1+\alpha-\beta}}$$
$$= C_{\alpha,\beta} \int_0^{|\eta|} \frac{(|\eta|-t)^\alpha}{||v|-t|^{\alpha-\beta}} dt \equiv C_{\alpha,\beta} I, \quad v<0.$$

Similarly

$$\int_{-\infty}^{+\infty} |W^{-\alpha} \log |h_\alpha(u+iv,\zeta)||du$$
$$\le C_{\alpha,\beta} \int_0^{|\eta|} \frac{(|\eta|-t)^\alpha}{(|v|+t)^{\alpha-\beta}} dt < C_{\alpha,\beta}|\eta|^{1+\beta} \int_0^1 (1-x)^\alpha x^{\beta-\alpha} dx \equiv C'_{\alpha,\beta}|\eta|^{1+\beta}.$$

Hence

$$\sup_{v<0}\int_{-\infty}^{+\infty}|W^{-\alpha}\log|H_\alpha(u+iv,\{w_k\})||du \le C'_{\alpha,\beta}\sum_k |v_k|^{1+\alpha} < +\infty,$$

and it is easy to verify that $H_\alpha \in \mathfrak{N}_\beta^m \subset N_\beta\{G^-\}$. For evaluation of $I$ in the above estimate for $g_\alpha$, first assume $|v| \ge |\eta|$. Then obviously

$$I = \int_0^{|\eta|}\frac{(|\eta|-t)^\alpha}{(|v|-t)^{\alpha-\beta}}dt < \int_0^{|\eta|}(|\eta|-t)^\beta dt = \frac{|\eta|^{1+\beta}}{1+\beta}.$$

If $|v| < |\eta|$ and $\alpha \le 0$, then

$$\begin{aligned} I &< \int_0^{|v|}(|v|-t)^\beta dt + \int_{|v|}^{(|v|+|\eta|)/2}(t-|v|)^\beta dt + \int_{(|v|+|\eta|)/2}^{|\eta|}(|\eta|-t)^\beta dt \\ &= \frac{1}{1+\beta}\left[|v|^{1+\beta} + 2\left(\frac{|\eta|-|v|}{2}\right)^{1+\beta}\right] < \frac{3}{1+\beta}|\eta|^{1+\beta}. \end{aligned}$$

Now let $|v| < |\eta|$ and $\alpha > 0$. Then integration by parts gives

$$\begin{aligned} I &< \frac{1}{1+\beta-\alpha}\Bigg\{|\eta|^{1+\beta} + \alpha\int_{|v|}^{(|v|+|\eta|)/2}(|\eta|-t)^\beta dt \\ &\quad + \alpha(|\eta|-|v|)^{1+\beta-\alpha}\int_{(|v|+|\eta|)/2}^{|\eta|}(|\eta|-t)^\alpha dt\Bigg\} < \frac{(3+\beta)|\eta|^{1+\beta}}{(1+\beta)(1+\beta-\alpha)}. \end{aligned}$$

From these estimates it follows that

$$\sup_{v<0}\int_{-\infty}^{+\infty}|W^{-\alpha}\log|G_\alpha(u+iv,\{w_k\})||du < +\infty.$$

Hence $G_\alpha \in \mathfrak{N}_\beta^m \subset N_\beta\{G^-\}$, and it remains to see that $B_\alpha \in N_\beta\{G^-\}$ by the inclusions $G_\alpha \in N_\beta\{G^-\}$ and $H_\alpha \in N_\beta\{G^-\}$.

**Lemma 2.6.** *Let* $\alpha \in (0,+\infty)$ *and let the sequence* $\{w_k\} \equiv \{u_k + iv_k\}$ *be such that*

$$\sum_k |v_k|^\alpha < +\infty.$$

*Then*

$$B_\alpha(w,\{w_k\}) = B_{\alpha-1}(w,\{w_k\})\,[H_{\alpha-1}(w,\{w_k\})]^{-2}, \tag{2.15}$$

*where* $H_{\alpha-1}$ *is a function holomorphic in* $G^-$*, having the form (2.13′). If* $0<\alpha\le 1$*, then* $H_{\alpha-1}$ *is a bounded function in* $G^-$.

*If* $\alpha \in (1, +\infty)$ *and the sequence* $\{w_k\} \equiv \{u_k + iv_k\}$ *satisfies*

$$\sum_k |v_k|^{\alpha-1} < +\infty,$$

*then*

$$B_\alpha(w, \{w_k\}) = B_{\alpha-1}(w, \{w_k\}) \left[H_{\alpha-2}(w, \{w_k\})\right]^2 \left[R_{\alpha-1}(w, \{w_k\})\right]^2, \quad (2.16)$$

*where* $R_{\alpha-1}$ *is holomorphic in* $G^-$ *and bounded for* $1 < \alpha \leq 2$.

**Proof** follows from formulas (2.12), (2.13), (2.13′) by successive integration by parts. Besides, it is clear that $|H_{\alpha-1}| \leq 1$ in $G^-$ for $0 < \alpha \leq 1$. In addition, one can show that

$$R_{\alpha-1}(w, \{w_k\}) = \exp\left\{-\frac{1}{1+\alpha} \sum_k \frac{|v_k|^{\alpha-1}}{[i(w-u_k)]^{\alpha-1}}\right\}.$$

Using this, one can prove that $|R_{\alpha-1}| \leq 1$ in $G^-$ for $1 < \alpha \leq 2$.

*2.3.* **Proof of Theorem 1.2.** 1°. Let (1.11) be valid with a fixed $\gamma \in (0, 1)$. First consider the case $\gamma - 1 \leq \alpha < \gamma$. If we denote $\beta = \gamma - 1$, then $-1 < \beta < 0$ and $\beta \leq \alpha < \beta + 1$. Therefore $B_\alpha \in N_\beta\{G^-\} \equiv N_{\gamma-1}\{G^-\}$ by Lemma 2.5, and our assertion follows from Theorem 1.1.

Next, let $\gamma \leq \alpha < \gamma + 1$. Then the representation (2.15) of Lemma 2.6 is valid, and $\gamma - 1 \leq \alpha - 1 < \gamma$. Therefore $B_{\alpha-1}$, $H_{\alpha-1} \in N_{\gamma-1}\{G^-\}$ by Lemma 2.5. Hence $B_\alpha$ is of the same class, and our assertion follows from Theorem 1.1. At last, let $\alpha = \gamma + 1$. Then, by (2.15) and (2.16)

$$B_\alpha \equiv B_{\gamma+1} = B_\gamma [H_\gamma]^{-2}, \quad H_\gamma = [H_{\gamma-1} R_{\gamma-1}]^{-1}, \quad (2.17)$$

where, by proven above, $B_\gamma$ has non-zero, finite boundary values at all $u \in (-\infty, +\infty)$, with possible exception of a set of zero $\gamma$-capacity. By Lemma 2.5, $H_{\gamma-1} \in N_{\gamma-1}\{G^-\}$. Therefore, by Theorem 1.1 also the set of those $u$, where $H_{\gamma-1}$ has not non-zero, finite boundary values is of zero $\gamma$-capacity. Thus, if we also prove that

$$\lim_{v \to -0} H_\gamma(u + iv, \{w_k\}) \neq 0 \quad (2.18)$$

out of a set of zero $\gamma$-capacity, then, by Lindelöf's theorem, non-zero, finite boundary values of $R_{\gamma-1}$ will exist out of such a set. By Lemma 2.1, $H_\gamma$ will have the same boundary property. And returning to the first equality in (2.17), we shall arrive at the desired assertion for $B_{\gamma+1}$.

Thus, for completion of the proof of assertion 1° it suffices to show that (2.18) exists as a finite limit outside a set of zero $\gamma$-capacity. One can prove this quite similar to Lemma 2.3. The following estimates will be the only difference:

$$\left|\left[\log H_\gamma(u+iv,\{w_k\})\right]'\right| \le (\gamma+1)\sum_k \left|\int_0^{|v_k|} \frac{(|v_k|-t)^\gamma dt}{[i(w-u_k)+t]^{2+\gamma}}\right|$$

$$< 2^{(1+\gamma)/2}\sum_k \frac{|v_k|^\gamma}{(|u-u_k|+|v|)^{\gamma+1}},$$

$$\int_{-A}^{A} d\mu_0(u)\int_{-\infty}^{0}\left|\left[\log H_\gamma(u+iv,\{w_k\})\right]'\right| dv$$

$$< 2^{(1+\gamma)/2}\gamma^{-1}\sum_k |v_k|^\gamma \int_{-A}^{A}\frac{d\mu_0(u)}{|u-u_k|^\gamma} \le 2^{(1+\gamma)/2}\gamma^{-1}S_1\sum_k |v_k|^\gamma < +\infty.$$

2°. Let (1.12) be satisfied. The desired assertion is well-known for $\alpha = 0$. If $0 < \alpha \le 1$, then (2.15) implies that $B_\alpha$ is a function of bounded type in $G^-$, and hence our assertion follows. If $1 < \alpha \le 2$, we use the representation (2.16), where $H_{\alpha-2}$ and $R_{\alpha-1}$ are holomorphic and bounded in $G^-$ and $B_{\alpha-1}$ is of bounded type in $G^-$ by proven above. Thus, also $B_\alpha$ is of bounded type in $G^-$, and hence our assertion follows. The proof is complete.

It remains only to prove the inequality (1.15) for the Blaschke type products in the half-plane. To this end, we use formula (2.14) one more time. In view of formulas (1.1′)–(1.2) of Ch. 2 we get

$$\varphi(w) \equiv W^{-\alpha_2}\log\left|\frac{b_{\alpha_2}(w+\xi,\zeta)}{b_{\alpha_1}(w+\xi,\zeta)}\right|$$

$$= \frac{\Gamma(1+\alpha_1-\alpha_2)}{\Gamma(1+\alpha_1)}\mathrm{Re}\int_0^{|\eta|}(|\eta|-t)^{\alpha_1}\left[\frac{1}{(iw-t)^{1+\alpha_1-\alpha_2}}+\frac{1}{(iw+t)^{1+\alpha_1-\alpha_2}}\right]dt$$

$$-\frac{1}{\Gamma(1+\alpha_2)}\mathrm{Re}\int_0^{|\eta|}(|\eta|-t)^{\alpha_2}\left[\frac{1}{iw-t}+\frac{1}{iw+t}\right]dt$$

for any $\alpha_1$, $\alpha_2$ $(-1 < \alpha_1 < \alpha_2 \le 0)$, $w \in G^-$ and $\zeta = \xi + i\eta \in G^-$. Hence it is evident that the function $\varphi$ is harmonic in $G^-$ and continuous up to the real axis. But a simple calculation shows that for $w = u \in (-\infty, +\infty)$ the integrand of one of the integrals in the representation of $\varphi$ is a positive function, while the integrand in the other integral is identically zero. Hence $\varphi(u) > 0$. Further, $\varphi(w) > 0$ $(w \in G^-)$ by the Phragmén–Lindelöf principle, and applying Theorem 1.4 of Ch. 1 we find that

$$B_{\alpha_1}(w,\{w_k\}) = B_{\alpha_2}(w,\{w_k\})\exp\left\{-\int_{-\infty}^{+\infty}\frac{d\sigma(t)}{[i(w-t)]^{1+\alpha_2}}\right\}, \quad w \in G^-,$$

where $\sigma(t)$ is a nondecreasing function satisfying (2.2) with $\beta = \alpha_2$. Hence (1.15) follows.

## 3. PROOF OF THEOREM 1.3

*3.1.* First note that in our case there is no similarity of formula (2.14) in the disc. Therefore, it is not evident how to prove the similarity of the inequality (1.15) for Blaschke–M.M.Djrbashian products. Nevertheless, for our aims it is enough to replace (2.14) by the following

**Lemma 3.1.** *For any $\beta \in (-1, +\infty)$ and $\alpha > \beta - 1$*

$$\int_0^1 \frac{(1-t)^\beta}{(1-tz)^{2+\alpha}} dt = \frac{J_{\alpha,\beta}(z)}{(1-z)^{1+\alpha-\beta}}, \quad z \in \overline{\mathbb{C}},$$

*where $J_{\alpha,\beta}(z)$ is such that*

$$M_{\alpha,\beta}(R_0) = \sup_{|z|<R_0} |J_{\alpha,\beta}(z)| < +\infty, \quad 0 < R_0 < +\infty.$$

**Proof.** The change of variable $t = 1-(1-z)x/z$ gives our representation, where

$$J_{\alpha,\beta}(z) = \frac{1}{z^{1+\beta}} \int_0^{z/(1-z)} \frac{x^\beta dx}{(1+x)^{2+\alpha}}$$

is a function holomorphic in the $z$-plane cut along $1 \le z < +\infty$ ($\operatorname{Im} z = 0$). Besides, here occurs the well-known arbitrariness in choice of integration contour. Since $|z/(1-z)+1| = 1/|1-z| > (1+R_0)^{-1} > 0$, one can choose this contour to be distant from $-1$ not less than $(1+R_0)^{-1}$. Concretely, we choose the integration contour in the following way. If the point $w = z/(1-z)$ is out of $\Delta_{R_0} = \{w : |\arg w| > \pi - \arcsin(1+R_0)^{-1}, \operatorname{Re} w < -1 + (1+R_0)^2\}$, then we choose the contour to be the intercept $[0, w]$. If $w \in \Delta_{R_0}$ and $|w| \le 2$, then we choose the union of the intercept $\left[0, |w|e^{i\varphi(w)}\right]$ (where $\varphi(w) = [\pi - \arcsin(1+R_0)^{-1}]\operatorname{sign}(\arg w)$) and the arc $[|w|e^{i\varphi(w)}, w]$ of a circle of radius $|w|$. Further, if $w \in \Delta_{R_0}$ and $|w| > 2$, then we unite the intercept $[0, 2e^{i\varphi(w)}]$, the arc $[2e^{i\varphi(w)}, 2e^{i\arg w}]$ of a circle with the radius 2 and the intercept $[2e^{i\arg w}, w]$.

Let $|z| \le 1/3$. Then $w \notin \Delta_{R_0}$ and $|w| = |z/(1-z)| \le 3|z|/2$. Therefore

$$\sup_{|z|\le 1/3} |J_{\alpha,\beta}(z)| < \frac{(1+R_0)^{2+\alpha}}{|z|^{1+\beta}} \int_0^{3|z|/2} x^\beta dx = \frac{(1+R_0)^{2+\alpha}}{1+\beta} \left(\frac{3}{2}\right)^{1+\beta}. \tag{3.1}$$

If $1/3 < |z| < R_0$ and $w \notin \Delta_{R_0}$, $|w| \le 2$. Then

$$|J_{\alpha,\beta}(z)| \le 3^{1+\beta}(1+R_0)^{2+\alpha} \int_0^2 x^\beta dx = \frac{6^{1+\beta}}{1+\beta}(1+R_0)^{2+\alpha}.$$

If $1/3 < |z| < R_0$ and $w \notin \Delta_{R_0}$ but $|w| > 2$, then obviously

$$|J_{\alpha,\beta}(z)| \le 3^{1+\beta}(1+R_0)^{2+\alpha} \int_0^2 x^\beta dx + 3^{1+\beta} \int_2^{+\infty} \frac{x^\beta dx}{(x-1)^{2+\alpha}} < +\infty.$$

Thus, if $z/(1-z)$ remains out of $\Delta_{R_0}$, then

$$\sup_{1/3<|z|<R_0} |J_{\alpha,\beta}(z)| < +\infty. \tag{3.1$'$}$$

Now let $1/3 < |z| < R_0$ and let $w \in \Delta_{R_0}$, $|w| \le 2$. Then

$$\begin{aligned}|J_{\alpha,\beta}(z)| &< \frac{6^{1+\beta}}{1+\beta}(1+R_0)^{2+\alpha} + 3^{1+\beta}(1+R_0)^{2+\alpha}\left| i|w|^{1+\beta}\int_{\varphi(w)}^{\arg w} e^{i(2+\beta)\vartheta}d\vartheta\right| \\ &< 6^{1+\beta}(1+R_0)^{2+\alpha}[(1+\beta)^{-1}+\pi] < +\infty.\end{aligned}$$

Further, if $1/3 < |z| < R_0$ and $w \in \Delta_{R_0}$ but $|w| > 2$, then

$$\begin{aligned}|J_{\alpha,\beta}(z)| &< 6^{1+\beta}(1+R_0)^{2+\alpha}\left[\frac{1}{1+\beta}+\pi\right] + 3\left|\int_{2e^{i\arg(w)}}^{w}\frac{x^\beta dx}{(1+x)^{2+\alpha}}\right| \\ &< 6^{1+\beta}(1+R_0)^{2+\alpha}\left[\frac{1}{1+\beta}+\pi\right] + 3^{1+\beta}\int_2^{+\infty}\frac{x^\beta dx}{(1+x)^{2+\alpha}} < +\infty.\end{aligned}$$

Thus, the relation (3.1′) is true also when the point $w = z/(1-z)$ lies in $\Delta_{R_0}$. Together with (3.1), this completes the proof.

*3.2.* We shall use also the below similarities of Lemmas 2.5 and 2.6.

**Lemma 3.2.** 1°. *Let $\alpha \in (-1, +\infty)$ be any number, and let $\{z_k\} \subset \mathbb{D}$ ($z_k \ne 0$, $k \ge 1$) be a sequence satisfying (1.6). Then*

$$B_\alpha(z,\{z_k\}) = C_\alpha G_\alpha(z,\{z_k\})H_\alpha(z,\{z_k\}), \tag{3.2}$$

*where $C_\alpha$ is a constant depending on $\alpha$ and $\{z_k\}$ and*

$$G_\alpha(z,\{z_k\}) = \prod_k g_\alpha(z,z_k) \equiv \prod_k \exp\left\{-\int_{|z_k|}^1 \frac{(1-x)^\alpha}{\left(1-\frac{zx}{z_k}\right)^{1+\alpha}}\frac{dx}{x}\right\}, \tag{3.3}$$

$$H_\alpha(z,\{z_k\}) = \prod_k h_\alpha(z,z_k) \equiv \prod_k \exp\left\{-\int_{|z_k|}^1 \frac{(1-x)^\alpha}{\left(1-\frac{z\bar{z}_k}{x}\right)^{1+\alpha}}\frac{dx}{x}\right\} \tag{3.3$'$}$$

*are holomorphic functions in $\mathbb{D}$, and $H_\alpha(z,\{z_k\}) \ne 0$, $z \in \mathbb{D}$.*

2°. *If $-1 < \beta \le 0$, $\beta \le \alpha < 1+\beta$ and the sequence $\{z_k\} \subset \mathbb{D}$ ($z_k \ne 0$, $k \ge 1$) satisfies (1.13) with $\gamma = 1+\beta$, then the functions $B_\alpha(z,\{z_k\})$, $G_\alpha(z,\{z_k\})$ and $H_\alpha(z,\{z_k\})$ belong to the class $N_\beta$ of M.M.Djrbashian.*

**Proof.** 1°. It is easy to verify that the products (3.3) and (3.3′) are absolutely and uniformly convergent inside $\mathbb{D}$. Hence, (3.2) directly follows from

(1.7)–(1.8). It remains to observe that the integral in the exponent of the representation of each factor of $H_\alpha$ is a holomorphic function in $\mathbb{D}$.

2°. First observe that under our requirements there exists $\rho_0 \in (0,1)$ for which $|z_k| \geq \rho_0$ $(k \geq 1)$. Further, assuming $\zeta = \rho e^{i\varphi}$ $(\rho_0 \leq \rho < 1)$ and $z = re^{i\vartheta}$ $(0 < r < 1)$ we use the following formula from [19] ((1.22), p. 573): *if* $f'(x) \in L_1(0,l)$ $(0 < l < +\infty)$, *then for any* $\alpha \in (0,1)$

$$D^\alpha f(x) = \frac{f(0)}{\Gamma(1-\alpha)} x^{-\alpha} + \frac{1}{\Gamma(1-\alpha)} \int_0^x (x-t)^{-\alpha} f'(t)dt, \quad a.e. \quad x \in (0,l).$$

Then for any $\beta \in (-1,0]$ and $\alpha \in (-1,+\infty)$ we get

$$\begin{aligned} r^{-\beta} D^{-\beta} \log g_\alpha(z,\zeta) &= r^{-\beta} D^{-(1+\beta)} \frac{d}{dr} \log g_\alpha(z,\zeta) + \frac{1}{\Gamma(1+\beta)} \log g_\alpha(z,\zeta) \\ &= -\frac{1+\alpha}{\Gamma(1+\beta)} \frac{r}{\rho} e^{i(\varphi-\vartheta)} \int_\rho^1 (1-x)^\alpha dx \int_0^1 \frac{(1-t)^\beta dt}{\left(1 - \frac{rx}{\rho} te^{i(\varphi-\vartheta)}\right)^{2+\alpha}} \\ &\quad - \frac{1+\alpha}{\Gamma(1+\beta)} \int_\rho^1 \frac{(1-x)^\alpha}{x} dx. \end{aligned}$$

Similarly, we find

$$\begin{aligned} r^{-\beta} D^{-\beta} \log h_\alpha(z,\zeta) = &-\frac{1+\alpha}{\Gamma(1+\beta)} \int_\rho^1 \frac{(1-x)^\alpha}{x} dx \\ &- \frac{1+\alpha}{\Gamma(1+\beta)} r\rho e^{i(\varphi-\vartheta)} \int_\rho^1 \frac{(1-x)^\alpha}{x^2} dx \int_0^1 \frac{(1-t)^\beta dt}{\left(1 - \frac{r\rho}{x} te^{i(\varphi-\vartheta)}\right)^{2+\alpha}}. \end{aligned}$$

For $\alpha = \beta$, one can show that

$$\begin{aligned} \left| r^{-\alpha} D^{-\alpha} \log h_\alpha(re^{i\vartheta}, \rho e^{i\varphi}) \right| \leq &\frac{1+2\alpha}{\Gamma(2+\alpha)} \int_\rho^1 \frac{(1-x)^\alpha}{x} dx \\ &+ \frac{1}{2\Gamma(2+\alpha)} \int_\rho^1 \frac{1 - \left(\frac{r\rho}{x}\right)^2}{\left|1 - \frac{r\rho}{x} e^{i(\vartheta-\varphi)}\right|^2} \frac{(1-x)^\alpha}{x} dx. \end{aligned}$$

Using this estimate, we obtain that for $0 \leq r < 1$

$$\frac{1}{2\pi} \int_{-\pi}^{\pi} \left| r^{-\alpha} D^{-\alpha} \log \left| H_\alpha(re^{i\vartheta}, \{z_k\}) \right| \right| \leq \frac{4}{\rho_0 \Gamma(2+\alpha)} \sum_k (1 - |z_k|)^{1+\alpha} < +\infty,$$

which provides $H_\alpha \in N_\alpha$. Hence $G_\alpha \in N_\alpha$ in view of $B_\alpha \in N_\alpha$ (see [19], Ch. IX). Now let $\beta < \alpha < 1 + \beta$. Then

$$\begin{aligned} \left| r^{-\beta} D^{-\beta} \log h_\alpha(re^{i\vartheta}, \rho e^{i\varphi}) \right| \leq &\frac{(1-\rho)^{1+\alpha}}{\rho_0 (1+\alpha) \Gamma(1+\beta)} \\ &+ \frac{1+\alpha}{\Gamma(1+\beta)} \int_\rho^1 (1-x)^\alpha dx \int_0^1 \frac{(1-t)^\beta dt}{\left|x - tr\rho e^{i(\vartheta-\varphi)}\right|^{2+\alpha}}, \end{aligned}$$

and by Lemma 3.1

$$\left|r^{-\beta}D^{-\beta}\log g_\alpha(re^{i\vartheta},\rho e^{i\varphi})\right| \le \frac{(1-\rho)^{1+\alpha}}{\rho_0(1+\alpha)\Gamma(1+\beta)} + \frac{1+\alpha}{\Gamma(1+\beta)}M_{\alpha,\beta}\left(\frac{1}{\rho_0}\right)\int_\rho^1 \frac{(1-x)^\alpha dx}{\left|\rho - rxe^{i(\vartheta-\varphi)}\right|^{1+\alpha-\beta}}. \quad (3.4)$$

Observe that for $\vartheta-\varphi=\lambda$, $|\lambda|<\pi$

$$\left|x - r\rho te^{i\lambda}\right|^2 = (x-r\rho t)^2 + 4xr\rho t\sin^2\frac{\lambda}{2} \ge (x-r\rho t)^2 + 4xr\rho t\frac{\lambda^2}{\pi^2}.$$

Therefore

$$\left|x-r\rho te^{i\lambda}\right|^{-(2+\alpha)} \le 2^{1+\alpha/2}\left[x - r\rho t + 2\sqrt{xr\rho t}\frac{|\lambda|}{\pi}\right]^{-(2+\alpha)}$$

and

$$\frac{1}{2\pi}\int_{-\pi}^{\pi}\left|r^{-\beta}D^{-\beta}\log h_\alpha(re^{i\vartheta},\rho e^{i\varphi})\right| d\vartheta \le \frac{(1-\rho)^{1+\alpha}}{\rho_0(1+\alpha)\Gamma(1+\beta)}$$
$$+ \frac{2^{2+\alpha/2}(1+\alpha)}{\pi\Gamma(1+\beta)}\int_\rho^1(1-x)^\alpha dx\int_0^1(1-t)^\beta dt\int_0^\pi \frac{d\lambda}{\left(x-r\rho t+\frac{2}{\pi}\sqrt{xr\rho t}\lambda\right)^{2+\alpha}}$$
$$< \frac{(1-\rho)^{1+\alpha}}{\rho_0(1+\alpha)\Gamma(1+\beta)} + \frac{2^{2+\alpha/2}}{\pi\Gamma(1+\beta)}\int_\rho^1(1-x)^\alpha dx\int_0^1\frac{(1-t)^\beta dt}{(x-\rho t)^{1+\alpha}}.$$

The change of variable $t = 1-(x/\rho-1)\tau$ gives

$$\int_0^1 \frac{(1-t)^\beta dt}{(x-\rho t)^{1+\alpha}} < \frac{\rho_0^{-(1+\beta)}}{(x-\rho)^{\alpha-\beta}}\int_0^{+\infty}\frac{\tau^\beta d\tau}{(1+\tau)^{1+\alpha}} \equiv \frac{C_1}{(x-\rho)^{\alpha-\beta}}.$$

Consequently, by $x = 1-(1-\rho)y$ we get

$$\frac{1}{2\pi}\int_{-\pi}^{\pi}\left|r^{-\beta}D^{-\beta}\log h_\alpha(re^{i\vartheta},\rho e^{i\varphi})\right| d\vartheta$$
$$< \frac{(1-\rho)^{1+\beta}}{\rho_0(1+\alpha)\Gamma(1+\beta)} + \frac{2^{2+\alpha/2}C_1}{\pi\Gamma(1+\beta)}\int_0^1\frac{y^\alpha dy}{(1-y)^{\alpha-\beta}}(1-\rho)^{1+\beta}.$$

Hence, for $0\le r<1$

$$\frac{1}{2\pi}\int_{-\pi}^{\pi}\left|r^{-\beta}D^{-\beta}\log H_\alpha(re^{i\vartheta},\{z_k\})\right| d\vartheta$$
$$< \left[\frac{1}{\rho_0(1+\alpha)\Gamma(1+\beta)} + \frac{2^{2+\alpha/2}C_1}{\pi\Gamma(1+\beta)}\int_0^1\frac{y^\alpha dy}{(1-y)^{\alpha-\beta}}\right]\sum_k(1-|z_k|)^{1+\beta} < +\infty.$$

Thus, $H_\alpha \in N_\beta$. For proving that also $G_\alpha \in N_\beta$, observe that in view of (3.4) the repetition of the above argument gives

$$\frac{1}{2\pi}\int_{-\pi}^{\pi}\left|r^{-\beta}D^{-\beta}\log g_\alpha(re^{i\vartheta},\rho e^{i\varphi})\right|d\vartheta$$
$$< \frac{(1-\rho)^{1+\beta}}{\rho_0(1+\alpha)\Gamma(1+\beta)} + \frac{1+\alpha}{\pi\Gamma(1+\beta)}M_{\alpha,\beta}\left(\frac{1}{\rho_0}\right)2^{2+\alpha/2}I_{\alpha,\beta}, \tag{3.5}$$

where

$$I_{\alpha,\beta} = \int_\rho^1 \frac{(1-x)^\alpha dx}{|\rho - rx|^{\alpha-\beta}}. \tag{3.6}$$

For evaluating of the last integral, first assume $r \le \rho$. Then

$$I_{\alpha,\beta} \le \int_\rho^1 \frac{(1-x)^\alpha dx}{(\rho - rx)^{\alpha-\beta}} \le \frac{(1-\rho)^{1+\beta}}{\rho_0^{\alpha-\beta}(1+\beta)}, \quad 0 \le r \le \rho. \tag{3.7}$$

Next, assume $\rho < r < 1$ and $\alpha \le 0$. Then

$$I_{\alpha,\beta} = \int_\rho^{\rho/r} \frac{(1-x)^\alpha dx}{(\rho - rx)^{\alpha-\beta}} + \int_{\rho/r}^1 \frac{(1-x)^\alpha dx}{(rx-\rho)^{\alpha-\beta}} \equiv J_1 + J_2.$$

One can verify that

$$J_1 < \frac{1}{r^{\alpha-\beta}}\int_\rho^{\rho/r}\left(x-\frac{\rho}{r}\right)^\beta dx < \frac{(1-\rho)^{1+\beta}}{\rho_0^{\alpha-\beta}(1+\beta)},$$
$$J_2 < \frac{1}{r^{\alpha-\beta}}\int_{\rho/r}^{(1+\rho/r)/2}\left(x-\frac{\rho}{r}\right)^\beta dx + \frac{1}{r^{\alpha-\beta}}\int_{(1+\rho/r)/2}^1 (1-x)^\beta dx < \frac{2(1-\rho)^{1+\beta}}{\rho_0^{1+\alpha}(1+\beta)}.$$

Therefore

$$I_{\alpha,\beta} \le \left[\frac{1}{\rho_0^{\alpha-\beta}} + \frac{2}{\rho_0^{1+\alpha}}\right]\frac{(1-\rho)^{1+\beta}}{1+\beta}, \quad \rho < r < 1, \quad \alpha \le 0. \tag{3.8}$$

Now let $\rho < e < 1$ and $0 < \alpha < 1+\beta$. Then integration by parts gives

$$I_{\alpha,\beta} = \frac{r^{-(\alpha-\beta)}}{1+\beta-\alpha}\left\{-\int_\rho^{\rho/r}(1-x)^\alpha d\left(\frac{\rho}{r}-x\right)^{1+\beta-\alpha}\right.$$
$$\left.+\int_{\rho/r}^{(1+\rho/r)/2}(1-x)^\alpha d\left(x-\frac{\rho}{r}\right)^{1+\beta-\alpha}\right\} + \frac{1}{r^{\alpha-\beta}}\int_{(1+\rho/r)/2}^1 \frac{(1-x)^\alpha dx}{\left(x-\frac{\varrho}{r}\right)^{\alpha-\beta}}$$
$$< \left[\frac{1}{\rho_0^{\alpha-\beta}} + \frac{1}{2^{1+\beta}\rho_0^{1+\alpha}} + \frac{1}{\rho_0^{1+\alpha}(1+\beta)}\right]\frac{(1-\rho)^{1+\beta}}{1+\beta-\alpha}.$$

Hence, by (3.5)–(3.8)

$$\frac{1}{2\pi}\int_{-\pi}^{\pi}\left|r^{-\beta}D^{-\beta}\log G_\alpha(re^{i\vartheta},\{z_k\})\right|d\vartheta < C_2\sum_k(1-|z_k|)^{1+\beta} < +\infty$$

for $0 \le r < 1$, where $C_2$ is a constant depending only on $\alpha$, $\beta$, and $\rho_0$. Thus, $G_\alpha \in N_\beta$, which provides $B_\alpha \in N_\beta$ by already proven $H_\alpha \in N_\beta$ and (3.2).

**Lemma 3.3.** 1°. *Let $\alpha \in (0,+\infty)$ be any number, and let $\{z_k\} \subset \mathbb{D}$ ($z_k \neq 0$, $k \ge 1$) be a sequence satisfying*

$$\sum_k(1-|z_k|)^\alpha < +\infty.$$

*Then*

$$B_\alpha(z,\{z_k\}) = C'_\alpha B_{\alpha-1}(z,\{z_k\})\left[\widetilde{H}^*_{\alpha-1}(z,\{z_k\})\right]^{-2}, \tag{3.9}$$

*where $C'_\alpha = \exp\left\{-\alpha^{-1}\sum_k(1-|z_k|)^\alpha\right\}$ is a constant and the function*

$$H^*_{\alpha-1}(z,\{z_k\}) = \prod_k \exp\left\{\int_{|z_k|}^1 \frac{(1-x)^{\alpha-1}}{\left(1-\frac{z\bar{z}_k}{x}\right)^\alpha}dx\right\} \neq 0 \tag{3.10}$$

*is holomorphic in $\mathbb{D}$ and bounded if $0 < \alpha \le 1$.*

2°. *If $\alpha \in (1,+\infty)$ and the sequence $\{z_k\}$ satisfies*

$$\sum_k(1-|z_k|)^{\alpha-1} < +\infty,$$

*then*

$$B_\alpha(z,\{z_k\}) = C'_\alpha B_{\alpha-1}(z,\{z_k\})\frac{\left[H^{***}_{\alpha-2}(z,\{z_k\})\right]^4}{\left[H^{**}_{\alpha-2}(z,\{z_k\})\right]^2\left[R_{\alpha-1}(z,\{z_k\})\right]^2}, \tag{3.11}$$

*where*

$$H^{**}_{\alpha-2}(z,\{z_k\}) = \prod_k \exp\left\{-\frac{1}{\alpha-1}\int_{|z_k|}^1 \frac{(1-x)^{\alpha-2}}{\left(1-\frac{z\bar{z}_k}{x}\right)^{\alpha-1}}dx\right\}, \tag{3.12}$$

$$H^{***}_{\alpha-2}(z,\{z_k\}) = \prod_k \exp\left\{-\int_{|z_k|}^1 \frac{(1-x)^{\alpha-2}}{\left(1-\frac{z\bar{z}_k}{x}\right)^{\alpha-1}}x\,dx\right\}, \tag{3.12'}$$

$$R_{\alpha-1}(z,\{z_k\}) = \exp\left\{-\frac{1}{\alpha-1}\sum_k |z_k|\frac{(1-|z_k|)^{\alpha-1}}{\left(1-\frac{z\bar{z}_k}{|z_k|}\right)^{\alpha-1}}\right\} \tag{3.12''}$$

*are holomorphic, non-vanishing functions in $\mathbb{D}$, bounded if $1 < \alpha \le 2$.*

**Proof.** 1°. The representation (3.9)–(3.10) is proved in [6]. The required properties of $H^*_{\alpha-1}$ are obvious. 2°. One can easily show (see [6]) that for any $z, \zeta \in \mathbb{D}$

$$h_\alpha(z,\zeta) = h_{\alpha-1}(z,\zeta)\exp\left\{2\int_{|\zeta|}^1 \frac{(1-x)^{\alpha-1}}{\left(1-\frac{z\bar\zeta}{x}\right)^\alpha}dx - \frac{1}{\alpha}\frac{(1-|\zeta|)^\alpha}{\left(1-\frac{z\bar\zeta}{|\zeta|}\right)^\alpha}\right\}.$$

On the other hand, if $h^*_{\alpha-1}(z,\zeta)$ is a factor of the product (3.10), then obviously

$$\varphi_{\alpha-1}(z,\zeta) \equiv \frac{h^*_{\alpha-1}(z,\zeta)}{h_{\alpha-1}(z,\zeta)} = \int_{|\zeta|}^1 \frac{(1-x)^\alpha}{\left(1-\frac{z\bar\zeta}{x}\right)^\alpha}\frac{dx}{x}.$$

Further, one can verify that

$$\frac{\varphi_{\alpha-1}(z,\zeta)}{\varphi_{\alpha-2}(z,\zeta)} = \exp\left\{\frac{1}{\alpha-1}\frac{(1-|\zeta|)^\alpha}{\left(1-\frac{z\bar\zeta}{|\zeta|}\right)^{\alpha-1}} - \frac{1+2(\alpha-1)}{\alpha-1}\int_\rho^1 \frac{(1-x)^{\alpha-1}}{\left(1-\frac{z\bar\zeta}{x}\right)^{\alpha-1}}dx\right\}$$

The representation (3.11) is a consequence of these equalities. For $1 < \alpha \le 2$ the boundedness of functions (3.12), (3.12′) and (3.12″) in $\mathbb{D}$ is obvious.

**Proof of Theorem 1.3.** 1°. Let (1.13) be true for some $\gamma \in (0,1)$. First, assume that $\gamma - 1 \le \alpha < \gamma$. Denoting $\beta = \gamma - 1$ we have $-1 < \beta < 0$ and $\beta \le \alpha < \beta + 1$. Consequently, $B_\alpha \in N_\beta$ by Lemma 3.2, and our assertion holds in view of Theorem 2 (Introduction) of M.M.Djrbashian–V.S.Zakarian.

Next, let $\gamma \le \alpha < \gamma + 1$. Then the representation (3.9)–(3.10) of Lemma 3.3 holds, and $\gamma - 1 \le \alpha - 1 < \gamma$. Therefore, in this representation $B_{\alpha-1} \in N_\beta$ by Lemma 3.2. In essence, the proof of $H^*_{\alpha-1} \in N_\beta$ coincides with that of the inclusion $H_{\alpha-1} \in N_\beta$. The desired statement follows from the above mentioned boundary property of functions of $N_\beta$.

Now consider the case $\alpha = \gamma + 1$. According to formulas (3.9) and (3.11), $B_\alpha \equiv B_{\gamma+1} = C'_{\gamma+1}B_\gamma[H^*_\gamma]^{-2}$, where $H^*_\gamma = H^{**}_{\gamma-1}[H^{***}_{\gamma-1}]^{-2}R_\gamma$. As it is proved above, the multiplier $B_\gamma$ has non-zero, finite boundary values at all points of the unit circle, with possible exception of a set of zero $\gamma$-capacity. As it was mentioned, $H^{**}_{\gamma-1} \in N_{\gamma-1}$, and one can easily verify that also $H^{***}_{\gamma-1} \in N_{\gamma-1}$. Hence these functions possess the required boundary property in view of the common boundary property of functions from $N_{\gamma-1}$. Therefore, if we prove that the limit

$$\lim_{r\to 1-0} H^*_{\gamma-1}(re^{i\vartheta}, \{z_k\}) \neq 0 \tag{3.13}$$

exists and is finite out of a set of zero $\gamma$-capacity, then by Lindelöf's theorem $R_\gamma$ and successively $H^*_\gamma$ and $B_{\gamma+1}$ will have non-zero, finite boundary values out of such a set. And this will complete the proof of 1°. The proof of (3.13) is quite similar to that of Lemma 2.3. The following estimates appear to be the only difference.

Let $h_\gamma^*$ be a factor of the product (3.10), let $z = re^{i\vartheta}$ $(0 \le r < 1)$ and let $\zeta = \rho e^{i\varphi}$ $(0 < \rho_0 \le \rho < 1)$. Then one can verify that

$$\left|\frac{d}{dr}\log h_\gamma^*(re^{i\vartheta},\zeta)\right| < 2^{2+\gamma/2}\int_\rho^1 \frac{(1-x)^\gamma dx}{\left(x - r\rho + \frac{2}{\pi}\sqrt{r\rho}|\vartheta-\varphi|\right)^{2+\gamma}}$$
$$= \frac{2^{2+\gamma/2}}{1+\gamma}\frac{(1-\rho)^{\gamma+1}}{\left(1-r\rho+\frac{2}{\pi}\sqrt{r\rho}|\vartheta-\varphi|\right)\left[\rho(1-r)+\frac{2}{\pi}\sqrt{r\rho}|\vartheta-\varphi|\right]^{\gamma+1}}$$
$$< 2^{2+\gamma/2}\frac{(1-\rho)^\gamma}{\left[\rho(1-r)+\frac{2}{\pi}r\rho|\vartheta-\varphi|\right]^{\gamma+1}}.$$

Consequently, for $z_k = |z_k|e^{i\varphi_k}$, $|z_k| \ge \rho_0 > \rho$

$$\int_0^1 \left|\frac{d}{dr}\log H_\gamma^*(re^{i\vartheta},\{z_k\})\right| dr$$
$$\le 2^{2+\gamma/2}\sum_k (1-|z_k|)^\gamma \int_0^1 \frac{dr}{\left[|z_k|(1-r)+\frac{2}{\pi}r|z_k||\vartheta-\varphi_k|\right]^{\gamma+1}}$$
$$< 2^{2+\gamma/2}\sum_k (1-|z_k|)^\gamma\left[\frac{2^\gamma}{\rho_0^{\gamma+1}} + \frac{1}{\rho_0^{\gamma+1}}\int_{1/2}^1 \frac{dr}{(1+\pi^{-1}|\vartheta-\varphi_k|-r)^{\gamma+1}}\right]$$
$$< \frac{2^{2+\gamma/2}}{\rho_0^{\gamma+1}}\sum_k (1-|z_k|)^\gamma\left[2^\gamma + \frac{\pi^\gamma}{\gamma}\frac{1}{|\vartheta-\varphi_k|^\gamma}\right].$$

At last, if $S_2$ is the number from (1.10), then

$$\int_{-\pi}^{\pi} d\mu_0(\vartheta)\int_0^1\left|\frac{d}{dr}\log H_\gamma^*(re^{i\vartheta},\{z_k\})\right| dr$$
$$< \frac{2^{2+\gamma/2}}{\rho_0^{\gamma+1}}\sum_k (1-|z_k|)^\gamma\left[2^\gamma + \frac{\pi^\gamma}{\gamma}S_2\right] < +\infty,$$

since for any $\varphi \in (-\pi,\pi]$

$$\int_{-\pi}^{\pi}\frac{d\mu_0(\vartheta)}{|\vartheta-\varphi|^\gamma} \le \sup_{z\in\mathbb{D}}\int_{-\pi}^{\pi}\frac{d\mu_0(\vartheta)}{|z-e^{i\vartheta}|^\gamma} \equiv S_2 < +\infty.$$

2°. If (1.14) is true, then our assertion is well-known for $\alpha = 0$. If $0 < \alpha \le 1$ then by (3.9) $B_\alpha$ is a function of bounded type in $\mathbb{D}$, which provides the validity of our assertion. If $1 < \alpha \le 2$, then we use the representation (3.11), where the functions $H^{**}_{\alpha-2}$, $H^{***}_{\alpha-2}$, and $R_{\alpha-1}$ are holomorphic and bounded in $\mathbb{D}$. And $B_{\alpha-1}$ is of bounded type in $\mathbb{D}$ by the previous case. Thus, also $B_\alpha$ is of bounded type in $\mathbb{D}$. This completes the proof.

One can be convinced that the requirements of Theorems 1.2 and 1.3 for $\alpha > \gamma+1$ (in 1°) or for $\alpha > 2$ (in 2°) do not provide the inclusion of the considered Blaschke type products in the suitable classes of functions of $\alpha$-bounded type. Therefore, the method used in this chapter is not applicable for investigation of boundary properties of these products in the mentioned cases.

NOTES. Theorem 1.1 is a distinctive similarity of a result proved by M.M.Djrbashian and V.S.Zakarian [28, 29, 30] for the classes $N_\alpha$ $(-1 < \alpha < 0)$ in the disc. Similar to the case of unit disc, here also a contraction of Nevanlinna's class leads to delicate boundary properties. For $-1 < \alpha < 0$ and $\gamma = 1 + \alpha$ the assertion 1° of Theorem 1.3 was proved earlier by M.M.Djrbashian and V.S.Zakarian [28, 29]. In another particular case $0 < \alpha < 1$ and $\gamma = \alpha$ the same assertion, in essence, was proved by D.T.Baghdasarian and I.V.Ohanyan [6]. For all $-1 < \alpha \leq 1$ the assertions 1° of Theorems 1.2 and 1.3 can be considered as generalizations (for the Blaschke type products in the half-plane and in the disc) of the well known result of Frostman [33], which coincides with the case $\alpha = 0$ of assertions 1° of Theorems 1.2 and 1.3. Further, by assertions 2° of Theorems 1.2 and 1.3 a well-known property of the classical Blaschke product is extended to Blaschke type products.

The author's united work with G.V.Mikaelyan [71] contains similar results on the boundary properties of the Blaschke–M.M.Djrbashian products of the version of 1945 [16, 17] (formulas (7), (8) in Introduction), as well as for the Blaschke type products considered in [69] (formulas (2.6), (2.7) in Ch. 4) and used in factorizations of Ch. 4. The structure of that products permits to prove in [71] the similarities of results of this chapter for all $\alpha \in (-1, +\infty)$.

# CHAPTER 6

# UNIFORM APPROXIMATIONS

## 1. MAIN RESULTS

In this section we prove the following three theorems which are related to approximations and some other problems in the classes $N_\alpha\{G^-\}$ of functions of $\alpha$-bounded type (see Definition 1.1 in Ch. 5).

**Theorem 1.1.** *The class $N_\alpha\{G^-\}$ $(-1 < \alpha < +\infty)$ coincides with the set of functions which are representable in $G^-$ in the form*

$$F(w) = \frac{B_\alpha(w,\{a_m\})}{B_\alpha(w,\{b_n\})} \exp\Bigg\{ c_0 + c_1 w - \frac{\Gamma(2+\alpha)}{2\pi} \lim_{\eta\to -0} |\eta|^{-1-\alpha} \int_{-\infty}^{+\infty} \log b_\alpha(w, t+i\eta)\, d\mu(t) + iC \Bigg\}, \quad (1.1)$$

*where the limit is uniform inside $G^-$, $B_\alpha$ are convergent Blaschke type products (of Ch. 2), and the numbers $c_0$, $c_1$, $C$, the sequence $\{b_n\}$ and the function $\mu(t)$ are those in factorization (1.1), (1.2), (1.3) of Ch. 5.*

**Theorem 1.2.** *Let $F(w)$ be a holomorphic function from $N_\alpha\{G^-\}$ $(-1 < \alpha < +\infty)$, representable by the formula (1.1), (1.2), (1.3) of Ch. 5, where $c_0 = c_1 = C = 0$ and $\mu(t)$ is a nondecreasing function.*

*Then for any natural numbers $\kappa$ and $N$ there exists a triangular matrix of pairwise different complex numbers $\{w_l^{(k)}(\kappa,N)\} \subset G^-$ $(k = 1,2,\ldots,\ l = 1,2,\ldots,k)$ such that:*

*1°. For any $k \geq 1$*

$$\operatorname{Im} w_l^{(k)}(\kappa,N) = \eta(\kappa,N), \quad l = 1,2,\ldots,k,$$

*and*

$$\sup_{k\geq 1,\kappa\geq 1} \left\{ k\, |\eta_k(\kappa,N)|^{1+\alpha} \right\} \leq \frac{\Gamma(2+\alpha)}{2\pi} \left[ \bigvee_{-N}^{N} \mu + 1 \right].$$

*2°. From the associated with the mentioned matrix finite Blaschke type products*

$$B_\alpha(w, \{w_l^{(k)}(\kappa,N)\}_1^k) = \prod_{l=1}^{k} b_\alpha(w, w_l^{(k)}(\kappa,N))$$

*one can choose a sequence*

$$\left\{B_\alpha(w, \{w_l^{(k_j)}(\kappa_j, N_j)\}_1^{k_j})\right\}_{j=1}^{\infty}$$

*such that*

$$F(w) = B_\alpha(w, \{a_m\}) \lim_{j\to\infty} B_\alpha(w, \{w_l^{(k_j)}(\kappa_j, N_j)\}_1^{k_j})$$

*uniformly inside* $G^-$.

**Remark 1.1.** This theorem is similar to a well-known result due to Schur [93], stating that any bounded holomorphic function in $|z| < 1$ (or in a half-plane) can be uniformly approximated by a sequence of finite Blaschke products, although, for $\alpha = 0$ Theorem 1.2 somewhat differs from Schur's theorem.

**Remark 1.2.** Any function of $N_\alpha\{G^-\}$ (see (1.1) in Ch. 5) is representable as the product of $\exp\{c_0+c_1w+iC\}$ and a quotient of two functions satisfying the conditions of Theorem 1.2. Therefore, Theorem 1.2 implies a similar assertion on uniform approximation of arbitrary meromorphic functions from $N_\alpha\{G^-\}$ $(-1 < \alpha < +\infty)$ by means of finite Blaschke type products.

**Theorem 1.3.** *Let* $F(w) \in N_\alpha\{G^-\}$ $(-1 < \alpha < +\infty)$, *and let* $F(w) \not\equiv 0$.

1°. *If*

$$\liminf_{v\to -0} \int_{-\infty}^{+\infty} \left|W^{-\alpha} \log|F(u+iv)|\right| \frac{du}{1+u^2} = 0, \tag{1.2}$$

*then*

$$F(w) = e^{c_0+c_1w+iC} \frac{B_\alpha(w, \{a_m\})}{B_\alpha(w, \{b_n\})}, \quad w \in G^-, \tag{1.3}$$

*where* $c_0$, $c_1$, *and* $C$ *are those in (1.1) of Ch. 5, and* $B_\alpha$ *are some convergent Blaschke type products.*

2°. *Conversely, if* $F(w)$ *is representable in the form (1.3), then*

$$\lim_{v\to -0} \int_{-\infty}^{+\infty} \left|W^{-\alpha} \log|F(u+iv)|\right| du = 0, \tag{1.4}$$

**Remark 1.3.** Being a generalization of Akutowicz' theorem [3] for the classes $N_\alpha\{G^-\}$, Theorem 1.3 becomes a somewhat different assertion for $\alpha = 0$. This difference of cases $\alpha = 0$ of Theorems 1.2 and 1.3 from the well known results is a consequence of the fact that $N_0\{G^-\}$ is a *subclass* of the set of all functions of bounded type in $G^-$.

## 2. PROOFS OF THEOREMS 1.1, 1.2, 1.3

*2.1.* The following lemma is to be used for proving Theorem 1.1.

**Lemma 2.1.** *For any $\alpha \in (-1, +\infty)$ and $t + i\eta \in G^-$*

$$\Omega_\alpha(w, t+i\eta) = \frac{2|\eta|^{1+\alpha}}{(1+\alpha)[i(w-t)]^{1+\alpha}} + R_\alpha(w, t+i\eta), \quad w \in G^-, \tag{2.1}$$

*where $R_\alpha$ is such that for any compact $\mathbf{K} \subset G^-$ and any $w \in \mathbf{K}$ and $t \in (-\infty, +\infty)$*

$$\big|R_\alpha(w, t+i\eta)\big| \le \frac{C_\alpha(\mathbf{K})|\eta|^{2+\alpha}}{1+|t|^{2+\alpha}}, \quad |\eta| < \frac{1}{2}\min_{w\in\mathbf{K}} |\mathrm{Im}\ w|, \tag{2.2}$$

*where $C_\alpha(\mathbf{K})$ is a constant depending only on $\alpha$ and $\mathbf{K}$.*

**Proof.** In view of the proof of Lemma 1.2 in Ch. 2

$$R_\alpha(w, t+i\eta) = \int_0^{|\eta|} \Big\{[i(w-t)-\sigma]^{-2-\alpha} - [i(w-t)+\sigma]^{-2-\alpha}\Big\}(|\eta|-\sigma)^{1+\alpha} d\sigma.$$

Hence (2.2) follows by the estimate

$$2|i(w-t)\pm\sigma| \ge \begin{cases} |t| & \text{for} \quad |t| \ge 2\kappa+\rho \\ \rho & \text{for} \quad |t| \le 2\kappa+\rho \end{cases} \qquad (0 < \sigma < +|\eta|),$$

where $\kappa = \max_{w\in\mathbf{K}} |\mathrm{Re}\ w|$ and $\rho = \min_{w\in\mathbf{K}} |\mathrm{Im}\ w|$.

**Proof of Theorem 1.1.** First we show that any $F(w) \in N_\alpha\{G^-\}$ $(\alpha > -1)$ is representable in the form (1.1). To this end, observe that by (2.1)

$$\frac{1}{[i(w-t)]^{1+\alpha}} = -\frac{1+\alpha}{2}|\eta|^{-1-\alpha}\log b_\alpha(w, t+i\eta) - \frac{1+\alpha}{2}|\eta|^{-1-\alpha}R_\alpha(w, t+i\eta)$$

for any $\eta < 0$. Inserting this into formula (1.1) of Ch. 5, one can write the exponential factor of that formula in the form

$$\exp\left\{-\frac{\Gamma(2+\alpha)}{2\pi}|\eta|^{-1-\alpha}\int_{-\infty}^{+\infty}\log b_\alpha(w, t+i\eta)d\mu(t) \right.$$
$$\left. - \frac{\Gamma(2+\alpha)}{2\pi}|\eta|^{-1-\alpha}\int_{-\infty}^{+\infty} R_\alpha(w, t+i\eta)d\mu(t) + iC\right\}. \tag{2.3}$$

Consequently, by (2.2) and (1.2) of Ch. 5

$$\lim_{\eta\to -0} |\eta|^{-1-\alpha}\int_{-\infty}^{+\infty} R_\alpha(w, t+i\eta)d\mu(t) = 0 \tag{2.4}$$

uniformly in respect to $w \in \mathbf{K}$, whatever be the compact $\mathbf{K} \subset G^-$.

Thus, we proved the representation (1.1), where the passage is uniform by $w$ inside $G^-$. Conversely, if (1.1) is true, then $F(w) \in N_\alpha\{G^-\}$ by (2.3) and (2.4).

*2.2.* **Proof of Theorem 1.3.** Our assertion $2^\circ$ is already proved (see the relation (3.12) in Ch. 2). For proving $1^\circ$, note that from formula (1.1) of Ch. 5 one can easily derive that for any $w = v + iv \in G^-$

$$W^{-\alpha} \log |F(w)| = W^{-\alpha} \log \left| \frac{B_\alpha(w, \{a_m\})}{B_\alpha(w, \{b_n\})} \right| + \frac{|v|}{\pi} \int_{-\infty}^{+\infty} \frac{d\mu(t)}{(u-t)^2 + v^2}. \tag{2.5}$$

Denoting

$$\Phi(w) = \frac{|v|}{\pi} \int_{-\infty}^{+\infty} \frac{d\mu(t)}{(u-t)^2 + v^2}, \quad w = v + iv \in G^-, \tag{2.6}$$

by (1.2) and (1.4) we find

$$\lim_{v \to -0} \int_{-\infty}^{+\infty} \frac{|\Phi(u+iv)|}{1+u^2} = 0.$$

Observe that for any $x \in (-\infty, +\infty)$ and $y > 0$

$$\frac{y}{\pi} \frac{C(x,y)}{(x-u)^2 + v^2} \leq \frac{1}{1+u^2}, \quad -\infty < u < +\infty,$$

where $C(x, y) > 0$ depends on $x$ and $y$. Hence, for fixed $x$ and $y$

$$\lim_{v \to -0} \frac{y}{\pi} \int_{-\infty}^{+\infty} \frac{|\Phi(u+iv)|}{(x-u)^2 + y^2} = 0. \tag{2.7}$$

Further, set

$$\varphi_{1,2}(w) = \frac{|v|}{\pi} \int_{-\infty}^{+\infty} \frac{d\mu_{1,2}(t)}{(u-t)^2 + v^2}, \quad w = u + iv \in G^-, \tag{2.8}$$

$$I_{1,2} = \frac{y}{\pi} \int_{-\infty}^{+\infty} \frac{\varphi_{1,2}(u+iv)}{(x-u)^2 + y^2} du, \quad v < 0 \tag{2.9}$$

(where $\mu_{1,2}(t)$ is assumed to satisfy (1.2) of Ch. 5). Then one can see that

$$I_{1,2} = \frac{y + |v|}{\pi} \int_{-\infty}^{+\infty} \frac{d\mu_{1,2}(t)}{(x-t)^2 + (y + |v|)^2}.$$

Now observe that for any $v < 0$

$$\frac{y+|v|}{\pi}\frac{C'(x,y)}{(x-t)^2+(y+|v|)^2} \leq \frac{1}{1+t^2}, \quad -\infty < t < +\infty,$$

where $C(x,y) > 0$ depends on $x$ and $y$. Therefore, by Lebesgue's theorem

$$\lim_{v\to-0} I_{1,2} = \frac{y}{\pi}\int_{-\infty}^{+\infty}\frac{d\mu_{1,2}(t)}{(x-t)^2+y^2} \tag{2.10}$$

for any fixed $x$ and $y$. But in view of (2.8) and (2.9)

$$\frac{y}{\pi}\int_{-\infty}^{+\infty}\frac{\Phi(u+iv)}{(x-u)^2+y^2}du \geq |I_1 - I_2|.$$

Together with (2.7) and (2.10), this leads to the conclusion that the quantity

$$\frac{y}{\pi}\int_{-\infty}^{+\infty}\frac{d\mu(t)}{(x-t)^2+y^2} = \frac{y}{\pi}\int_{-\infty}^{+\infty}\frac{d\mu_1(t)}{(x-t)^2+y^2} - \frac{y}{\pi}\int_{-\infty}^{+\infty}\frac{d\mu_2(t)}{(x-t)^2+y^2}$$

is equal to zero. Hence $\mu(t) \equiv \text{const}$ by arbitrariness of $x \in (-\infty, +\infty)$ and $y > 0$, and consequently the representation (1.1) of Ch. 5 can be written in the form (1.3).

*2.3.* For proving Theorem 1.2 we shall use

**Lemma 2.2.** *Let $\alpha \in (-1, +\infty)$, let $N \geq 1$ be a natural number and let*

$$f_0(w,N) = \exp\left\{-\frac{\Gamma(1+\alpha)}{\pi}e^{-i\frac{\pi}{2}(1+\alpha)}\int_{-N}^{N}\frac{d\mu(t)}{(w-t)^{1+\alpha}}\right\}, \quad w \in G^-,$$

*where $\mu(t)$ is a continuous, nondecreasing function on $[-N,N]$.*

*Then there exists a triangular matrix of pairwise different complex numbers $\{w_l^{(k)}(\mu,N)\}_{l=1}^{k} \subset G^-$ $(k = 1,2,\dots)$ such that:*

1°. *For any $k \geq 1$*

$$\operatorname{Im} w_l^{(k)}(\mu,N) = \eta_k(\mu,N) \quad (l = 1,2,\dots,k), \tag{2.11}$$

*and*

$$|\eta_k(\mu,N)|^{1+\alpha} = \frac{1}{k}\frac{\Gamma(2+\alpha)}{2\pi}\bigvee_{-N}^{N}\mu. \tag{2.12}$$

2°. *Uniformly inside $G^-$*

$$f_0(w,N) = \lim_{k\to\infty} B_\alpha(w, \{w_l^{(k)}(\mu,N)\}_{l=1}^{k}). \tag{2.13}$$

**Proof.** For any $k \geq 1$ we split $[-N, N]$ as $-N < t_1^{(k)} < t_2^{(k)} < \cdots < t_k^{(k)} < t_{k+1}^{(k)} = N$ in a way to provide

$$\mu(t_{l+1}^{(k)}) - \mu(t_l^{(k)}) = \frac{1}{k} \bigvee_{-N}^{N} \mu \quad (l = 1, 2, \ldots, k).$$

Also, we assume that

$$\eta_k \equiv \eta_k(\mu, N) = -\left[\frac{\Gamma(2+\alpha)}{2\pi} \bigvee_{-N}^{N} \mu\right]^{\frac{1}{1+\alpha}},$$

$$w_l^{(k)}(\mu, N) = t_l^{(k)} + i\eta_k \quad (k = 1, 2, \ldots; l = 1, 2, \ldots, k).$$

One can be convinced that in these assumptions 1° is true. Besides, by (1.1′)–(1.2) of Ch. 2 and (2.1)

$$\log B_\alpha(w, \{w_l^{(k)}(\mu, N)\}_{l=1}^k) = \sum_{l=1}^{k} \log b_\alpha(w, w_l^{(k)}(\mu, N))$$

$$= -\frac{\Gamma(1+\alpha)}{\pi} \sum_{l=1}^{k} \frac{\mu(t_{l+1}^{(k)}) - \mu(t_l^{(k)})}{\left[i(w - t_l^{(k)})\right]^{1+\alpha}} - \sum_{l=1}^{k} R_\alpha(w, t_l^{(k)} + i\eta_k).$$

But, if $\mathbf{K} \subset G^-$ is any compact, then by (2.2)

$$\left|\sum_{l=1}^{k} R_\alpha(w, t_l^{(k)} + i\eta_k)\right| \leq C_\alpha(\mathbf{K}) |\eta_k|^{2+\alpha} = o(1) \quad \text{as} \quad k \to \infty$$

uniformly in respect to $w \in \mathbf{K}$. In addition, uniformly by $w \in \mathbf{K}$

$$\lim_{k \to \infty} \sum_{l=1}^{k} \frac{\mu(t_{l+1}^{(k)}) - \mu(t_l^{(k)})}{\left[i(w - t_l^{(k)})\right]^{1+\alpha}} = \int_{-N}^{N} \frac{d\mu(t)}{[i(w-t)]^{1+\alpha}}.$$

Hence (2.13) follows.

*2.4.* **Proof of Theorem 1.2.** Assuming that $\alpha \in (-1, +\infty)$, set

$$f(w) \equiv \exp\left\{-\frac{\Gamma(1+\alpha)}{\pi} \int_{-\infty}^{+\infty} \frac{d\mu(t)}{[i(w-t)]^{1+\alpha}}\right\},$$

$$f(w, N) \equiv \exp\left\{-\frac{\Gamma(1+\alpha)}{\pi} \int_{-N}^{N} \frac{d\mu(t)}{[i(w-t)]^{1+\alpha}}\right\},$$

where $\mu(t)$ is a nondecreasing function satisfying the condition (1.2) of Ch. 5. Let $\{\mathbf{K}_j\}_1^\infty$ be a family of compacts exhausting $G^-$, i.e.

$$\mathbf{K}_j \subseteq \mathbf{K}_{j+1} \quad (j = 1, 2, \dots) \quad \text{and} \quad \bigcup_{j=1}^{\infty} \mathbf{K}_j = G^-.$$

Further, for any $j \geq 1$ choose $N_j$ $(N_{j+1} > N_j)$ enough great to provide

$$|f(w) - f(w, N_j)| < \frac{1}{3j}, \quad w \in \mathbf{K}_j. \tag{2.15}$$

Consider the following sequence of functions, which are continuous and increasing on $[-N, N]$ $(N \geq 1)$:

$$\mu_\kappa(t) = \kappa \int_t^{t+\frac{1}{\kappa}} \varphi_\kappa(x)dx + \frac{t}{2\kappa N}, \quad \kappa = 1, 2, \dots,$$

where

$$\varphi_\kappa(x) = \begin{cases} \mu(N) & \text{if } \; x \geq N \\ \mu(x) & \text{if } \; -N + \frac{1}{\kappa} < x < N \\ \mu(-N) & \text{if } \; -N \leq x \leq -N + \frac{1}{\kappa}. \end{cases}$$

One can verify, that $\lim\limits_{\kappa \to \infty} \mu_\kappa(t) = \mu(t)$ for any $t \in [-N, N]$ and

$$|\mu_\kappa(t)| \leq |\mu(-N)| + \bigvee_{-N}^{N} \mu + 1, \quad \bigvee_{-N}^{N} \mu_\kappa \leq \bigvee_{-N}^{N} \mu + 1$$

for any $\kappa \geq 1$. Therefore, by Helly's theorem on the passage in the Stieltjes integral,

$$\lim_{\kappa \to \infty} \int_{-N}^{N} \frac{d\mu_\kappa(t)}{[i(w-t)]^{1+\alpha}} = \int_{-N}^{N} \frac{d\mu(t)}{[i(w-t)]^{1+\alpha}}, \quad w \in G^-. \tag{2.16}$$

One can be convinced that this passage is uniform by $w$ inside $G^-$.

Now for any $\kappa \geq 1$ and $j \geq 1$ set

$$f_\kappa(w, N_j) \equiv \exp\left\{ -\frac{\Gamma(1+\alpha)}{\pi} \int_{-N_j}^{N_j} \frac{d\mu_\kappa(t)}{[i(w-t)]^{1+\alpha}} \right\}, \quad w \in G^-.$$

Then, by (2.16) one can choose a sequence of natural numbers $\{\kappa_j\}_1^\infty$ to provide that for any $j \geq 1$

$$\left|f(w, N_j) - f_{\kappa_j}(w, N_j)\right| < \frac{1}{3j}, \quad w \in \mathbf{K}_j. \tag{2.17}$$

By Lemma 2.2, for each $j \geq 1$ there exists a triangular matrix of pairwise different numbers $\{w_l^{(k)}(\kappa_j, N_j)\}$ $(k = 1, 2, \dots; l = 1, 2, \dots, k)$ satisfying (2.11) and (2.12) and such that (2.13) is uniform in respect to $w$ inside $G^-$. Consequently, one can extract a sequence of natural numbers $\{k_j\}_1^\infty$ such that

$$\left| B_\alpha(w, \{w_l^{(k_j)}(\kappa_j, N_j)\}_1^{(k_j)}) - f_{\kappa_j}(w, N_j) \right| < \frac{1}{3j}, \quad w \in \mathbf{K}_j$$

for any $j \geq 1$. Hence by (2.15) and (2.17) we come to the desired assertion.

# CHAPTER 7

# SUBHARMONIC FUNCTIONS WITH NONNEGATIVE HARMONIC MAJORANTS

## 1. MAIN RESULTS

*1.1.* It is well known, that Nevanlinna's factorization of the class $N$ of meromorphic functions of bounded type in the unit disc [82] (see also [84], Ch. VII) leads to the following complete characterization of the growth of functions $u(z)$ which are subharmonic in $|z| < 1$ and have nonnegative harmonic majorants:

$$\sup_{0<r<1} \int_0^{2\pi} u^+(re^{i\vartheta})d\vartheta = \liminf_{r\to 1-0} \int_0^{2\pi} u^+(re^{i\vartheta})d\vartheta < +\infty \tag{1.1}$$

(here and everywhere $u^+ = \max\{u, 0\}$, $u^- = u^+ - u$). In contrast with this, the problem of finding a complete characterization of the growth of functions which are of the same type in a half-plane was solved only partially, since the conditions

$$\liminf_{R\to+\infty} \frac{1}{R} \int_0^{\pi} u^+(R\,e^{i\vartheta}) \sin\vartheta d\vartheta < +\infty \tag{1.2}$$

and

$$\int_{-\infty}^{+\infty} u^+(x)\frac{dx}{1+x^2} < +\infty \tag{1.3}$$

arising from Nevanlinna factorization in the upper half-plane $G^+ = \{z : \operatorname{Im} z > 0\}$ [83] completely characterize the growth of such functions only in the particular case when they are subharmonic in $\overline{G^+} = \{z : \operatorname{Im} z \geq 0\}$ (see also [10], Sec. 6.3–6.5, where it is assumed that $u(z)$ can somehow be continuously extended to the real axis).

*1.2.* The main results of this chapter are the following three theorems particularly containing a complete characterization of the growth of functions subharmonic in $G^+$, having there nonnegative harmonic majorants. Besides, these theorems somehow improve Nevanlinna's uniqueness theorem ([84], Ch. III, Sec. 38) and also the Phragmén-Lindelöf type theorem which follows from a result due to M.Heins and L.Ahlfors [1], by replacing the condition

$$\limsup_{z\to t,\ \operatorname{Im} z>0} u(z) \leq 0, \quad -\infty < t < +\infty \tag{1.4}$$

by a less restrictive one.

**Theorem 1.1.** *Let $S(\Omega)$ be the class of functions $u(z) \not\equiv -\infty$ subharmonic in $G^+$ and satisfying (1.2) and*

$$\lim_{R\to+\infty} \liminf_{y\to+0} \int_{-R}^{R} u^+(x+iy)\Omega(x)\,dx < +\infty, \tag{1.5}$$

*where $\Omega(x)$ is a continuous function such that $\Omega(x) \geq C(1+x^2)^{-1}$ $(-\infty < x < +\infty)$ for a constant $C > 0$. Then:*

$1^\circ$. *$S(\Omega)$ coincides with the set of functions representable in the form*

$$u(z) = \iint_{G^+} \log\left|\frac{z-\zeta}{z-\overline{\zeta}}\right| d\nu(\zeta) + hy + \frac{y}{\pi}\int_{-\infty}^{+\infty} \frac{d\mu(t)}{(x-t)^2+y^2} \tag{1.6}$$

*$(z = x+iy \in G^+)$, where $\nu(\zeta)$ is a nonnegative Borel measure for which*

$$\iint_{G^+} \frac{\operatorname{Im}\zeta}{1+|\zeta|^2} d\nu(\zeta) < +\infty, \tag{1.7}$$

*$h$ is a real number, and $\mu(t)$ is a function representable as the difference $\mu(t) = \mu_+(t) - \mu_-(t)$ of two nondecreasing functions such that*

$$\int_{-\infty}^{+\infty} \Omega(t)\,d\mu_+(t) < +\infty \quad \text{and} \quad \int_{-\infty}^{+\infty} \frac{d\mu_-(t)}{1+t^2} < +\infty. \tag{1.8}$$

$2^\circ$. *If the representation (1.6)–(1.8), is valid, then $\mu_\pm(t)$ can be deduced by*

$$\mu_\pm(t) = \lim_{y\to+0} \int_0^t u^\pm(x+iy)\,dx, \quad -\infty < t < +\infty. \tag{1.9}$$

*With this choice of $\mu_\pm(t)$*

$$\bigvee_a^b \mu = \bigvee_a^b \mu_+ + \bigvee_a^b \mu_-, \quad \forall[a,b] \subset (-\infty,+\infty), \tag{1.10}$$

*and*

$$\lim_{y\to+0} \int_a^b u^\pm(x+iy)g(x)\,dx = \int_a^b g(x)\,d\mu_\pm(x) \tag{1.11}$$

*for any function $g(x)$ continuous in $[a,b]$. Besides,*

$$\lim_{R\to+\infty} \lim_{y\to+0} \int_{-R}^{R} u^+(x+iy)\Omega(x)\,dx = \int_{-\infty}^{+\infty} \Omega(x)\,d\mu_+(x), \tag{1.12}$$

$$\lim_{y\to+0} \int_{-\infty}^{+\infty} u^-(x+iy)\frac{dx}{1+x^2} = \int_{-\infty}^{+\infty} \frac{d\mu_-(x)}{1+x^2}, \tag{1.13}$$

*and the relation*

$$\lim_{y\to+0}\int_{-\infty}^{+\infty} u^+(x+iy)\Omega(x)\,dx = \int_{-\infty}^{+\infty}\Omega(x)\,d\mu_+(x) \tag{1.14}$$

*is true, if* $\Omega(x)=(1+|x|)^{-\gamma}$ $(1<\gamma\le 2)$ *or* $\Omega(x)=(1+|x|)^{-\gamma}$ $(-1<\gamma\le 1)$ *and* $h\le 0$.

3°. *If the representation (1.6)–(1.8) is valid, then*

$$h^{\pm} = \frac{2}{\pi}\lim_{R\to+\infty}\frac{1}{R}\int_0^{\pi} u^{\pm}(R\,e^{i\vartheta})\sin\vartheta d\vartheta, \tag{1.15}$$

*for any* $\vartheta\in(0,\pi)$

$$h\sin\vartheta = \limsup_{R\to+\infty} R^{-1}u(R\,e^{i\vartheta}),\quad h^+\sin\vartheta = \limsup_{R\to+\infty} R^{-1}u^+(R\,e^{i\vartheta}), \tag{1.16}$$

*and for any* $\vartheta\in(0,\pi)$, *except at most for a set of outer capacity zero,*

$$h^{\pm}\sin\vartheta = \lim_{R\to+\infty} R^{-1}u^{\pm}(R\,e^{i\vartheta}). \tag{1.17}$$

**Remark 1.1.** It is well known that the Nevanlinna class of functions (which are subharmonic in $G^+$ and have there nonnegative harmonic majorants) coincides with the set of functions representable in the form (1.6)–(1.8) where $\Omega(x) = (1+x^2)^{-1}$, i.e. coincides with $S((1+x^2)^{-1})$. Thus, the pair of conditions (1.2) and (1.5) presents a complete characterization of the growth of such functions when $\Omega(x) = (1+x^2)^{-1}$. Besides, if $u(x)$ is subharmonic in $\overline{G^+}$, then it is obvious that the condition (1.5) with $\Omega(x)=(1+x^2)^{-1}$ is equivalent to (1.3). The relations (1.5)–(1.7) are well known even in the most general case when $\Omega(x) = (1+x^2)^{-1}$ (these relations are true for $h$, $h^+$ and $h^-$, since $u(z)$ and $u^+(z)$ have the same least nonnegative harmonic majorant).

**Remark 1.2.** Using a result from [101] one can verify that the subset of functions of $S(1)$ $(\Omega(x)\equiv 1)$, for which $h\le 0$, coincides with the class of those functions subharmonic in $G^+$, for which

$$\sup_{y>0}\int_{-\infty}^{+\infty} u^+(x+iy)\,dx < +\infty. \tag{1.18}$$

**Theorem 1.2.** *Let* $u(z)$ *be subharmonic in* $G^+$, *and let there exists a sequence* $R_n\uparrow+\infty$ *such that for any* $n\ge 1$

$$\int_0^{\pi} u^+(R_n\,e^{i\vartheta})\sin\vartheta d\vartheta < +\infty \quad\text{and}\quad \liminf_{y\to+0}\int_{-R_n}^{R_n} u^+(x+iy)\,dx < +\infty. \tag{1.19}$$

*If for some $R_o > 0$*

$$\liminf_{R\to+\infty}\left\{\frac{1}{R}\int_0^\pi u(Re^{i\vartheta})\sin\vartheta d\vartheta + \liminf_{y\to+0}\int_{-R}^{R} u(x+iy)g_{R,R_o}(x)\,dx\right\} = -\infty, \quad (1.20)$$

*where*

$$g_{R,R_o}(x) = \begin{cases} 2^{-1}(x^{-2} - R^{-2}) & \text{for} \quad R_o \le |x| \le R, \\ 2^{-1}(R_o^{-2} - R^{-2}) & \text{for} \quad |x| < R_o, \end{cases}$$

*then $u(z) \equiv -\infty$.*

**Remark 1.3.** Theorem 1.2 is an improvement of Nevanlinna's uniqueness theorem, since Nevanlinna's conditions (1.4) and

$$\liminf_{R\to+\infty} R^{-1}M(R) = -\infty \quad (M(R) = \sup_{0<\vartheta<\pi} u(Re^{i\vartheta}))$$

are more restrictive than the conditions (1.19)–(1.20).

**Theorem 1.3.** *Let $u(z)$ be subharmonic in $G^+$, and let*

$$\liminf_{y\to+0}\int_{-R}^{R} u^+(x+iy)\,dx = 0 \quad (1.21)$$

*for any $R > 0$. Then:*

$$1^o. \quad \lim_{R\to+\infty} R^{-1}M(R) = \lim_{R\to+\infty} R^{-1}[M(R)]^+ = \frac{2}{\pi}\lim_{R\to+\infty}\frac{1}{R}\int_0^\pi u^+(Re^{i\vartheta})\sin\vartheta d\vartheta = \beta \in [0,+\infty]. \quad (1.22)$$

*Besides, if $\alpha = \sup_{\text{Im } z>0} (\text{Im } z)^{-1}u(z)$, then*

$$\alpha^+ = \sup_{\text{Im } z>0} (\text{Im } z)^{-1}u^+(z) = \beta. \quad (1.23)$$

*$2^o$. If $\beta = \alpha^+ < +\infty$, then (1.15), (1.16) and (1.17) hold for $h^+ = \beta = \alpha^+$.*

**Remark 1.4.** The above theorem is an improvement of the Phragmén–Lindelöf type theorem which follows from a result of M.Heins and L.Ahlfors [1] since their theorem implies the same assertions under the condition (1.4) which is more restrictive than (1.20).

*1.3.* The techniques used to prove Theorem 1.1 can be applied to prove also a theorem on some weighted $H^p$ classes in $G^+$. Consider the class $H^p(\Omega(x)\,dx)$ $(0 < p < +\infty)$ of functions $f(z)$ holomorphic in $G^+$, for which

$$\liminf_{R\to+\infty} \frac{1}{R} \int_0^{\pi} |f(Re^{i\vartheta})|^p \sin\vartheta d\vartheta < +\infty \tag{1.24}$$

and

$$\lim_{R\to+\infty} \liminf_{y\to+0} \int_{-R}^{R} |f(x+iy)|^p \Omega(x)\,dx < +\infty. \tag{1.25}$$

By Theorem 1.1, $H^p((1+x^2)^{-1}dx)$ coincides with the set of functions $f(z)$ holomorphic in $G^+$, for which $|f(z)|^p$ has a harmonic majorant in $G^+$, i.e. it coincides with the conformal mapping of Hardy's $H^p$ class for the disc. Besides, it is obvious that, by Remark 1.2 $H^p(dx)$ coincides with the class $H^p$ introduced by Hille and Tamarkin by means of the condition (1.18), where $u(z) = |f(z)|^p$. The following theorem relates to a more general case.

**Theorem 1.4.** *Let $0 < p < +\infty$, and let $\Omega(x)$ be a measurable function such that almost everywhere $\Omega(x) \geq C(1+x^2)^{-1}$ for a constant $C > 0$ and such that for any $R > 0$, $\Omega(x)$ is uniformly bounded almost everywhere in $(-R, R)$. Then:*

$1^o$. *The class $H^p(\Omega(x)\,dx)$ coincides with the subset of those functions of the conformal mapping of Hardy's class, for which $f(x) \in L_p(\Omega(x)\,dx)$.*

$2^o$. *The condition (1.24) can be replaced by*

$$\liminf_{R\to+\infty} \frac{1}{R} \int_0^{\pi} \log^+ |f(Re^{i\vartheta})| \sin\vartheta d\vartheta = 0 \tag{1.26}$$

*without changing $H^p(\Omega(x)\,dx)$. Besides, if $f(z) \in H^p(\Omega(x)\,dx)$, then*

$$\lim_{R\to+\infty} \frac{1}{R} \int_0^{\pi} |f(Re^{i\vartheta})|^p \sin\vartheta d\vartheta = 0. \tag{1.27}$$

$3^o$. *If $\Omega(x)$ satisfies the additional condition*

$$\int_{-\infty}^{+\infty} \frac{\log^+ \Omega(x)}{1+x^2} dx < +\infty, \tag{1.28}$$

*then*

$$H^p(\Omega(x)\,dx) = [\Omega(z)]^{-1/p} H^p(dx), \tag{1.29}$$

*where*

$$\Omega(z) = \exp\left\{ \frac{1}{\pi i} \int_{-\infty}^{+\infty} \frac{1+tz}{t-z} \frac{\log \Omega(t)}{1+t^2} dt \right\}, \quad z \in G^+.$$

*Particularly*

$$H^p((1+|x|)^{-\gamma}dx) = (z+i)^{\gamma/p}H^p(dx), \quad -\infty < \gamma \le 2. \tag{1.30}$$

## 2. REPRESENTATION IN THE SEMI-DISC

The main tool used in this chapter is a theorem on necessary and sufficient growth conditions under which a function subharmonic in a semi-disc larger than $G_R^+ = \{z : \text{Im } z > 0, \ |z| < R\}$ has a nonnegative harmonic majorant in $G_R^+$. Before stating our theorem we consider the function

$$\omega_\rho(\zeta, z) = \left\{\left(\frac{R_\rho+\zeta-i\rho}{R_\rho-\zeta+i\rho}\right)^{\pi/\alpha} - \left(\frac{R_\rho+z-i\rho}{R_\rho-z+i\rho}\right)^{\pi/\alpha}\right\} \times \left\{\left(\frac{R_\rho+\zeta-i\rho}{R_\rho-\zeta+i\rho}\right)^{\pi/\alpha} - \left(\frac{R_\rho+\overline{z}+i\rho}{R_\rho-\overline{z}-i\rho}\right)^{\pi/\alpha}\right\}^{-1},$$

where $0 \le \rho < R$, $R_\rho = \sqrt{R^2-\rho^2}$, and $\alpha = \arccos(\rho/R)$ and prove some lemmas. Observe that $\omega_\rho(\zeta, z)$ gives a conformal one–to–one mapping of the segment $G_{R,\rho}^+ = \{\zeta : \text{Im } \zeta > \rho, \ |\zeta| < R\}$ onto the unit disc and $\omega_\rho(z, z) = 0$. Therefore $g_{R,\rho}(\zeta, z) = -\log|\omega_\rho(\zeta, z)|$ ($z, \zeta \in G_{R,\rho}^+$) is the Green's function of $G_{R,\rho}^+$. If $u(z)$ is a function subharmonic in a semi-disc $G_{R^*}^+$ ($R^* > R$), then by Riesz' theorem in $G_{R,\rho}^+$

$$u(z) = -\iint\limits_{G_{R,\rho}^+} g_{R,\rho}(\zeta, z)\, d\nu(\zeta) + \int_\beta^{\pi-\beta} u(Re^{i\vartheta})\varphi_{R,\rho}(\vartheta, z)d\vartheta + \int_{-R_\rho}^{R_\rho} u(t+i\rho)\psi_{R,\rho}(t, z)dt, \tag{2.1}$$

where $\nu(\zeta)$ is a nonnegative Borel measure in $G_{R^*}^+$, finite in any domain $D$ compactly contained in $G_{R^*}^+$, and $\varphi_{R,\rho}$, $\psi_{R,\rho}$ are the expressions for the Poisson kernel of $G_{R,\rho}^+$, written on the arc $\{\zeta = Re^{i\vartheta} : \beta < \vartheta < \pi-\beta\}$ ($\beta = \arcsin(\rho/R) = \pi/2 - \alpha$) and on the interval $\{\zeta = t+i\rho : -R_\rho < t < R_\rho\}$. Using the well known formula

$$\frac{\partial g_{R,\rho}(\zeta, z)}{\partial n}|d\zeta| = \frac{1}{i}\left[\frac{\partial}{\partial\zeta}\omega_\rho(\zeta, z)\right]\frac{d\zeta}{\omega_\rho(\zeta, z)},$$

where $\partial n$ is the differentiation along the inner normal, one can calculate

$$\varphi_{R,\rho}(\vartheta, z) = \frac{2i}{\alpha} R_\rho R e^{i\vartheta} [R_\rho^2 - (Re^{i\vartheta} - i\rho)^2]^{\pi/\alpha - 1} \left\{ \mathrm{Im} \left( \frac{R_\rho + z - i\rho}{R_\rho - z + i\rho} \right)^{\pi/\alpha} \right\}$$

$$\times \left\{ (R_\rho + Re^{i\vartheta} - i\rho)^{\pi/\alpha} - (R_\rho - Re^{i\vartheta} + i\rho)^{\pi/\alpha} \left( \frac{R_\rho + z - i\rho}{R_\rho - z + i\rho} \right)^{\pi/\alpha} \right\}^{-1}$$

$$\times \left\{ (R_\rho + Re^{i\vartheta} - i\rho)^{\pi/\alpha} - (R_\rho - Re^{i\vartheta} + i\rho)^{\pi/\alpha} \left( \frac{R_\rho + \overline{z} + i\rho}{R_\rho - \overline{z} - i\rho} \right)^{\pi/\alpha} \right\}^{-1} \quad (2.2)$$

$$\psi_{R,\rho}(t, z) = \frac{2}{\alpha} R_\rho (R_\rho^2 - t^2)^{\pi/\alpha - 1} \left\{ \mathrm{Im} \left( \frac{R_\rho + z - i\rho}{R_\rho - z + i\rho} \right)^{\pi/\alpha} \right\}$$

$$\times \left| (R_\rho + t)^{\pi/\alpha} - (R_\rho - t)^{\pi/\alpha} \left( \frac{R_\rho + z - i\rho}{R_\rho - z + i\rho} \right)^{\pi/\alpha} \right|^{-2} \quad (2.2')$$

and state that $\varphi_{R,\rho}(\vartheta, z) \geq 0$ $(\beta \leq \vartheta \leq \pi - \beta)$ and $\psi_{R,\rho}(t, z) \geq 0$ $(-R_\rho \leq t \leq R_\rho)$. The below lemmas relate to the behavior of these functions when $\rho \to +0$.

**Lemma 2.1.** *If $R > 0$ and $z \in G_R^+$ are fixed numbers and $\rho \geq 0$ is sufficiently small, then*

$$C_1 \leq \left( \sin \frac{\pi(\vartheta - \beta)}{\pi - 2\beta} \right)^{1 - \pi/\alpha} \varphi_{R,\rho}(\vartheta, z) \leq C_2, \quad \beta \leq \vartheta \leq \pi - \beta, \quad (2.3)$$

*where $C_{1,2} > 0$ are constants depending only on $z$ and $R$. Besides, uniformly in $(0, \pi)$*

$$\lim_{\rho \to +0} \varphi_{R,\rho}(\vartheta, z) = \frac{2\mathrm{Im}\, z}{\pi} \frac{R(R^2 - |z|^2) \sin \vartheta}{|Re^{i\vartheta} - z|^2 |Re^{-i\vartheta} - z|^2} \equiv \varphi_R(\vartheta, z). \quad (2.4)$$

**Proof.** First we shall prove the inequalities (2.3) for $\vartheta$ close to the ends of $[\beta, \pi - \beta]$. Next we shall prove the relation (2.4). Then, estimating $\varphi_R(\vartheta, z)$, we shall extend (2.3) to all $\vartheta \in [\beta, \pi - \beta]$. So, we start from the obvious relation

$$\lim_{\rho \to +0} \arg \left( \frac{R_\rho + z - i\rho}{R_\rho - z + i\rho} \right) = \arctan \frac{2Ry}{R^2 - |z|^2} \equiv \eta \in \left(0, \frac{\pi}{2}\right), \quad y = \mathrm{Im}\, z,$$

using which one can verify that for sufficiently small $\rho > 0$

$$\lambda \frac{4R^2}{\pi} \left( \frac{y}{5R} \right)^3 < \left| \frac{2i}{\alpha} R_\rho R e^{i\vartheta} \mathrm{Im} \left( \frac{R_\rho + z - i\rho}{R_\rho - z + i\rho} \right)^{\pi/\alpha} \right| < \frac{6R^2}{\pi} \left( \frac{5R}{y} \right)^3, \quad (2.5)$$

where $\lambda = \sin\eta$, if $0 < \eta \le \pi/4$, and $\lambda = \sin(\pi/2 - \eta)$, if $\pi/4 < \eta < \pi/2$. To estimate the denominator in (2.2), first observe that

$$|R_\rho + Re^{i\vartheta} - i\rho|^{\pi/\alpha} + |R_\rho - Re^{i\vartheta} + i\rho|^{\pi/\alpha}\left|\frac{R_\rho + z - i\rho}{R_\rho - z + i\rho}\right|^{\pi/\alpha} < \left(\frac{5R}{2}\right)^3\left[1 + \left(\frac{5R}{y}\right)^3\right]$$

for sufficiently small $\rho > 0$. Next, observe that

$$|R_\rho + Re^{i\vartheta} - i\rho|^{\pi/\alpha} - |R_\rho - Re^{i\vartheta} + i\rho|^{\pi/\alpha}\left|\frac{R_\rho + z - i\rho}{R_\rho - z + i\rho}\right|^{\pi/\alpha} \ge |R_\rho + R\cos\delta|^{\pi/\alpha} - \left(4R\sin\frac{\delta}{2}\right)^2\left(\frac{5R}{y}\right)^3 > R^2$$

for sufficiently small $\rho$, $\delta > 0$ $(\delta > \beta)$ and for any $\vartheta \in [\beta, \delta]$. Now, using (2.5) and the last two inequalities (where the quantities estimated are even functions of $\vartheta - \pi/2$), we obtain that for sufficiently small $\rho$, $\delta > 0$ $(\delta > \beta)$ and for any $\vartheta \in [\beta, \delta] \cup [\pi - \delta, \pi - \beta]$

$$a_1 \le |R_\rho^2 - (Re^{i\vartheta} - i\rho)^2|^{1-\pi/\alpha}\varphi_{R,\rho}(\vartheta, z) \le a_2,$$

where $a_{1,2} > 0$ are constants depending only on $z$ and $R$. Since

$$|R_\rho^2 - (Re^{i\vartheta} - i\rho)^2| = 2R^2\sin[(\vartheta - \beta)/2]\sin[(\pi - \beta - 0)/2],$$

we conclude that the estimates (2.3) are true for sufficiently small $\rho$, $\delta > 0$ $(\delta > \beta)$ and for any $\vartheta \in [\beta, \delta] \cup [\pi - \delta, \pi - \beta]$. Now observe that (2.4) can be easily derived for any $\vartheta \in (0, \pi)$. To prove that this relation is true uniformly in $[\delta, \pi - \delta]$ for any $\delta \in (0, \pi/2)$, it is enough to observe that the limits of both the numerator and denominator of $\varphi_{R,\rho}(\vartheta, z)$ are uniformly separated from zero. Finally, it follows from (2.4) that for any $\vartheta \in [\delta, \pi - \delta]$ $(0 < \delta < \pi/2)$ we have $0 < a_1^* \le \varphi_R(\vartheta, z) \le a_2^* < +\infty$, where the constant $a_1^*$ depends only on $z$, $R$ and $\delta$, and the constant $a_2^*$ depends only on $z$ and $R$. Using the uniformity of (2.4) and the estimates (2.3) (which were proved for $\vartheta \in [\beta, \delta] \cup [\pi - \delta, \pi - \beta]$), we conclude that these estimates hold for any $\vartheta \in [\beta, \pi - \beta]$.

**Lemma 2.2.** *If $R > 0$ and $z \in G_R^+$ are fixed numbers and $\rho \ge 0$ is sufficiently small, then*

$$C_1^* \le (R_\rho^2 - t^2)^{1-\pi/\alpha}\psi_{R,\rho}(t, z) \le C_2^*, \quad -R_\rho \le t \le R_\rho, \tag{2.6}$$

*where* $C^*_{1,2} > 0$ *are constants depending only on* $z$ *and* $R$. *Besides, uniformly in* $(-R, R)$

$$\lim_{\rho\to+0}\psi_{R,\rho}(t,z) = \frac{\operatorname{Im} z}{\pi}\left\{\frac{1}{|t-z|^2} - \frac{R^2}{|R^2-tz|^2}\right\}$$
$$= \frac{\operatorname{Im} z}{\pi}\frac{(R^2-t^2)(R^2-|z|^2)}{|t-z|^2|R^2-tz|^2} \equiv \psi_R(t,z). \tag{2.7}$$

**Proof** is similar to that of the previous lemma.

**Theorem 2.1.** *Let* $u(z) \not\equiv -\infty$ *be a function subharmonic in the semi-disc* $G^+_{R^*}$ $(0 < R^* \leq +\infty)$. *Then* $u(z)$ *has a nonnegative harmonic majorant in* $G^+_R$ $(0 < R < R^*)$ *if and only if the following conditions are satisfied:*

$$\int_0^\pi u^+(R\,e^{i\vartheta})\sin\vartheta d\vartheta < +\infty, \tag{2.8}$$

$$\liminf_{y\to+0}\int_{-R_y}^{R_y}(R_y^2 - x^2)^{\pi/\alpha-1}u^+(u+iy)dx < +\infty, \tag{2.9}$$

*where* $R_y = \sqrt{R^2-y^2}$ *and* $\alpha = \arccos(y/R)$. *If it is so, then*

$$u(z) = \iint_{G^+_R}\log\left|\frac{z-\zeta}{z-\overline{\zeta}}\frac{R^2-\zeta z}{R^2-\overline{\zeta}z}\right|d\nu(\zeta)$$
$$+\frac{2y}{\pi}R(R^2-|z|^2)\int_0^\pi\frac{u(R\,e^{i\vartheta})\sin\vartheta}{|Re^{i\vartheta}-z|^2|R\,e^{-i\vartheta}-z|^2}d\vartheta$$
$$+\frac{y}{\pi}\int_{-R}^{R}\left\{\frac{1}{|t-z|^2}-\frac{R^2}{|R^2-tz|^2}\right\}d\mu(t), \quad z = x+iy \in G^+_R, \tag{2.10}$$

*where* $\nu(\zeta)$ *is a nonnegative Borel measure in* $G^+_{R^*}$, *such that*

$$\iint_{G^+_R}(R-|\zeta|)\operatorname{Im}\zeta\, d\nu(\zeta) < +\infty, \tag{2.11}$$

*and* $\mu(t)$ *is a function representable as the difference* $\mu(t) = \mu_+(t) - \mu_-(t)$ *of two nondecreasing functions such that*

$$\int_{-R}^{R}(R^2-t^2)d\mu_\pm(t) < +\infty. \tag{2.12}$$

*The functions* $\mu_\pm(t)$ *can be chosen to be those deduced from*

$$\mu_\pm(t) = \lim_{y\to+0}\int_0^t u^\pm(x+iy)dx, \quad -R < t < R. \tag{2.13}$$

*Under this choice*

$$\bigvee_a^b \mu = \bigvee_a^b \mu_+ + \bigvee_a^b \mu_-, \quad \forall[a,b] \subset (-R,R), \tag{2.14}$$

*and*

$$\lim_{y\to+0} \int_a^b u^{\pm}(x+iy)g(x)dx = \int_a^b g(x)d\mu_{\pm}(x) \tag{2.15}$$

*for any function* $g(x)$ *continuous in* $[a,b]$. *Besides, for any* $R_0$ $(0 < R_0 \leq R)$

$$\int_0^{\pi} |u(R_0\, e^{i\vartheta})| \sin\vartheta d\vartheta < +\infty. \tag{2.16}$$

**Proof.** Choose $0 < y_0 < R$ such that $u(iy_0) \neq -\infty$, then take $z = iy_0$ and suppose $\rho > 0$ in (2.1) is small enough to ensure the validity of the estimates (2.3) and (2.6). Using these estimates we obtain

$$\begin{aligned}
\iint_{G^+_{R,\rho}} g_{R,\rho}(\zeta, z)d\nu(\zeta) &+ C_1 \int_{\beta}^{\pi-\beta} u^-(Re^{i\vartheta}) \left(\sin\frac{\pi(\vartheta-\beta)}{\pi-2\beta}\right)^{\pi/\alpha-1} d\vartheta \\
&+ C_1^* \int_{-R_\rho}^{R_\rho} (R_\rho^2 - t^2)^{\pi/\alpha-1} u^-(t+i\rho)dt \\
\leq u^-(iy_0) &+ C_2 \int_{\beta}^{\pi-\beta} u^+(Re^{i\vartheta}) \left(\sin\frac{\pi(\vartheta-\beta)}{\pi-2\beta}\right)^{\pi/\alpha-1} d\vartheta \\
&+ C_2^* \int_{-R_\rho}^{R_\rho} (R_\rho^2 - t^2)^{\pi/\alpha-1} u^+(t+i\rho)dt.
\end{aligned} \tag{2.17}$$

Further, assuming that $\lambda_m \downarrow 0$ is any sequence such that the last integral is uniformly bounded for $\rho = \lambda_m$ $(m \geq 1)$, we conclude that the whole right-hand side of (2.17) is uniformly bounded for such $\rho$. Thus

$$\begin{aligned}
&\sup_{m\geq 1} \iint_{G^+_{R,\lambda_m}} g_{R,\lambda_m}(\zeta, iy_0)d\nu(\zeta) < +\infty, \\
&\sup_{m\geq 1} \int_{\beta_m}^{\pi-\beta_m} u^-(Re^{i\vartheta}) \left(\sin\frac{\pi(\vartheta-\beta_m)}{\pi-2\beta_m}\right)^{\pi/\alpha_m-1} d\vartheta < +\infty, \\
&\sup_{m\geq 1} \int_{-R_{\lambda_m}}^{R_{\lambda_m}} (R_{\lambda_m}^2 - t^2)^{\pi/\alpha_m-1} u^{\pm}(t+i\lambda_m)dt < +\infty,
\end{aligned} \tag{2.18}$$

where $\alpha_m = \arccos(\lambda_m/R)$, $\beta_m = \pi/2 - \alpha_m$ and $R_{\lambda_m} = \sqrt{R^2 - \lambda_m^2}$. It is not difficult to verify that the first of these relations is equivalent to (2.11) while the second, together with (2.8), gives (2.16) for $R_0 = R$. Now let $\delta_n \downarrow 0$ $(0 < \delta_n < R)$ be any sequence. Then, by (2.18), the nondecreasing functions

$$\mu_m^{(\pm)}(t) = \int_0^t u^{\pm}(x + i\lambda_m)dx$$

are uniformly bounded in any segment $[-(R - \delta_n), (R - \delta_n)]$ $(n \geq 1)$, if $m \geq N(n) \geq 1$. Consequently, by Helly's theorem, there exists a subsequence $\{\lambda_m^{(n)}\} \subseteq \{\lambda_m\}$ such that the relations (2.13) are true when $t \in [-(R - \delta_n), (R - \delta_n)]$ and $y = \lambda_m^{(n)} \downarrow 0$. Therefore, the relations (2.13) are true for any $t \in (-R, R)$, if $y$ takes values from the diagonal sequence $\{\rho_n\} = \{\lambda_n^{(n)}\}$. At the same time, Helly's theorem on passage to a limit leads to relations (2.15) for $y = \rho_n \downarrow 0$. This implies the validity of the equality (2.14). Consider the nondecreasing functions

$$\Lambda_n^{(\pm)}(t) = \int_0^t (R_{\rho_n}^2 - x^2)^{\pi/\alpha_n - 1} d\mu_n^{(\pm)}(x), \quad -R_{\rho_n} \leq t \leq R_{\rho_n} \tag{2.19}$$

and put

$$\Lambda_n^{(\pm)}(t) = \Lambda_n^{(\pm)}(R_{\rho_n}) \quad (R_{\rho_n} < t \leq R)$$

and

$$\Lambda_n^{(\pm)}(t) = \Lambda_n^{(\pm)}(-R_{\rho_n}) \quad (-R \leq t < -R_{\rho_n}).$$

Then (2.18) shows that these functions are uniformly bounded in $[-R, R]$. Hence, by Helly's theorem, there exists a subsequence of $\{\rho_n\}$ (which for convenience we again denote $\{\rho_n\}$) such that

$$\Lambda_n^{(\pm)}(t) \to \Lambda_{\pm}(t) \quad (-R \leq t \leq R) \quad \text{as} \quad n \to \infty,$$

where $\Lambda_{\pm}(t)$ are some nondecreasing and bounded functions. It is clear that

$$\Lambda_{\pm}(t) = \int_0^t (R^2 - x^2) d\mu_{\pm}(x), \quad -R \leq t \leq R, \tag{2.20}$$

and so the relations (2.12) are true. On the other hand, Helly's other theorem (on the passage to the limit in Stiltyes integrals) implies the relations (2.15) in any $[a, b] \subseteq [-R, R]$ for the measures $d\Lambda_n^{(\pm)}(t)$. Therefore,

$$\lim_{n\to\infty} \bigvee_a^b \Lambda_n^{(\pm)} = \bigvee_a^b \Lambda_{\pm}, \tag{2.21}$$

and, since the function $\psi_R(t, z)/(R^2 - t^2)$ is continuous in $[-R, R]$,

$$\lim_{n\to\infty} \int_{-R}^{R} \frac{\psi_R(t, z)}{R^2 - t^2} d\Lambda_n^{(\pm)}(t) = \int_{-R}^{R} \frac{\psi_R(t, z)}{R^2 - t^2} d\Lambda_{\pm}(t). \tag{2.22}$$

Let

$$\int_{-R}^{R} \frac{\psi_{R,\rho_n}(t,z)}{(R_{\rho_n}^2 - t^2)^{\pi/\alpha_n - 1}} d\Lambda_n^{(\pm)}(t) = \int_{-R}^{R} \frac{\psi_R(t,z)}{R^2 - t^2} d\Lambda_n^{(\pm)}(t) + Q_n^{(\pm)}. \tag{2.23}$$

Then

$$|Q_n^{(\pm)}| \le \int_{R-\delta<|t|<R} \left[ \frac{\psi_{R,\rho_n}(t,z)}{(R_{\rho_n}^2 - t^2)^{\pi/\alpha_n - 1}} + \frac{\psi_R(t,z)}{R^2 - t^2} \right] d\Lambda_n^{(\pm)}(t)$$
$$+ \int_{-(R-\delta)}^{(R-\delta)} \left| \frac{\psi_{R,\rho_n}(t,z)}{(R_{\rho_n}^2 - t^2)^{\pi/\alpha_n - 1}} - \frac{\psi_R(t,z)}{R^2 - t^2} \right| d\Lambda_n^{(\pm)}(t) \equiv J_n' + J_n'' \tag{2.24}$$

for any $\delta$ $(0 < \delta < R)$. Assuming that $\varepsilon > 0$ is arbitrary, choose $\delta > 0$ such that

$$\left( \bigvee_{-R}^{-(R-\delta)} + \bigvee_{R-\delta}^{R} \right) \Lambda_n^{(\pm)} < \frac{\varepsilon}{8C_2^*},$$

where the constant $C_2^*$ is that of (2.6). Then, by (2.21)

$$\left( \bigvee_{-R}^{-(R-\delta)} + \bigvee_{R-\delta}^{R} \right) \Lambda_\pm < \frac{\varepsilon}{4C_2^*},$$

if $n \ge 1$ is sufficiently large. Therefore, $J_n' < \varepsilon/2$ by (2.6). On the other hand, it follows from (2.7) that the integrand of $J_n''$ tends to zero uniformly in $[-(R-\delta), (R-\delta)]$ as $n \to \infty$. Therefore, by (2.21), $J_n'' < \varepsilon/2$ for sufficiently large $n \ge 1$. Consequently, the relations (2.19)–(2.24) give

$$\lim_{n\to\infty} \int_{-R_{\rho_n}}^{R_{\rho_n}} u^\pm(t + i\rho_n)\psi_{R,\rho_n}(t,z)dt = \int_{-R}^{R} \psi_R(t,z)d\mu_\pm(t). \tag{2.25}$$

It follows from (2.3) that $0 \le \varphi_{R,\rho}(\vartheta, z) \le C_2 \sin\vartheta$ for sufficiently small $\rho > 0$. Hence, using the relation (2.16) (still proved only for $R_0 = R$), we obtain

$$\lim_{\rho\to+0} \int_{\beta}^{\pi-\beta} u^\pm(Re^{i\vartheta})\varphi_{R,\rho}(\vartheta,z)d\vartheta$$
$$= \frac{2y}{\pi} R(R^2 - |z|^2) \int_0^{\pi} \frac{u^\pm(R\,e^{i\vartheta}) \sin\vartheta}{|Re^{i\vartheta} - z|^2 |R\,e^{-i\vartheta} - z|^2} d\vartheta. \tag{2.26}$$

For the next passage, observe that the function $\Phi(z) = g_R(\zeta, z) - g_{R,\rho}(\zeta, z)$ is harmonic in the closure of $G_{R,\rho}^+$ whatever be $\zeta \in G_{R,\rho}^+$. Besides, $\Phi(Re^{i\vartheta}) = 0$ $(\beta \le \vartheta \le \pi - \beta)$ and $\Phi(t + i\rho) = g_R(\zeta, t + i\rho) > 0$ $(-R_\rho < t < R_\rho)$. Therefore,

$\Phi(z) > 0$ in $G^+_{R,\rho}$, and $g_R(\zeta, z) > g_{R,\rho}(\zeta, z)$ for any $z, \zeta \in G^+_{R,\rho}$. It is easy to prove that the condition (2.11) is sufficient for the convergence of the integral

$$\iint_{G^+_R} g_R(\zeta, z)d\nu(\zeta) \equiv \iint_{G^+_R} \log\left|\frac{z-\zeta}{z-\overline{\zeta}}\frac{R^2-\zeta z}{R^2-\overline{\zeta}z}\right| d\nu(\zeta).$$

Using this, one can prove that

$$\lim_{\rho\to+0} \iint_{G^+_{R,\rho}} g_{R,\rho}(\zeta, z)d\nu(\zeta) = \iint_{G^+_R} g_R(\zeta, z)d\nu(\zeta). \tag{2.27}$$

The relations (2.25)–(2.27) permit to obtain the representation (2.10)–(2.12) by letting $\rho = \rho_n \downarrow 0$ in (2.1). On the other hand, the representation (2.10)–(2.12) is necessary and sufficient for the existence of a nonnegative harmonic majorant of $u(z)$ in $G^+_R$, since using a conformal mapping this representation is derived from the similar representation of subharmonic functions which are of the same type in the unit disc. Thus, $u(z)$ has a nonnegative harmonic majorant in $G^+_R$, and, as its majorant is the same in any half-disc $G^+_{R_o}$ $(0 < R_o < R)$, the relation (2.16) is true for any $R_o$ $(0 < R_o \le R)$. Finally, the relations (2.13) and (2.15) without the indices $\pm$ can be easily verified using directly the representation (2.10). But also $u^+(z)$ is a subharmonic function having the same nonnegative harmonic majorant. This proves the relations (2.13) and (2.15) with the index $+$, hence follows their validity with the index $-$.

## 3. PROOFS OF THEOREMS 1.1 – 1.4

**Proof of Theorem 1.1.** Let $u(z) \not\equiv -\infty$ be a function of $S(\Omega)$, and let $R_k \uparrow \infty$ be a sequence on which the lower limit in (1.2) is attained. Then the hypotheses of Theorem 2.1 are true for any $R = R_k$, and hence the representation (2.10) is valid for any $R = R_k$. From (2.10) we subtract the same representation written for a smaller half-disc $G^+_{R_o}$ $(0 < R_o < R)$, put $z = iy$, divide the obtained equality by $2y$ and let $y \to +0$. This gives the following Carleman type formula:

$$\begin{aligned}\iint_{G^+_R\setminus G^+_{R_o}} \left(\frac{1}{|\zeta|^2} - \frac{1}{R^2}\right)\operatorname{Im}\zeta\, d\nu(\zeta) + \left(\frac{1}{R_o^2} - \frac{1}{R^2}\right)\iint_{G^+_{R_o}} \operatorname{Im}\zeta\, d\nu(\zeta)\\ = \frac{1}{\pi R}\int_0^\pi u(Re^{i\vartheta})\sin\vartheta d\vartheta + \int_{R_o+0<|t|<R}\left(\frac{1}{t^2} - \frac{1}{R^2}\right)d\mu(t)\\ + \left(\frac{1}{R_o^2} - \frac{1}{R^2}\right)\int_{-R_o}^{R_o} d\mu(t) - \frac{1}{\pi R_o}\int_0^\pi u(R_o e^{i\vartheta})\sin\vartheta d\vartheta.\end{aligned} \tag{3.1}$$

The right-hand side of this formula remains bounded from above as $R = R_k \to +\infty$. This follows from (1.2), (2.16) and from the relation

$$\int_{-\infty}^{+\infty} \Omega(x)d\mu_+(x) = \lim_{R\to+\infty}\liminf_{y\to+0}\int_{-R}^{R} u^+(x+iy)\Omega(x)\,dx < +\infty \tag{3.2}$$

which is a consequence of (1.5) and (2.15). Hence, also the left-hand side of (3.1) is bounded. Using this we arrive at (1.7). The proof of convergence of the second integral of (1.8) as well as the proof of the representation (1.6) are similar to the proofs of the corresponding assertions of Nevanlinna's theorem [83] (see also [10], Sec. 6.3 – 6.5). Now let $u(z)$ be a function representable in the form (1.6)–(1.8). Then, obviously

$$u^+(z) \le h^+ y + \frac{y}{\pi}\int_{-\infty}^{+\infty} \frac{d\mu_+(t)}{(x-t)^2+y^2} \equiv U(z), \quad z = x+iy \in G^+.$$

Hence (1.2) follows. Further, it is easy to show that for any $R > 0$ there exists a constant $C > 0$ such that

$$\frac{y}{\pi}\int_{-R}^{R} \frac{\Omega(x)dx}{(x-t)^2+y^2} \le \frac{C}{1+t^2}, \quad 0 < y < 1, \quad -\infty < t < +\infty.$$

Hence, using Lebesgue's theorem on dominated convergence we arrive at relation (1.12) for $U(z)$. This implies (1.5), and so, $u(z) \in S(\Omega)$. Consequently, the measures $\mu_\pm(t)$ in (1.6) can be recovered by the relations (1.9), and (1.10), (1.11) follow from (2.14), (2.15) of Theorem 2.1. Further,

$$\lim_{R\to+\infty} \limsup_{y\to+0} \int_{-R}^{R} u^+(x+iy)\Omega(x)\,dx \le \int_{-\infty}^{+\infty} \Omega(x)d\mu_+(x)$$

since (1.2) is true for $U(z)$. This inequality and (3.2) imply (1.12). On the other hand, one can prove that

$$\frac{y}{\pi}\int_{-\infty}^{+\infty} \frac{1}{(x-t)^2+y^2}\frac{dx}{(1+|x|)^\gamma} \le \frac{C}{(1+|t|)^\gamma}, \quad -\infty < t < +\infty,$$

for any $0 < y < M < +\infty$ and $-1 < \gamma \le 2$, where the constant $C > 0$ depends only on $M$ and $\gamma$. Therefore, if $\Omega(t) = (1+|t|)^{-\gamma}$ $(1 < \gamma \le 2)$ or $\Omega(t) = (1+|t|)^{-\gamma}$ $(-1 < \gamma \le 1)$ and $h \le 0$, then by Lebesgue's theorem

$$\limsup_{y\to+0} \int_{-\infty}^{+\infty} u^+(x+iy)\Omega(x)\,dx \le \int_{-\infty}^{+\infty} \Omega(x)d\mu_+(x).$$

The converse inequality for the lower limit is true by (3.2). Hence (1.14) holds. To prove (1.13), observe that the function

$$\iint_{G^+} \log\left|\frac{z-\zeta}{z-\bar\zeta}\right| d\nu(\zeta) - u(z)$$

is from $S((1+|x|)^{-2})$. Therefore, the relation (1.14) with $\Omega(x) = (1+|x|)^{-2}$ is true for this function. Using (1.14) we come to (1.13).

**Remark 3.1.** By a similar argument, one can prove that the set $S(\Omega_1, \Omega_2)$ of functions which are subharmonic in $G^+$ and satisfy the conditions (1.2), (1.5) with $\Omega \equiv \Omega_1$ and also the condition

$$\lim_{R\to+\infty} \liminf_{y\to+0} \int_{-R}^{R} u^-(x+iy)\Omega_2(x)\,dx < +\infty$$

(where $\Omega_{1,2}(x)$ are continuous and such that $\Omega_{1,2} \geq C_{1,2}(1+x^2)^{-1}$ for some $C_{1,2} > 0$) coincides with the set of functions representable in the form (1.6)–(1.7), where $\mu_\pm(t)$ are such that

$$\int_{-\infty}^{+\infty} \Omega_1(x)d\mu_+(x) < +\infty \quad \text{and} \quad \int_{-\infty}^{+\infty} \Omega_2(x)d\mu_-(x) < +\infty.$$

Besides, by the results of [103] the subclass of $S(1,1)$ ($\Omega_1 \equiv \Omega_2 \equiv 1$) for which $h = 0$ and

$$\iint_{G+} \operatorname{Im} \zeta \; d\nu(\zeta) < +\infty$$

coincides with the set of those functions $u(z)$ subharmonic in $G^+$, for which

$$\sup_{y>0} \int_{-\infty}^{+\infty} |u(x+iy)|dx < +\infty.$$

**Proof of Theorem 1.2.** Assuming $u(z) \not\equiv -\infty$, observe that the conditions of Theorem 2.1 are satisfied for any $R = R_n$. Therefore, they are satisfied for any $R > 0$, and (2.15), (2.10)–(2.12) and (3.1) are true for any $R$, $R_o$ ($0 < R_o < R < +\infty$), if $\mu_\pm(t)$ are recovered by (2.13). By (3.1) and (2.15)

$$\frac{1}{R_o}\int_0^\pi u(R_o e^{i\vartheta}) \sin\vartheta d\vartheta \leq \frac{1}{R}\int_0^\pi u(R e^{i\vartheta}) \sin\vartheta d\vartheta + \lim_{y\to+0} \int_{-R}^{R} g_{R,R_o}(x)u(x+iy)dx,$$

where the left-hand side integral is absolutely convergent in accordance with (2.16). So, the hypothesis $u(z) \not\equiv -\infty$ contradicts to the condition (1.20).

**Proof of Theorem 1.3.** 1°. Let $\limsup R^{-1}M(R) = +\infty$ as $R \to +\infty$. In this case, for any $K > 0$ we can find $z = x+iy \in G^+$ such that $u(z) > K|z| \geq Ky$. Hence $\alpha = \alpha^+ = +\infty$. If we have also $\liminf R^{-1}M(R) < +\infty$ as $R \to +\infty$, then by Theorem 1.1, $u(z)$ is representable in the form (1.6)–(1.8), where $\mu_+(t) \equiv 0$. Thus, $u(z)$ satisfies the condition (1.4), and, by the result of M.Heins and L.Ahlfors, there exists $\lim R^{-1}M(R)$ as $R \to +\infty$. This is a contradiction. Therefore, if $\limsup R^{-1}M(R) = +\infty$ as $R \to +\infty$, then there exists $\lim R^{-1}M(R) = +\infty$ as $R \to +\infty$, and $\alpha^+ = \beta = +\infty$. In our case

$$\lim_{R\to+\infty} \frac{1}{R}\int_0^\pi u^+(R e^{i\vartheta}) \sin\vartheta d\vartheta = +\infty$$

since otherwise $u(z)$ would be representable in the form (1.6) – (1.8), where $\mu_+(t) \equiv 0$, and obviously $\alpha \le h < +\infty$ which is a contradiction. Now consider the case when $\limsup R^{-1}M(R) < +\infty$ as $R \to +\infty$. In this case $u(z)$ is representable in the form (1.6)–(1.8) with $\mu_+(t) \equiv 0$, besides, (1.4) is true and there exists

$$\lim R^{-1}M(R) = \beta = \alpha^+ \quad \text{as} \quad R \to +\infty,$$

according to the result of M.Heins and L.Ahlfors. Besides, from the representation (1.6)–(1.8) immediately follows that $\alpha \le h$. On the other hand, $\alpha \ge \limsup y^{-1}u(iy)$ as $y \to +\infty$ by the first of relations (1.16). Thus, $\alpha^+ = h^+ = \beta$ and, additionally, (1.15) is true.

2°. If $\beta = \alpha^+ < +\infty$, then $u(z)$ is representable in the form (1.6)–(1.8). For such functions the relations (1.15)–(1.17) with $h = \beta = \alpha^+$ hold.

**Proof of Theorem 1.4.** 1°. Let $f(z) \in H^p(\Omega(x)dx)$. Then $|f(z)|^p$ has a harmonic majorant in $G^+$ by Theorem 1.1. Thus, $f(z)$ belongs to the conformal mapping of Hardy's class, and it has nontangential boundary values $f(x)$ almost everywhere on $(-\infty, +\infty)$. Hence

$$\int_{-R}^{R} |f(x)|^p \Omega(x)dx \le \liminf_{y \to +0} \int_{-R}^{R} |f(x+iy)|^p \Omega(x)dx < +\infty, \quad R > 0, \tag{3.3}$$

by Fatou's lemma, and $f(x) \in L_p(\Omega(x)dx)$. Now let $f(z)$ be from the conformal mapping of Hardy's class $H^p$ $(|z| < 1)$, and let $f(x) \in L_p(\Omega(x)dx)$. Then, using the factorization of $f(z)$ one can obtain

$$|f(z)|^p \le \frac{y}{\pi} \int_{-\infty}^{+\infty} \frac{|f(t)|^p dt}{(x-t)^2 + y^2}, \qquad z = x + iy \in G^+. \tag{3.4}$$

On the other hand, it can be easily verified that for almost all $t \in (-\infty, +\infty)$

$$\frac{y}{\pi} \int_{-R}^{R} \frac{\Omega(x)dx}{(x-t)^2 + y^2} \le C^* \Omega(t), \quad 0 < y < 1, \quad R > 0,$$

for a constant $C^* > 0$ depending only on $R$ and $\Omega(x)$. Consequently,

$$\limsup_{y \to +0} \int_{-R}^{R} |f(x+iy)|^p \Omega(x)dx \le \int_{-R}^{R} |f(t)|^p \Omega(t)dt \tag{3.5}$$

by Lebesgue's theorem. The relation (1.27) follows from (3.4).

2°. If $f(z) \in H^p(\Omega(x)dx)$, then $f(z)$ is from the conformal mapping of Hardy's class. Using the factorization of this function we obtain (1.26). Now let (1.24) be replaced by (1.26). Then evidently $\log^+ |f(z)|$ has a harmonic majorant in $G^+$, i.e. $f(z)$ is of bounded type in $G^+$. Therefore, $|f(Re^{i\vartheta})|^p$ is continuous in $0 \le \vartheta \le \pi$ for almost all $R > 0$. Besides,

$$\liminf_{y\to+0} \int_{-R}^{R} |f(x+iy)|^p \frac{dx}{1+x^2} < +\infty$$

for any $R > 0$. Therefore, by Theorem 2.1, $|f(z)|^p$ has a harmonic majorant in any $G_R^+$ $(R > 0)$, i.e. in any $G_R^+$ the function $f(z)$ is from the conformal mapping of Hardy's class. The transformation of the corresponding factorization by means of the conformal mapping of $|z| < 1$ onto $G_R^+$ gives

$$\begin{aligned}\log|f(z)| = &\sum_{z_k\in G_R^+} \log\left|\frac{z-z_k}{z-\overline{z}_k}\frac{R^2-zz_k}{R^2-z\overline{z}_k}\right| \\ &+ \frac{2y}{\pi}R(R^2-|z|^2)\int_0^{\pi} \frac{\log|f(Re^{i\vartheta})|\sin\vartheta}{|Re^{i\vartheta}-z|^2|Re^{-i\vartheta}-z|^2}d\vartheta \\ &+ \frac{y}{\pi}\int_{-R}^{R}\left\{\frac{1}{|t-z|^2} - \frac{R^2}{|R^2-tz|^2}\right\}d\mu(t), \quad z = x+iy \in G_R^+. \end{aligned} \tag{3.6}$$

Here $z_k$ are the zeros of $f(z)$ and $d\mu(t) = \log|f(t)|dt - d\omega(t)$, where $\omega(t)$ is a nondecreasing function such that $\omega'(t) = 0$ almost everywhere in $(-R, R)$. But we have already proved that $\log^+|f(z)|$ has a harmonic majorant in $G^+$. Therefore it is obvious that the passage $R \to +\infty$ in (3.6) leads to the factorization

$$f(z) = \prod_{z_k\in G^+} \frac{z-z_k}{z-\overline{z_k}}\frac{|1+z_k^2|}{1+z_k^2}\exp\left\{\frac{1}{\pi i}\int_{-\infty}^{+\infty}\frac{1+tz}{t-z}\frac{d\mu(t)}{1+t^2} + ihz + iC\right\} \tag{3.7}$$

$(z \in G^+)$, where $d\mu(t) = \log|f(t)|dt - d\omega(t)$, $\operatorname{Im} C = 0$ and

$$h = \frac{2}{\pi}\lim_{R\to+\infty}\frac{2}{R}\int_0^{\pi}\log|f(Re^{i\vartheta})|\sin\vartheta d\vartheta \le 0$$

according to (1.15) and (1.26). Consequently, $f(z)$ is from the conformal mapping of Hardy's class, and (3.4) is true. Hence (1.27) follows.

$3^o$. Let $f(z) \in H^p(\Omega(x)dx)$, where $\Omega(x)$ satisfies the additional condition (1.28). Then it is clear that $\Omega(z)$ and

$$[\Omega(z)]^{1/p} \equiv \exp\left\{\frac{1}{p\pi i}\int_{-\infty}^{+\infty}\frac{1+tz}{t-z}\frac{\log\Omega(t)}{1+t^2}dt\right\}$$

both are non-vanishing holomorphic functions in $G^+$. Using (3.7) one can easily obtain

$$|f(z)[\Omega(z)]^{1/p}|^p \le \frac{y}{\pi}\int_{-\infty}^{+\infty}\frac{|f(t)|^p\Omega(t)}{(x-t)^2+y^2}dt, \quad z = x+iy \in G^+,$$

where $|f(t)|^p\Omega(t) \in L_1(dt)$. Hence $f(z)[\Omega(z)]^{1/p} \in H^p(dx)$. Thus

$$H^p(\Omega(x)dx)[\Omega(z)]^{1/p} \subseteq H^p(dx).$$

On the other hand, the assumption $f(z) \in H^p(dx)$ gives $f(x)[\Omega(x)]^{-1/p} \in L_p(\Omega(x)dx)$. Therefore,

$$f(z)[\Omega(z)]^{-1/p} \in H^p(\Omega(x)dx)$$

by assertion 1°, and

$$H^p(dx)[\Omega(z)]^{-1/p} \subseteq H^p(\Omega(x)dx).$$

The equality (1.30) is for the particular case of (1.29), when $\Omega(x) = (1+x^2)^{-\gamma/2}$.

# CHAPTER 8

# WEIGHTED CLASSES OF SUBHARMONIC FUNCTIONS

## 1. GREEN TYPE POTENTIALS

*1.1.* We start by the following auxiliary

**Lemma 1.1.** *Let $U(w)$ be subharmonic in $G^-$, and let $U(w) \in M_p$ ($\equiv M_p\{G^-\}$) for a natural $p$. Then for any $\alpha \in (0,p]$:*

1°. *$W^{-\alpha}\, u(z)$ is continuous and subharmonic in $G^-$,*

2°. $\dfrac{\partial^p}{\partial v^p} W^{-(p-\alpha)}\, W^{-\alpha}\, u(z) = U(w),\ w = u + iv \in G^-.$

**Proof.** 1°. Since $U(w) \in M_p$, the integral $W^{-\alpha}\, u(z)$ is absolutely convergent for any $w \in G^-$. By the same reason, for any fixed $w_0 = u_0 + iv_0 \in G^-$ the integral

$$\int_M^{+\infty} \sigma^{\alpha-1}\, |U(w - i\sigma)|\, d\sigma, \quad |w - w_0| < \frac{|v_0|}{2},$$

can be made arbitrary small by taking $M > 0$ large enough. On the other hand, one can verify that for any finite $M > 0$ the integral

$$\int_0^M \sigma^{\alpha-1}\, |U(w - i\sigma)|\, d\sigma$$

is a function continuous at $w = w_0$. To this end, one can use, for example, the Riesz representation of $U(w)$ in any disc in $G^-$, which contains the rectangle $\{w - i\sigma :\ |w - w_0| < |v_0|/2\},\ 0 < \sigma < M$. Hence it follows that $W^{-\alpha}\, u(z)$ is continuous in $G^-$. The inequality between $W^{-\alpha}\, u(z)$ and its mean value is easily obtained by changing the order of integration. 2° follows from Lemma 1.2 of Ch. 1.

*1.2.* The Green type potentials considered below are constructed by means of the Blaschke type products considered in Ch. 2.

**Lemma 1.2.** *Let $\nu(\zeta)$ be a nonnegative Borel measure in $G^-$, such that*

$$\iint_{G^-} |\eta|^{1+\alpha}\, d\nu(\zeta) < +\infty \quad (\zeta = \xi + i\eta) \tag{1.1}$$

*for some $\alpha > 0$. Then the Green type potential*

$$I_\alpha(w) = -\iint_{G^-} \log|b_\alpha(w,\zeta)|\,d\nu(\zeta) \tag{1.2}$$

*is a superharmonic function in $G^-$.*

**Proof.** Let $\rho_0 < 0$ be a fixed number, and let $w \in G^-_{\rho_0} = \{w : \operatorname{Im} w < \rho_0\}$. Using the representation (2.2) of Ch. 2, one can verify that for any $\zeta \in G^- \setminus G^-_{\rho_0/2}$ we have $|\log|b_\alpha(w,\zeta)|| \le C_{\alpha,\rho_0}|\eta|^{1+\alpha}$, where $C_{\alpha,\rho_0} > 0$ is a constant depending only on $\alpha$ and $\rho_0$. Hence

$$I_{\alpha,\rho_0}(w) = -\iint_{G^-\setminus G^-_{\rho_0/2}} \log|b_\alpha(w,\zeta)|\,d\nu(\zeta)$$

is a harmonic function in $G^-_{\rho_0}$. For consideration of the remaining part of $I_\alpha(w)$, recall from Ch. 2 that the function

$$U_\alpha(w,\zeta) = \log|b_\alpha(w,\zeta)| - \log|b_0(w,\zeta)| \tag{1.3}$$

is harmonic in $G^-$ and representable in the form

$$\begin{aligned} U_\alpha(w,\zeta) =& C_\alpha + \operatorname{Re}\int_{|\eta|}^{+\infty}\left[\frac{\tau^\alpha}{(iw-i\zeta+\tau)^{1+\alpha}} - \frac{1}{iw-i\zeta+\tau}\right]d\tau \\ &- \operatorname{Re}\int_0^{|\eta|}\frac{\tau^\alpha\,d\tau}{(iw-i\bar\zeta-\tau)^{1+\alpha}} + \operatorname{Re}\int_0^{|\eta|}\frac{d\tau}{iw-i\bar\zeta-\tau}, \end{aligned} \tag{1.4}$$

where

$$C_\alpha = \int_0^{+\infty}\left[1-\left(\frac{\sigma}{1+\sigma}\right)^\alpha\right]\frac{d\sigma}{1+\sigma} > 0$$

is a constant while the other terms are functions harmonic in $G^-$. One can see that for any $w \in G^-_{\rho_0}$ and $\zeta \in G^-_{\rho_0/2}$

$$\left|C_\alpha - \int_0^{|\eta|}\frac{\tau^\alpha\,d\tau}{(iw-i\bar\zeta-\tau)^{1+\alpha}} + \int_0^{|\eta|}\frac{d\tau}{iw-i\bar\zeta-\tau}\right| \le C'_{\alpha,\rho_0}|\eta|^{1+\alpha}, \tag{1.5}$$

where $C'_{\alpha,\rho_0} > 0$ is a constant depending only on $\alpha$ and $\rho_0$. To estimate the remaining term of (1.4), denote

$$\begin{aligned} J &= \int_{|\eta|}^{+\infty}\left|\frac{\tau^\alpha}{(iw-i\zeta+\tau)^{1+\alpha}} - \frac{1}{iw-i\zeta+\tau}\right|d\tau \\ &= \int_0^{+\infty}\left|1-\left(\frac{\sigma+|\frac{\eta}{v}|}{i\frac{u-\xi}{|v|}+1+\sigma}\right)^\alpha\right|\frac{d\sigma}{\left|i\frac{u-\xi}{|v|}+1+\sigma\right|}, \qquad w = u+iv, \end{aligned} \tag{1.6}$$

and consider separately the cases $|\eta| > 2|v|$ and $|\eta| \le 2|v|$ ($\zeta = \xi + i\eta \in G^-_{\rho_0/2}$, $w = u + iv \in \overline{G^-_{\rho_0}}$). If $|\eta| > 2|v|$, then

$$\begin{aligned} J =& \left(\int_0^{|\frac{\eta}{v}|-2} + \int_{|\frac{\eta}{v}|-2}^{+\infty}\right)\left|1-\left(\frac{\sigma+|\frac{\eta}{v}|}{i\frac{u-\xi}{|v|}+1+\sigma}\right)^\alpha\right|\frac{d\sigma}{\left|i\frac{u-\xi}{|v|}+1+\sigma\right|} \\ &\equiv J_1 + J_2, \end{aligned} \tag{1.7}$$

where obviously

$$J_1 < 2^\alpha \left|\frac{\eta}{v}\right|^\alpha \int_0^{+\infty} \frac{d\sigma}{(1+\sigma)^{1+\alpha}} + \log\left|\frac{\eta}{v}\right| < C''_{\alpha,\rho_0} |\eta|^{1+\alpha}. \tag{1.8}$$

On the other hand, putting

$$\frac{\sigma + \left|\frac{\eta}{v}\right|}{i\frac{u-\xi}{|v|} + 1 + \sigma} = 1 - \Delta, \quad \text{where} \quad \Delta = \frac{i\frac{u-\xi}{|v|} + 1 - \left|\frac{\eta}{v}\right|}{i\frac{u-\xi}{|v|} + 1 + \sigma},$$

observe that $|\Delta| < 1$ if and only if $\sigma > |\eta/v| - 2$.

When $|\Delta| < 1$ we have $|(1-\Delta)^\alpha - 1|/|\Delta| \le M_\alpha < +\infty$. Therefore

$$\begin{aligned} J_2 &\le 2M_\alpha \left(\left|\frac{u-\xi}{v}\right| + \left|\frac{\eta}{v}\right| - 1\right) \int_{\left|\frac{\eta}{v}\right|-2}^{+\infty} \left(\left|\frac{u-\xi}{v}\right| + 1 + \sigma\right)^{-2} d\sigma \\ &< 2M_\alpha \le C'''_{\alpha,\rho_0} |\eta|^{1+\alpha}. \end{aligned} \tag{1.9}$$

If $|\rho_0|/2 < |\eta| \le 2|v|$, then $|\Delta| < 1$ for any $\sigma > 0$. Hence

$$\begin{aligned} J &\le 2M_\alpha \left(\left|\frac{u-\xi}{v}\right| + \left|1 - \left|\frac{\eta}{v}\right|\right|\right) \int_0^{+\infty} \left(\left|\frac{u-\xi}{v}\right| + 1 + \sigma\right)^{-2} d\sigma \\ &\le 2M_\alpha \le C'''_{\alpha,\rho_0} |\eta|^{1+\alpha}. \end{aligned}$$

Using formulas (1.4)–(1.9) and the last estimate we obtain that $|U_\alpha(z\zeta)| \le C^*_{\alpha,\rho_0} |\eta|^{1+\alpha}$ for $w \in \overline{G^-_{\rho_0}}$, $\zeta \in G^-_{\rho_0/2}$, where the constant $C^*_{\alpha,\rho_0} > 0$ depends only on $\alpha$ and $\rho_0$. Therefore, by (1.3) and (1.1)

$$\begin{aligned} I^*_{\alpha,\rho_0}(w) &= -\iint_{G^-_{\rho_0/2}} \log|b_\alpha(w,\zeta)|\, d\nu(\zeta) \\ &= -\iint_{G^-_{\rho_0/2}} \log|b_0(w,\zeta)|\, d\nu(\zeta) - \iint_{G^-_{\rho_0/2}} U_\alpha(w,\zeta)\, d\nu(\zeta), \end{aligned}$$

where the last integral converges in $G^-_{\rho_0}$ absolutely and uniformly, while the whole $I^*_{\alpha,\rho_0}(w)$ is superharmonic in $G^-_{\rho_0}$, as a sum of an ordinary Green potential and a function harmonic in $G^-_{\rho_0}$. The desired assertion follows, since $I_\alpha(w) = I_{\alpha,\rho_0}(w) + I^*_{\alpha,\rho_0}(w)$ for any $\rho_0 < 0$.

**Lemma 1.3.** *Let $\nu(\zeta)$ be a nonnegative Borel measure in $G^-$, satisfying (1.1) for some $\alpha > 0$. Then the Green type potential (1.2) has the following properties:*

1°. *$I_\alpha(w) \in M_\beta$ for any $\beta \in (0, 1+\alpha)$.*

2°. *$W^{-\alpha} I_\alpha(w)$ is a continuous superharmonic function in $G^-$,*

$$W^{-\alpha} I_\alpha(w) = -\iint_{G^-} W^{-\alpha} \log|b_\alpha(w,\zeta)|\, d\nu(\zeta) \ge 0, \quad w \in G^-, \tag{1.10}$$

*where the integral is uniformly convergent in any half-plane $G^-_\rho$ ($\rho < 0$) and*

$$\int_1^{+\infty} W^{-\alpha} I_\alpha(-it)\,\frac{dt}{t} < +\infty. \tag{1.11}$$

3°. *$W^{-\alpha}I_\alpha(w)$ is representable as an ordinary Green potential:*

$$W^{-\alpha}I_\alpha(w) = -\iint_{G^-} \log|b_0(w,\zeta)|\,d\nu_\alpha(\zeta), \quad w \in G^-, \tag{1.12}$$

*where $\nu_\alpha(\zeta)$ is a nonnegative Borel measure satisfying (1.1) for $\alpha = 0$.*

**Proof.** 1°. Let $\rho_0 < 0$ and $w = u + iv \in \overline{G^-_{\rho_0}}$. Assuming $\alpha \le \beta < 1+\alpha$ set

$$\begin{aligned} I &= \int_0^{+\infty} \sigma^{\beta-1}\,d\sigma \iint_{G^-} |\log|b_\alpha(w-i\sigma,\zeta)||\,d\nu(\zeta) \\ &= \left(\iint_{G^-\setminus G^-_v} + \iint_{G^-_v}\right)\left(\int_0^{+\infty} \sigma^{\beta-1}|\log|b_\alpha(w-i\sigma,\zeta)||\,d\sigma\right) d\nu(\zeta) \\ &\equiv I_1 + I_2 \end{aligned} \tag{1.13}$$

and estimate $I_1$ and $I_2$ separately. By formula (1.1′) and (1.2) of Chapter 2

$$\begin{aligned} I_1 &\le \iint_{G^-\setminus G^-_v} d\nu(\zeta) \int_{-|\eta|}^{|\eta|} (|\eta|-|t|)^\alpha\,dt \int_0^{+\infty} \frac{\sigma^{\beta-1}\,d\sigma}{(|v|-t+\sigma)^{1+\alpha}} \\ &= \int_0^{+\infty} \frac{\sigma^{\beta-1}\,d\sigma}{(1+\sigma)^{1+\alpha}} \iint_{G^-\setminus G^-_v} |\eta|^\beta\,d\nu(\zeta) \int_{-1}^1 \frac{(1-|t|)^\alpha\,dt}{(|v/\eta|-t)^{1+\alpha-\beta}} \\ &\le C(\alpha,\beta,\rho_0) \iint_{G^-\setminus G^-_v} |\eta|^{1+\alpha}\,d\nu(\zeta), \end{aligned} \tag{1.14}$$

where $C(\alpha,\beta,\rho_0) > 0$ is a constant depending only on $\alpha$, $\beta$, and $\rho_0$. Putting

$$\begin{aligned} I_2 &= \iint_{G^-_v} d\nu(\zeta) \int_{|\eta|}^{+\infty} \sigma^{\beta-1}\,|\log|b_\alpha(w-i\sigma,\zeta)||\,d\sigma \\ &+ \iint_{G^-_v} d\nu(\zeta) \int_0^{|\eta|} \sigma^{\beta-1}\,|\log|b_\alpha(w-i\sigma,\zeta)||\,d\sigma \equiv J_1 + J_2 \end{aligned} \tag{1.15}$$

and again using formulas (1.1′) and (1.2) of Ch.2, we obtain

$$\begin{aligned} J_1 &\le \iint_{G^-_v} d\nu(\zeta) \int_{-|\eta|}^{|\eta|} (|\eta|-|t|)^\alpha\,dt \int_{|\eta|}^{+\infty} \frac{\sigma^{\beta-1}\,d\sigma}{(|v|-t+\sigma)^{1+\alpha}} \\ &\le C'(\alpha,\beta,\rho_0) \iint_{G^-_v} |\eta|^{1+\alpha}\,d\nu(\zeta). \end{aligned} \tag{1.16}$$

To estimate $J_2$ we use the representation

$$\begin{aligned} \log|b_\alpha(w,\zeta)| &= \sum_{k=0}^{p-1} \frac{1}{\alpha-k}\left(\frac{|\eta|}{i(w-\xi)}\right)^{\alpha-k} - \operatorname{Re}\int_0^{|\eta|} \frac{\tau^{\alpha-p}\,d\tau}{(iw-i\zeta+\tau)^{1+\alpha-p}} \\ &- \operatorname{Re}\int_0^{|\eta|} \frac{\tau^\alpha\,d\tau}{(iw-i\overline{\zeta}-\tau)^{1+\alpha}} \quad (\alpha \le p < 1+\alpha) \end{aligned}$$

which is derived from the representation (1.1′)–(1.2) of Ch.2 by integration by parts. Then we obtain

$$J_2 \leq \iint_{G_v^-} d\nu(\zeta) \int_0^{|\eta|} \left[ \sum_{k=0}^{p-1} \frac{1}{\alpha - k} \left( \frac{|\eta|}{|v| + \sigma} \right)^{\alpha - k} + \int_0^{|\eta|} \frac{\tau^{\alpha - p}\, d\tau}{\left| |v| - |\eta| + \sigma + \tau \right|^{1+\alpha-p}} + \int_0^{|\eta|} \frac{\tau^{\alpha}\, d\tau}{(|v| + |\eta| - \tau + \sigma)^{1+\alpha}} \right] \sigma^{\beta-1}\, d\sigma \equiv K_1 + K_2 + K_3. \tag{1.17}$$

For any $\gamma \in (0, \alpha]$

$$|\eta|^{\gamma} \int_0^{|\eta|} \frac{\sigma^{\beta-1}\, d\sigma}{(|v| + \sigma)^{\gamma}} \leq C''(\gamma, \beta, \rho_0)\, |\eta|^{1+\alpha}.$$

Therefore

$$K_1 \leq \sum_{k=0}^{p-1} \frac{C''(\alpha - k, \beta, \rho_0)}{\alpha - k} \iint_{G_v^-} |\eta|^{1+\alpha}\, d\nu(\zeta). \tag{1.18}$$

On the other hand

$$K_3 \leq C'''(\alpha, \beta, \rho_0) \iint_{G_v^-} |\eta|^{1+\alpha}\, d\nu(\zeta). \tag{1.19}$$

Now observe that

$$K_2 = \iint_{G_v^-} |\eta|^{\beta}\, L(\eta)\, d\nu(\zeta),$$

where

$$L(\eta) = \int_0^1 \sigma^{\beta-1}\, d\sigma \int_0^1 \frac{\tau^{\alpha-p}\, d\tau}{\left| |v/\eta| - 1 + \sigma + \tau \right|^{1+\alpha-p}}$$

and $||v/\eta| - 1 + \sigma| < 1$. One can easily show that $L(\eta) \leq C^{\mathrm{IV}}(\alpha, \beta) < +\infty$. Therefore, for $K_2$ an estimate similar to (1.18)–(1.19) holds and by (1.17)

$$J_2 \leq C^{\mathrm{V}}(\alpha, \beta, \rho_0) \iint_{G_v^-} |\eta|^{1+\alpha}\, d\nu(\zeta).$$

Thus, by (1.1) and (1.13)–(1.16)

$$I \leq C^*(\alpha, \beta, \rho_0) \iint_{G^-} |\eta|^{1+\alpha}\, d\nu(\zeta) < +\infty,$$

where $C^*(\alpha, \beta, \rho_0) > 0$ is a constant depending only on $\alpha$, $\beta$, and $\rho_0$. This obviously proves the inclusion $I_\alpha(w) \in M_\beta$ ($\alpha \leq \beta < 1 + \alpha$) and the equality (1.10), where the integral is absolutely and uniformly convergent in $G_\rho^-$. Besides, the mentioned inclusion is true for any $\beta \in (0, 1 + \alpha)$, since $M_{\beta_1} \subset M_{\beta_2}$ for any $\beta_1 > \beta_2$. The inequality in (1.10) is valid since the integral in the formula (2.2) of Ch. 2 is nonnegative. By Lemma 1.1, $W^{-\alpha} I_\alpha(w)$ is a continuous superharmonic function in $G^-$ since $I_\alpha(w) \in M_p$ ($\alpha \leq p < 1 + \alpha$). To prove representation (1.12), observe that by (1.10)

$$\begin{aligned}\varphi(v) =&\Gamma(1+\alpha)\int_{-\infty}^{+\infty}\left|W^{-\alpha}I_\alpha(u+iv)\right|\,du\\ =&\pi\iint_{G^-}d\nu(\zeta)\int_{-|\eta|}^{|\eta|}(|\eta|-|t|)^\alpha\,dt\,\frac{|v|-t}{\pi}\int_{-\infty}^{+\infty}\frac{dx}{(u-\xi)^2+(|v|-t)^2}\\ =&\pi\iint_{G^-}d\nu(\zeta)\int_{-|\eta|}^{|\eta|}(|\eta|-|t|)^\alpha\,\mathrm{sign}(|v|-t)\,dt\\ <&\frac{2\pi}{1+\alpha}\iint_{G^-}|\eta|^{1+\alpha}\,d\nu(\zeta)<+\infty\end{aligned}$$

for any $v<0$. On the other hand, $\varphi(v)\to 0$ as $v\to -0$. Therefore, the results of [103] and Ch. 7 lead to (1.12), the latter implies (1.11).

**Remark 1.1.** Using the formula (2.2) of Ch. 2, one can verify that for any $w,\zeta\in G^-$ and $\alpha>0$

$$W^{-\alpha}\log|b_\alpha(w,\zeta)|=\frac{1}{\Gamma(\alpha)}\int_0^{|\eta|}\tau^{\alpha-1}\log|b_0(w,\zeta+i\tau)|\,d\tau.$$

This formula permits to write explicitly the measure $\nu_\alpha(\zeta)$ of representation (1.12). Namely, one can prove that

$$\nu_\alpha(E)=W^{-\alpha}\nu(E)\equiv\frac{1}{\Gamma(\alpha)}\int_0^{+\infty}\sigma^{\alpha-1}\nu(E-i\sigma)\,d\sigma$$

for any Borel set $E$ such that $\overline{E}\subset G^-$ (see [66]).

## 2. SUBHARMONIC FUNCTIONS WITH NONPOSITIVE LIOUVILLE PRIMITIVES

**Definition 2.1.** *By $s_\alpha$ $(0\le\alpha<+\infty)$ we denote the class of functions $U(w)$ which are subharmonic in $G^-$ and satisfy the following conditions:*

(i) $U(w)\in M_p$ *for the integer* $p\in[\alpha,1+\alpha)$,

(ii) $W^{-\alpha}u(z)\le 0,\ w\in G^-$,

(iii) *the associated measure of $U(w)$ is supported in a neighborhood of the origin,*

(iv) *if $\alpha=p\ge 0$ is an integer, then*

$$\int_1^{+\infty}W^{-\alpha}U(-it)\,\frac{dt}{t}>-\infty.$$

*2.1.* Our aim is to find Riesz type representations for functions from $s_\alpha$.

**Lemma 2.1.** *If $U(w)\in s_\alpha$ $(\alpha\ge 0)$, then its associated measure $\nu(\zeta)$ satisfies (1.1) and*

$$U(w)=\iint_{G^-}\log|b_\alpha(w,\zeta)|\,d\nu(\zeta)+U_*(w),\quad w\in G^-,\tag{2.1}$$

*where $U_*(w)$ is a harmonic function of $s_\alpha$.*

**Proof.** Let $\alpha > 0$, and let $D$ be a bounded domain such that $\overline{D} \subset G^-$. Then

$$U_D(w) = U(w) - \iint_D \log |b_\alpha(w, \zeta)| \, d\nu(\zeta)$$

is a function subharmonic in $G^-$ and harmonic in $D$, and by Lemma 1.3 $U_D(w) \in M_p$. Therefore

$$W^{-\alpha} U_D(w) = W^{-\alpha} U(w) - \iint_D W^{-\alpha} \log |b_\alpha(w, \zeta)| \, d\nu(\zeta), \quad w \in G^-, \tag{2.2}$$

by Lemma 1.2, and consequently

$$W^{-\alpha} U_D(w) \leq - \iint_D W^{-\alpha} \log |b_\alpha(w, \zeta)| \, d\nu(\zeta) = V_D(w),$$

where $V_D(w)$ is a nonnegative superharmonic function, as proved in Lemma 1.3. The representation (1.12) yields

$$\sup_{v<0} \int_{-\infty}^{+\infty} \left[W^{-\alpha} U_D(u + iv)\right]^+ du \leq 2\pi \iint_{G^-} |\eta| \, d\nu_\alpha(\zeta) < +\infty.$$

In the same way we obtain

$$\lim_{v \to +0} \int_{-\infty}^{+\infty} \left[W^{-\alpha} U_D(u + iv)\right]^+ du = 0.$$

Consequently, $W^{-\alpha} U_D(w) \leq 0$ in $G^-$ by Theorem 1.1 of Ch. 7 (see also Remark 1.2 after that theorem). By (2.2) $V_D(w) \leq -W^{-\alpha} U(w)$ $(w \in G^-)$. But the measure $\nu(\zeta)$ vanishes when $|\zeta|$ is large enough, for example when $|\zeta| > R_*$. Therefore the representation (2.2) of Ch. 2 gives

$$+\infty > -W^{-\alpha} U(-2iR_*) \geq \frac{1}{\Gamma(1 + \alpha)} \iint_D d\nu(\zeta) \int_{-|\eta|}^{|\eta|} \frac{(|\eta| - |t|)^\alpha (2R_* - t)}{(2R_* - t)^2 + \xi^2} \, dt$$
$$\geq \frac{1}{5R_* \Gamma(2 + \alpha)} \iint_D |\eta|^{1+\alpha} \, d\nu(\zeta).$$

Exhausting $G^-$ by means of finite domains $D$ we arrive at (1.1). Further, defining $U_*(w)$ by (2.1) and using Lemma 1.3 we obtain $U_*(w) \in s_\alpha$. This completes the proof since our assertions are obvious when $\alpha = 0$.

*2.2.* We are ready to prove our main theorem on descriptive representations of classes $s_\alpha$.

**Theorem 2.1.** *The class $s_\alpha$ $(0 \le \alpha < +\infty)$ coincides with the set of functions representable in the form*

$$U(w) = \iint_{G^-} \log|b_\alpha(w,\zeta)|\, d\nu(\zeta) - \operatorname{Re} \frac{\Gamma(1+\alpha)}{\pi} e^{-i\frac{\pi}{2}(1+\alpha)} \int_{-\infty}^{+\infty} \frac{d\mu(t)}{(w-t)^{1+\alpha}}, \quad w \in G^-, \tag{2.3}$$

*where $\nu(\zeta)$ is a nonnegative Borel measure in $G^-$, supported in a neighborhood of the origin and such that*

$$\iint_{G^-} |\operatorname{Im}\, \zeta|^{1+\alpha}\, d\nu(\zeta) < +\infty, \tag{2.4}$$

*while $\mu(t)$ is a nondecreasing function for which*

$$\int_{-\infty}^{+\infty} \frac{d\mu(t)}{1+|t|^{1+\alpha-p}} < +\infty. \tag{2.5}$$

**Proof.** Let $u(z) \in s_\alpha$ $(0 \le \alpha < +\infty)$. By Lemma 2.1 the measure associated with $U(w)$ satisfies (2.4) and $U(w)$ is representable in the form (2.1), where $U_*(w) \in s_\alpha$ is a harmonic function. Using Theorem 3.5 of Ch. 3, we conclude that $U_*(w)$ can be written in the form of the last integral of (2.3), where $\mu(t)$ satisfies the mentioned conditions. Conversely, let $U(w)$ be representable in the form (2.3)–(2.5). Then using Theorem 3.5 of Ch. 3 we conclude that the last integral in (2.3) is a harmonic function of $s_\alpha$. At the same time, by Lemma 1.3 also the first term on the right–hand side of (2.3) is a function of $s_\alpha$. Hence $U(w) \in s_\alpha$.

**Remark 2.1.** Using Lemma 3.3 and formula (2.10) of Ch. 1, we find that the inclusion $U(w) \in s_\alpha$ $(0 \le \alpha < +\infty)$ implies that for $w = u + iv \in G^-$

$$W^{-\alpha} u(z) = \iint_{G^-} W^{-\alpha} \log|b_\alpha(w,\zeta)|\, d\nu(\zeta) + \frac{v}{\pi} \int_{-\infty}^{+\infty} \frac{d\mu(t)}{(u-t)^2+v^2}, \tag{2.6}$$

where $\nu(\zeta)$ and $\mu(t)$ are the same as in representation (2.3)–(2.5). By (1.12) for $w = u + iv \in G^-$

$$W^{-\alpha} u(z) = \iint_{G^-} \log|b_0(w,\zeta)|\, d\nu_\alpha(\zeta) + \frac{v}{\pi} \int_{-\infty}^{+\infty} \frac{d\mu(t)}{(u-t)^2+v^2}, \tag{2.7}$$

where $\nu_\alpha(\zeta)$ is some nonnegative Borel measure in $G^-$, satisfying (2.4) for $\alpha = 0$.

**Remark 2.2.** Consider the class $S_\alpha$ $(0 \le \alpha < +\infty)$ of those functions subharmonic in $G^-$, which are representable as differences of two functions of $s_\alpha$. Evidently $S_\alpha$ coincides with the set of those functions $U(w)$ subharmonic in $G^-$, whose Liouville $\alpha$–primitives (in imaginary variable) are representable in the form (2.7), with measures $\nu_\alpha(\zeta)$ from certain class, while $\mu(t)$ is the difference of any two nondecreasing functions satisfying (2.5). This representation describes a subclass of functions which are subharmonic and have nonnegative harmonic majorants in $G^-$.

## 3. THE GROWTH OF FUNCTIONS FROM $S_\alpha$

*3.1.* We aim at proving

**Theorem 3.1.** 1°. *The class $S_\alpha$ $(0 \leq \alpha < +\infty)$ coincides with the set of those functions $u(z) \in M_p$ $(\alpha \leq p < 1+\alpha)$ subharmonic in $G^-$, which satisfy the conditions*

$$\liminf_{R\to+\infty} \frac{1}{R} \int_0^\pi \left|W^{-\alpha} U(Re^{-i\vartheta})\right| \sin\vartheta \, d\vartheta = 0, \tag{3.1}$$

$$\lim_{R\to+\infty} \liminf_{y\to-0} \int_{-R}^{R} \left|W^{-\alpha} U(u+iv)\right| \frac{du}{1+|u|^{1+\alpha-p}} < +\infty, \tag{3.2}$$

*and associated measures of which are supported in the neighborhood of the origin.*

2°. *If $U(w) \in S_\alpha$ $(0 \leq \alpha < +\infty)$, then*

$$\lim_{R\to+\infty} \frac{1}{R} \int_0^\pi \left|W^{-\alpha} U(Re^{-i\vartheta})\right| \sin\vartheta \, d\vartheta = 0. \tag{3.3}$$

*The function $\mu(t)$ in the representations (2.3)–(2.4) and (2.6)–(2.7) can be found from the relations*

$$\mu_\pm(t) = \lim_{v\to-0} \int_0^t \left[W^{-\alpha} U(u+iv)\right]^\pm du, \quad -\mu(t) = \mu_+(t) - \mu_-(t), \tag{3.4}$$

*where the functions $\mu_\pm(t)$ are nondecreasing and satisfy (2.5). If $\mu_\pm(t)$ are recovered by (3.4), then for any $[a,b] \subset (-\infty,+\infty)$ and for any function $g(x)$ continuous in $[a,b]$*

$$\lim_{v\to-0} \int_a^b \left[W^{-\alpha} U(u+iv)\right]^\pm g(x)\, dx = \int_a^b g(u)\, d\mu_\pm(u). \tag{3.5}$$

*Besides,*

$$\lim_{v\to-0} \int_{-\infty}^{+\infty} \left[W^{-\alpha} U(u+iv)\right]^\pm \frac{du}{1+|u|^{1+\alpha-p}} = \int_{-\infty}^{+\infty} \frac{d\mu_\pm(u)}{1+|u|^{1+\alpha-p}}. \tag{3.6}$$

**Remark 2.3.** Evidently, $U(w) \in S_\alpha$ $(0 \leq \alpha < +\infty)$ implies $U(w) \in s_\alpha$ if and only if one of relations (3.4)–(3.6) gives $\mu_-(t) \equiv 0$ $(-\infty < t < +\infty)$. If both $\mu_\pm(t) \equiv 0$ then $U(w)$ is precisely a Green type potential.

*3.2.* To prove Theorem 3.1, we need the following lemma.

**Lemma 3.1.** *Let $\alpha \geq 0$ and let $U(w) \in M_p$ $(\alpha \leq p < 1+\alpha)$ be a function subharmonic in $G^-$ and satisfying (3.1) and (3.2). If the measure $\nu(\zeta)$ associated with $U(w)$ vanishes for $|\zeta| > R_*$, then $\nu(\zeta)$ satisfies (2.4).*

**Proof.** Let $\rho < 0$ be any number. Then the function

$$U_\rho(w) = U(w) - \iint_{G^-} \log|b_\alpha(w,s)|\, d\nu(s)$$

is harmonic in $G_\rho^-$ and is of $M_\rho$ by Lemma 1.3. By the same lemma,

$$W^{-\alpha} U_\rho(w) = W^{-\alpha} u(z) - \iint_{G_\rho^-} W^{-\alpha} \log|b_\alpha(w,s)|\, d\nu(s).$$

is harmonic in $G_\rho^-$. Writing the difference of two Poisson representations of $W^{-\alpha}U_\rho(w)$, in $G_\rho^-(R_0) = \{w \in G_\rho^- : |w| < R_0\}$ and in $G_\rho^-(R)$ $(R > R_0 > R_*)$, we arrive at the equality

$$\iint_{G_\rho^-} [I_{\rho,R}(w,s) - I_{\rho,R_0}(w,s)]\, d\nu(s) = \frac{1}{2\pi} \int_{\partial G_\rho^-(R)} W^{-\alpha}U(\zeta) \frac{\partial g_{\rho,R}(\zeta,w)}{\partial n} |d\zeta|$$
$$- \frac{1}{2\pi} \int_{\partial G_\rho^-(R_0)} W^{-\alpha}U(\zeta) \frac{\partial g_{\rho,R_0}(\zeta,w)}{\partial n} |d\zeta|, \quad w \in G_\rho^-(R_0), \qquad (3.7)$$

where

$$I_{\rho,R}(w,s) = \frac{1}{2\pi} \int_{\partial G_\rho^-(R)} W^{-\alpha} \log|b_\alpha(\zeta,s)| \frac{\partial g_{\rho,R}(\zeta,w)}{\partial n} |d\zeta| \quad (R = R_0, R)$$

and $g_{\rho,R_0}$, $g_{\rho,R}$ are the Green functions of $G_\rho^-(R_0)$ and $G_\rho^-(R)$. We conclude that $I_{\rho,R}(w,s) - I_{\rho,R_0}(w,s) \ge 0$ when $w \in G_\rho^-(R_0)$ and $s \in G_\rho^-$, since $W^{-\alpha}\log|b_\alpha(\zeta,s)|$ is subharmonic in $G^-$ for any fixed $s \in G^-$. Now assume that $R$ and $R_0$ $(R > R_0 > R_*)$ are such that

$$\int_0^\pi \left|W^{-\alpha}U(Re^{-i\vartheta})\right| \sin\vartheta\, d\vartheta < +\infty \quad (R = R_0, R).$$

By Lemmas 2.1 and 2.2 of Ch. 7, the right–hand side of (3.7) remains bounded for a sequence $\rho = \rho_n \uparrow 0$. Hence, by Fatou's lemma

$$\iint_{G_{\rho_0}^-} \liminf_{\rho\to -0} [I_{\rho,R}(w,s) - I_{\rho,R_0}(w,s)]\, d\nu(s) \le M(w) < +\infty,$$

where $M(w) > 0$ is a constant depending only on $w, R_0$ and $R$, while $\rho_0 < 0$ has sufficiently small modulus. Letting here $\rho_0 \to -0$, we obtain

$$\iint_{G^-} \liminf_{\rho\to -0} \left[I_{\rho,R}(w,s) - I_{\rho,R_0}(w,s)\right] d\nu(s) \le M(w) < +\infty. \qquad (3.8)$$

But for any $R > 0$ there exists

$$\lim_{\rho\to -0} I_{\rho,R}(w,s) = \frac{1}{2\pi i}\int_{\partial G^-(R)} W^{-\alpha}\log|b_\alpha(\zeta,s)|\left[\frac{1}{\zeta-w}-\frac{1}{\zeta-\overline{w}}\right.$$
$$\left.-\frac{w}{R^2-\zeta w}+\frac{\overline{w}}{R^2-\zeta\overline{w}}\right]d\zeta$$
$$= -\operatorname{Im}\frac{1}{\Gamma(1+\alpha)}\int_{-|\operatorname{Im} s|}^{|\operatorname{Im} s|}(|\operatorname{Im} s|-|t|)^\alpha J_R(t,w)\,dt,$$

where (see (2.2) of Ch. 2)

$$J_R(t,w) = \frac{1}{2\pi i}\int_{\partial G^-(R)}\frac{1}{\zeta-(\operatorname{Re} s+it)}\left[\frac{1}{\zeta-w}-\frac{1}{\zeta-\overline{w}}\right.$$
$$\left.-\frac{w}{R^2-\zeta w}+\frac{\overline{w}}{R^2-\zeta\overline{w}}\right]d\zeta.$$

Calculating this integral and returning to (3.8), we arrive at the estimate

$$\frac{2}{\Gamma(1+\alpha)}\iint_{\overline{G^-(R_*)}}\left\{\int_0^{|\operatorname{Im} s|}(|\operatorname{Im} s|-t)^\alpha\left[\frac{\frac{R_0^2}{|v_0|}+t}{\left(\frac{R_0^2}{|v_0|}+t\right)^2+(\operatorname{Re} s)^2}\right.\right.$$
$$\left.\left.-\frac{\frac{R^2}{|v_0|}+t}{\left(\frac{R^2}{|v_0|}+t\right)^2+(\operatorname{Re} s)^2}\right]dt\right\}d\nu(s)\le M(-iv_0)<+\infty,$$

where $M(-iv_0) > 0$ depends only on $v_0$ $(-R_* < v_0 < 0)$ and $R, R_0$. Taking $R$ sufficiently large, we obtain that $\nu(\zeta)$ satisfies (2.4).

*3.3.* **Proof of Theorem 3.1.** Let $U(w) \in M_\rho$ be a function subharmonic in $G^-$ and satisfying conditions (3.1), (3.2) for an $\alpha \in (p-1, p]$, and let the associated measure $\nu(\zeta)$ of $U(w)$ vanishes in a neighborhood of infinity. Then $\nu(\zeta)$ satisfies (2.4) according to Lemma 3.1. Therefore the function

$$U_*(w) = U(w) - \iint_{G^-}\log|b_\alpha(w,\zeta)|\,d\nu(\zeta) \tag{3.9}$$

is harmonic in $G^-$. By Lemma 1.3, $U_*(w) \in M_p$ and

$$W^{-\alpha}U_*(w) = W^{-\alpha}u(z) - \iint_{G^-}W^{-\alpha}\log|b_\alpha(w,\zeta)|\,d\nu(\zeta)$$

is harmonic in $G^-$. Besides, this function satisfies (3.1) and (3.2) since

$$W^{-\alpha}I_\alpha(w) = -\iint_{G^-}W^{-\alpha}\log|b_\alpha(w,\zeta)|\,d\nu(\zeta)$$

satisfies (3.1) and (3.2). Moreover, the representation (1.12) implies that

$$\lim_{R\to+\infty}\frac{1}{R}\int_0^{\pi} W^{-\alpha}I_\alpha(Re^{-i\vartheta})\sin\vartheta\,d\vartheta = 0 \tag{3.10}$$

and

$$\lim_{v\to-0}\int_{-\infty}^{+\infty} W^{-\alpha}I_\alpha(u+iv)\,du = 0. \tag{3.11}$$

Hence, by Theorem 1.1 of Ch. 7

$$W^{-\alpha}U_*(w) = \frac{v}{\pi}\int_{-\infty}^{+\infty}\frac{d\mu(t)}{(u-t)^2+v^2}, \quad w = u+iv \in G^-, \tag{3.12}$$

where $\mu(t)$ is the measure determined from relations (3.4) and satisfying (3.5), (3.6). In addition, (3.3) is obviously true. Applying $\frac{\partial^p}{\partial v^p}W^{-(p-\alpha)}$ to both sides of (3.12) and using the results of Ch. 1, we obtain a representation of the form (2.3). Therefore, Theorem 1.1 leads to the conclusion that $U(w) \in S_\alpha$. Conversely, by Remark 2.1, for any $U(w) \in S_\alpha$ the function $W^{-\alpha}u(z)$ is representable in the form (2.6). Now relations (3.3) and (3.6) follow from (3.10), (3.11) and Theorem 1.1 of Ch. 7.

NOTES. The inversion $z = w^{-1}$ transforms our requirement on boundedness of the support of associated measures in classes $s_\alpha$ and $S_\alpha$ to the requirement that the supports of measures are disjoint from the origin. This is a natural requirement for the classical Blaschke product with factors of the form $(1-z/\zeta)/(1-z/\bar{\zeta})$ in $G^+$ and its generalizations.

# CHAPTER 9

# FUNCTIONS OF $\alpha$-BOUNDED TYPE IN SPECTRAL THEORY OF NON-WEAK CONTRACTIONS

## 1. FACTORIZATION OF REGULARIZED DETERMINANTS

*1.1.* For any $p \geq 1$ we denote by $C_p$ the class of continuously invertible contractions $T$ in a separable Hilbert space $\mathfrak{H}$ for which the operator $D_T^2 = I - T^*T$ belongs to the Neuman–Schatten ideal $\mathfrak{S}_p$. The set $C_1$ coincides with the class of all invertible weak contractions [104]. We define the characteristic function $W_T$ of the operator $T$ as in [12]:

$$W_T(z)W_T(0) = [I - D_T(I - zT)^{-1}D_T] \,|\, \mathfrak{D}_T,$$
$$W_T(0) = (T^*T)^{1/2} \,|\, \mathfrak{D}_T, \quad \mathfrak{D}_T = \overline{D_T\mathfrak{H}}.$$

It is easy to verify that the operator-function $W_T^*(\overline{z})$ differs from the characteristic function $\Theta_T(z)$ of B.Sz.-Nagy and C.Foias [104] by a constant isometric factor. Let us recall from [104, 12] that $W_T(z)$ is holomorphic in $|z| < 1$, where its values are two-sided contractions in $\mathfrak{D}_T$, i.e. $W_T^*(z)W_T(z) \leq I$ and $W_T(z)W_T^*(z) \leq I$ in $|z| < 1$. Since

$$\begin{aligned} I - W_T^{-1}(0) &= W_T^{-1}(0)\left[(T^*T)^{1/2} - I\right] \,|\, \mathfrak{D}_T \\ &= W_T^{-1}(0)\left[I + (T^*T)^{1/2}\right]^{-1}(T^*T - I) \,|\, \mathfrak{D}_T \end{aligned}$$

and $I - W_T(z) = I - W_T^{-1}(0) + D_T(I - zT)^{-1}D_TW_T^{-1}(0)$, $D_T \in \mathfrak{S}_{2p}$, the operator $I - W_T(z)$ belongs to $\mathfrak{S}_p$ for any $z \notin \sigma(T^{-1})$. Hence for each $p \geq 1$ the regularized determinant

$$d_T(z) = \det\nolimits_p W_T(z) = \prod_k \lambda_k(z) \exp\left\{ \sum_{j=1}^{p-1} \frac{1}{j}[1 - \lambda_k(z)]^j \right\} \tag{1.1}$$

is holomorphic wherever the operator-function $W_T$ is holomorphic [37, Ch. IV]. In (1.1), $\{\lambda_k(z)\}$ is the set of eigenvalues of the operator $W_T(z)$. Further, the functions

$$\mathcal{D}_T = \det\nolimits_p W_T(z)W_T^*(\overline{z}) \tag{1.2}$$

will play an important role, and one can state that formulas (1.1) and (1.2) give a correspondence between the operators $T \in C_p$ (where $p$ is a natural number) and the functions $d_T$ and $\mathcal{D}_T$ which are holomorphic in $|z| < 1$.

*1.2.* The main result of this section states that $d_T$ and $\mathcal{D}_T$ both belong to M.M.Djrbashian's class $N_\alpha$ [19, Ch. IX]. As it is well known, the functions $d_T$ and $\mathcal{D}_T$ are bounded in $|z| < 1$ if $p = 1$. For the general case we have

**Theorem 1.1.** *If $p \geq 2$ is an integer and $T \in C_p$, then the holomorphic functions $d_T$ and $\mathcal{D}_T$ belong to $N_{p-1+\varepsilon}$ for any $\varepsilon > 0$.*

For proving this theorem we need

**Lemma 1.1.** *If $p \geq 2$ is an integer, then for $0 < r < 1$*

$$\frac{1}{2\pi} \int_{|z|=r} \|D_T(I - zT)^{-1} D_T\|_p^p |dz| \leq (1-r)^{-(p-1)} \|D_T^2\|_p^p, \tag{1.3}$$

*where $\|\cdot\|_p$ is the norm in $\mathfrak{S}_p$.*

**Proof.** First, suppose $p = 2^k$ ($k \geq 1$). By the elementary properties of eigenvalues ($\lambda_j$) and singular values ($s_j$) of compact operators [37, Ch. II]

$$\begin{aligned}
\|D_T(I - zT)^{-1} D_T\|_p^p &= \sum_j s_j^{p/2} \left(D_T(I - zT)^{-1} D_T^2 (I - \bar{z}T^*)^{-1} D_T\right) \\
&= \sum_j \lambda_j^{p/2} \left(D_T(I - zT)^{-1} D_T^2 (I - \bar{z}T^*)^{-1} D_T\right) \\
&= \sum_j \lambda_j^{p/2} \left(D_T^2(I - zT)^{-1} D_T^2 (I - \bar{z}T^*)^{-1}\right) \\
&\leq \sum_j s_j^{p/2} \left(D_T^2(I - zT)^{-1} D_T^2 (I - \bar{z}T^*)^{-1}\right) \\
&\leq \|(I - \bar{z}T^*)^{-1}\|^{p/2} \sum_j s_j^{p/2} \left(D_T^2(I - zT)^{-1} D_T^2\right) \\
&\leq (1-r)^{-p/2} \|D_T^2(I - zT)^{-1} D_T^2\|_{p/2}^{p/2} \quad (|z| = r).
\end{aligned}$$

Using these relations $k - 1$ times we get

$$\begin{aligned}
\|D_T(I - zT)^{-1} D_T\|_p^p &\leq (1-r)^{-p\sum_{j=1}^{k-1} 2^{-j}} \|D_T^{2^{k-1}} (I - zT)^{-1} D_T^{2^{k-1}}\|_2^2 \\
&= (1-r)^{-(p-2)} \|Q(I - zT)^{-1} Q\|_2^2,
\end{aligned}$$

where $Q = D_T^{p/2}$. Further,

$$\int_{|z|=r} \|D_T(I-zT)^{-1}D_T\|_p^p |dz|$$

$$\leq (1-r)^{-(p-2)} \int_{-\pi}^{\pi} \sum_{j,k\geq 0} \mathrm{Sp}\big(QT^jQ^2(T^*)^kQ\big) r^{j+k} e^{i(j-k)\vartheta} d\vartheta$$

$$= 2\pi(1-r)^{-(p-2)} \sum_{j=0}^{\infty} \mathrm{Sp}\big(QT^jQ^2(T^*)^kQ\big) r^{2j}$$

since

$$\|Q(I-zT)^{-1}Q\|_2^2 = \mathrm{Sp}\big(Q(I-zT)^{-1}Q^2(I-\overline{z}T^*)^{-1}Q\big)$$
$$= \sum_{j,k\geq 0} \mathrm{Sp}\big(QT^jQ^2(T^*)^jQ\big) z^j\overline{z}^k.$$

Recalling that $Q = D_T^{p/2}$, we get

$$\mathrm{Sp}\big(QT^jQ^2(T^*)^jQ\big) = \mathrm{Sp}\big(Q^2T^jQ^2(T^*)^j\big) \leq \|Q^2T^j\|_2 \|Q^2(T^*)^j\|_2$$
$$\leq \|Q^2\|_2^2 = \|D_T^p\|_2^2 = \|D_T^2\|_p^p.$$

Hence (1.3) holds for $p = 2^k$ $(k \geq 1)$. Note that the above argument remains valid if we replace $D_T$ by any normal operator $\Theta$ with the spectral decomposition

$$\Theta = \sum_i \varphi_i(\,\cdot\,, e_k)e_k, \quad \{\varphi_i\} \subset \mathbb{C}, \quad \sum_i |\varphi_i|^{2p} < +\infty.$$

We shall use this later. Returning to the proof of (1.3), for any $p \geq 2$ we shall use the well known Hadamard theorem on three lines and some techniques similar to one worked out in the proof of the Riesz-Thorin theorem on interpolation of operators [7]. First note that if the values of a function $F(\varphi)$ are in $\mathfrak{S}_p$, then

$$\sup_G \left| \int_{-\pi}^{\pi} \mathrm{Sp}\{F(\varphi)G(\varphi)\} d\varphi \right| = \left\{ \int_{-\pi}^{\pi} \|F(\varphi)\|_p^p d\varphi \right\}^{1/p}, \tag{1.4}$$

where the supremum is taken over all $\mathfrak{S}_q$-valued functions $G(\varphi)$ for which

$$\int_{-\pi}^{\pi} \|G(\varphi)\|_q^q \, d\varphi = 1, \qquad \frac{1}{p} + \frac{1}{q} = 1.$$

Of course, we presupposed that $F$ and $G$ satisfy the standard requirements providing the existence of the integrals in (1.4). It is easy to verify that (1.4) is an immediate consequence of the precise estimate $|\mathrm{Sp}\{FG\}| \leq \|F\|_p \|G\|_q$ (see [37]) and Hölder's inequality.

For interpolating the inequality (1.3) (already proved for $p = 2^{k-1}, 2^k$) for all $p \in [2^{k-1}, 2^k]$, denote $p_0 = 2^{k-1}$, $p_1 = 2^k$ and similar to the proof of the Riesz–Thorin theorem set

$$\frac{1}{p(z)} = \frac{1-z}{p_0} + \frac{z}{p_1}, \quad \frac{1}{q(z)} = \frac{1-z}{q_0} + \frac{z}{q_1}, \quad \frac{1}{p_j} + \frac{1}{q_j} = 1 \quad (j = 0, 1)$$

and for any $n \geq 1$ consider the operator-function

$$\Phi_n(z) = \sum_{k=1}^{n} s_k^{\frac{p}{p(z)}}(\,\cdot\,, e_k)e_k, \quad G_n(\varphi, z) = \sum_{j=1}^{n} a_j^{\frac{q}{q(z)}}(\varphi)(\,\cdot\,, v_j(\varphi))u_j(\varphi),$$

where $1/p + 1/q = 1$ and $\{e_k\}$, $\{u_j(\varphi)\}$, $\{v_j(\varphi)\}$ are some weak measurable orthonormal sets of vectors, $s_k > 0$ and $a_j(\varphi)$, $[a_j(\varphi)]^{-1}$ are nonnegative, bounded, measurable functions. In addition, we suppose that

$$\sum_{k=1}^{n} s_k^{2p} = 1 \quad \text{and} \quad \int_{-\pi}^{\pi} \sum_{j=1}^{n} a_j^q(\varphi)d\varphi = 1$$

for any fixed $n \geq 1$. Now introduce the entire function

$$\begin{aligned} f_r(z) &= \frac{1}{2\pi} \int_{-\pi}^{\pi} \mathrm{Sp}\{\Phi_n(z)(I - re^{i\varphi}T)^{-1}\Phi_n(z)G_n(\varphi, z)\}d\varphi \\ &= \frac{1}{2\pi} \sum_{k,m,j=1}^{n} s_k^{\frac{p}{p(z)}} s_m^{\frac{p}{p(z)}} \int_{-\pi}^{\pi} a_j^{\frac{q}{q(z)}}(\varphi)P_{k,m,j}(\varphi)d\varphi, \end{aligned}$$

where $P_{k,m,j}(\varphi) = (e_k, v_j(\varphi))(u_j(\varphi), e_m)((I - re^{i\varphi}T)^{-1}e_m, e_k)$. The inequality(1.3) is true for $p = p_0$ and for the normal operator $\Phi_n(z)$. Consequently

$$\begin{aligned} |f_r(iy)| &\leq (2\pi)^{\frac{1}{p_0}-1}\left\{\frac{1}{2\pi}\int_{-\pi}^{\pi} \|\Phi_n(iy)(I - re^{i\varphi}T)^{-1}\Phi_n(iy)\|_{p_0}^{p_0} d\varphi\right\}^{1/p_0} \\ &\quad \times \left\{\int_{-\pi}^{\pi} \|G_n(\varphi)\|_{q_0}^{q_0} d\varphi\right\}^{1/q_0} \\ &\leq (2\pi)^{-\frac{1}{q_0}}\|\Phi_n^2(iy)\|_{p_0}(1-r)^{-1+1/p_0}\left\{\int_{-\pi}^{\pi} \|G_n(\varphi, iy)\|_{q_0}^{q_0} d\varphi\right\}^{1/q_0} \\ &= (2\pi)^{-\frac{1}{q_0}}\left\{\sum_{k=1}^{n}\left|s_k^{\frac{2p}{p(iy)}}\right|^{p_0}\right\}^{1/p_0}(1-r)^{-\frac{1}{q_0}}\left\{\int_{-\pi}^{\pi}\sum_{j=1}^{n}\left|a_j^{\frac{q}{q(iy)}}(\varphi)\right|^{q_0} d\varphi\right\}^{1/q_0} \\ &= (2\pi)^{-\frac{1}{q_0}}\left\{\sum_{k=1}^{n} s_k^{2p}\right\}^{1/p_0}(1-r)^{-\frac{1}{q_0}}\left\{\int_{-\pi}^{\pi}\sum_{j=1}^{n} a_j^q(\varphi)d\varphi\right\}^{1/q_0} = [2\pi(1-r)]^{-\frac{1}{q_0}}. \end{aligned}$$

Taking $p = p_1$ we similarly come to the inequality $|f_r(1+iy)| \leq [2\pi(1-r)]^{-1/q_1}$. Consequently,

$$|f_r(\vartheta)| \leq [2\pi(1-r)]^{-\frac{1-\vartheta}{q_0}-\frac{\vartheta}{q_1}}, \quad 0 \leq \vartheta \leq 1.$$

Now let $p \in (2^{k-1}, 2^k)$ be arbitrary and let the corresponding $\vartheta \in (0,1)$ be chosen from

$$\frac{1}{p} = \frac{1-\vartheta}{p_0} + \frac{\vartheta}{p_1}, \quad \frac{1}{q} = \frac{1-\vartheta}{q_0} + \frac{\vartheta}{q_1} \quad \left(\frac{1}{p} + \frac{1}{q} = 1\right).$$

Then obviously

$$\Phi_n(\vartheta) = \sum_{k=1}^{n} s_k(\,\cdot\,, e_k)e_k, \quad G_n(\varphi, \vartheta) = \sum_{j=1}^{n} a_j(\varphi)(\,\cdot\,, v_j(\varphi))u_j(\varphi) \tag{1.6}$$

and therefore

$$\left|\frac{1}{2\pi}\int_{-\pi}^{\pi} \operatorname{Sp}\{\Phi_n(\vartheta)(I - re^{i\varphi}T)^{-1}\Phi_n(\vartheta)G_n(\varphi,\vartheta)\}d\varphi\right| \leq [2\pi(1-r)]^{-1/q}$$

for any $n \geq 1$ and any functions $\Theta_n(\vartheta)$, $G_n(\varphi, \vartheta)$ satisfying (1.5). Using this inequality and (1.4) instead of the first condition of (1.5), we get

$$\frac{1}{2\pi}\int_{-\pi}^{\pi} \|\Phi_n(\vartheta)(I - re^{i\varphi}T)^{-1}\Phi_n(\vartheta)\|_p^p d\varphi \leq \|\Phi_n^2(\vartheta)\|_p^p (1-r)^{-(p-1)}.$$

At last, taking as $\Phi_n(\vartheta)$ the $n$-th sum of the Schmidt series of the operator $D_T$ and letting $n \to \infty$ we come to (1.3).

**Proof of Theorem 1.1.** Using a simple inequality for regularized determinants [31, Ch. XI, Sec. 22], from (1.1) we get $\log^+ |d_T(z)| \leq m_p \|I - W_T(z)\|_p^p$ ($|z| < 1$). Therefore

$$\log^+ |d_T(z)| \leq 2^p M_p \Big\{\|I - W_T^{-1}(0)\|_p^p + \|W_T^{-1}(0)\|^p \|D_T(I - zT)^{-1}D_T\|_p^p\Big\}.$$

Consequently, by (1.3)

$$\begin{aligned}\int_{-\pi}^{\pi} D^{-\alpha}\log^+ |d_T(re^{i\vartheta})|d\vartheta &= D^{-\alpha}\left\{\int_{-\pi}^{\pi}\log^+ |d_T(re^{i\vartheta})|d\vartheta\right\} \\ &\leq D^{-\alpha}\Big[C_1 + C_2(1-r)^{-p+1}\Big] \leq C < +\infty\end{aligned}$$

for any $\alpha > p-1$. Hence $d_T \in A_\alpha^\circ$ $(\alpha > p-1)$. To prove that $\boldsymbol{\mathcal{D}}_T \in A_\alpha^\circ$ for any $\alpha > p-1$, observe that

$$\begin{aligned}\log^+ |\boldsymbol{\mathcal{D}}_T(z)| &\leq M_p \|I - W_T(z)W_T^*(\overline{z})\|_p^p \\ &\leq 2^p M_p\Big\{\|I - W_T(z)\|_p^p + \|I - W_T^*(\overline{z})\|_p^p\Big\} \\ &\leq C_1 + C_2\Big\{\|D_T(I - zT)^{-1}D_T\|_p^p + \|D_T(I - \overline{z}T^*)^{-1}D_T\|_p^p\Big\}.\end{aligned}$$

Application of (1.3) completes the proof.

**Corollary.** *If $p \geq 3$ is an integer, $T \in C_p$ and $I - T^*T \in \mathfrak{S}_\alpha$ for some $\alpha \in [p-1, p)$, then both $d_T$ and $\mathcal{D}_T$ belong to $A^\circ_{p-1}$.*

**Proof.** As Lemma 1.1 is true for any $p \geq 2$, which even may be not an integer, using the above notation we obtain

$$\begin{aligned}\|D_T(I - zT)^{-1}D_T\|_p^p &= \sum_j s_j^p = \sum_j s_j^{p-\alpha} s_j^\alpha \\ &\leq \|D_T(I - zT)^{-1}D_T\|^{p-\alpha} \sum_j s_j^\alpha \\ &\leq 2^{p-\alpha}\|D_T(I - zT)^{-1}D_T\|_\alpha^\alpha.\end{aligned}$$

Hence

$$\begin{aligned}\frac{1}{2\pi}\int_{|z|=r} \|D_T(I - zT)^{-1}D_T\|_p^p|dz| &\leq \frac{2^{p-\alpha}}{2\pi}\int_{|z|=r} \|D_T(I - zT)^{-1}D_T\|_\alpha^\alpha|dz| \\ &\leq 2^{p-\alpha}\|D_T^2\|_\alpha^\alpha(1-r)^{-(\alpha-1)}.\end{aligned}$$

Recalling the proof of Theorem 1.1 and using the fact that the class $A^\circ_\alpha$ enlarges as $\alpha$ increases, we complete the proof.

*1.3.* The following assertion is another corollary of Theorem 1.1.

**Proposition 1.1.** *Let $\{z_k\}$ be a sequence of numbers from $|z| < 1$, such that for a given integer $p \geq 2$*

$$\sum_k (1 - |z_k|)^p < +\infty. \tag{1.7}$$

*Then the products*

$$\begin{aligned} d_0(z) &= \prod_k b_{z_k}(z) \exp\left\{\sum_{k=1}^{p-1} \frac{1}{j}[1 - b_{z_k}(z)]^j\right\}, \\ \mathcal{D}_0(z) &= \prod_k B_k(z) \exp\left\{\sum_{k=1}^{p-1} \frac{1}{j}[1 - B_k(z)]^j\right\}, \end{aligned} \tag{1.8}$$

*where*

$$b_{z_k}(z) = \frac{z_k - z}{1 - \overline{z}_k z}\frac{|z_k|}{z_k} \quad \text{and} \quad B_k(z) = b_{z_k}(z)b_{\overline{z}_k}(z),$$

*are holomorphic functions in $|z| < 1$, which belong to $A^\circ_{p-1+\varepsilon}$ for any $\varepsilon > 0$.*

**Proof.** Consider the normal operator

$$T = \sum_k \overline{z}_k(\,\cdot\,, e_k)e_k.$$

in our Hilbert space $\mathfrak{H}$. Obviously, $T \in C_p$ by (1.7). Further, for $z \notin \sigma(T^{-1}) \cup \sigma(T^{*-1})$ the operators $W_T(z)$ and $W_T(z)W_T^*(\overline{z})$ are normal and the sequences $\{b_{z_k}(z)\}$ and $\{B_k(z)\}$ are their eigenvalues. Hence $d_0(z) = d_T(z)$, $\mathcal{D}_0(z) = \mathcal{D}_T(z)$, and our assertion follows from Theorem 1.1.

Now we shall check the precision of Theorem 1.1. First, it is easy to observe that the discrete spectrum of an operator $T \in C_p$ satisfies (1.7). Thus, the functions $d_T$ and $\mathcal{D}_T$ in general do not belong to $A_\alpha^\circ$ when $\alpha < p-1$. For checking this, observe that the set of zeros of $d_T$ is the discrete spectrum of $T^*$. Indeed, it follows from (1.1) that $d_T(z) = 0$ if and only if $z = 0$ is an eigenvalue of the operator $W_T(z)$, and this is equivalent to noninvertability of $W_T(z)$ since $I - W_T(z) \in \mathfrak{S}_p$. On the other hand, applying the equalities

$$W_T(z)W_T^*(\overline{z}^{-1}) = W_T^*(\overline{z}^{-1})W_T(z) = I, \qquad z \notin \sigma(T^*), \quad |z| < 1,$$

(see [104, 12]) we conclude that the operator $W_T(z)$ ($|z| < 1$) is invertible if and only if $z \notin \sigma(T^*)$. Similarly, $\{z_k\} \cup \{\overline{z}_k\}$ is the set of zeros of $\mathcal{D}_T$, where $\{z_k\}$ is the set of eigenvalues of $T$. Thus, if $d_T$ (or $\mathcal{D}_T$) belongs to $A_\alpha^\circ$ for some $\alpha < p-1$, then

$$\sum_k (1 - |z_k|)^{1+\alpha} < +\infty$$

according to the general assertion on the density of zeros in $A_\alpha^\circ$, and these series obviously does not converge for any $T \in C_p$.

It is not known weather the functions $d_T$ and $\mathcal{D}_T$ belong to $A_{p-1}^\circ$ or even to M.M.Djrbashian's more wider class $N_{p-1}$ for any $T \in C_p$. Also, it is not known weather the products $d_0$ and $\mathcal{D}_0$, constructed by a sequence satisfying (1.7), belong to $N_{p-1}$. But we can state that there exists a sequence $\{z_k\}_1^\infty$ satisfying (1.7) and

$$\sum_k (1 - |z_k|)^{p-\varepsilon} < +\infty$$

for any $\varepsilon > 0$, such that $d_0, \mathcal{D}_0 \in A_{p-1}^\circ$ for the corresponding products.

However, we mention that Theorem 1.1 is as much precise as it is necessary for the further assertions.

*1.4.* Finding factorizations for $d_T$ and $\mathcal{D}_T$, we shall pay the main attention to the function $\mathcal{D}_T$ since its factorization will play more important role in our further considerations. First we note that

$$0 \le \mathcal{D}_T(x) \le 1, \quad -1 < x < 1, \tag{1.9}$$

which immediately follows from $W_T(x)W_T^*(x) \le I$.

**Theorem 1.2.** *If $p \ge 2$ is an integer, $T \in C_p$ and $\{z_k\}$ is the discrete spectrum of $T$, then for any $\varepsilon > 0$*

$$\mathcal{D}_T(z) = \mathcal{D}_0(z) \exp\left\{ -\frac{1}{2\pi} \int_{-\pi}^{\pi} S_{p-1+\varepsilon}(e^{-i\vartheta}z) d\psi_\varepsilon(\vartheta) \right\}, \quad |z| < 1, \tag{1.10}$$

*where $\psi_\varepsilon$ is a real, continuous function of bounded variation in $[-\pi,\pi]$, $\mathcal{D}_0$ is defined by (1.8) and*

$$S_{p-1+\varepsilon}(\zeta)=\Gamma(p+\varepsilon)\left\{\frac{2}{(1-\zeta)^{p+\varepsilon}}-1\right\}.$$

*Each factor of this representation satisfies (1.9) and if it is assumed that $\psi_\varepsilon(-\pi)=0$, then the factorization (1.10) is unique for any $\varepsilon>0$.*

For proving this theorem we need the following

**Lemma 1.2.** *Let $A$ and $B$ be nonnegative contractions and let $A=I-\lambda P$, where $\lambda\in(0,1)$ and $P$ is a one-dimensional orthogonal projector. If $I-B\in\mathfrak{S}_p$ for some integer $p\geq 2$, then*

$$\operatorname{Sp}\left\{\sum_{k=1}^{p-1}\frac{1}{k}[(I-A)^k+(I-B)^k-(I-AB)^k]\right\}\geq 0. \tag{1.11}$$

**Proof.** First, consider the case when $A$ and $B$ are acting in a finite-dimensional space and $B$ is continuously invertible. Let $\{\lambda_k\}_1^n$, $\{\mu_k\}_1^n$, and $\{\nu_k\}_1^n$ be the sets of eigenvalues of the operators $A$, $B$, and $AB$ correspondingly, numerated in decreasing order (note that the spectrum of $AB$ is positive). It is well known [35, Addition] that

$$\prod_{j=1}^m\nu_j\leq\prod_{j=1}^m\lambda_j\mu_j\quad(1\leq m<n-1)\quad\text{and}\quad\prod_{j=1}^n\nu_j=\prod_{j=1}^n\lambda_j\mu_j.$$

In terms of these sequences, the inequality (1.11) takes the form

$$\sum_{j=1}^n\sum_{k=1}^{p-1}\frac{1}{k}\left\{(1-\lambda_j)^k+(1-\mu_j)^k-(1-\nu_j)^k\right\}\geq 0.$$

Observing that

$$\sum_{j=1}^n\sum_{k=1}^{\infty}\frac{1}{k}\left\{(1-\lambda_j)^k+(1-\mu_j)^k-(1-\nu_j)^k\right\}$$
$$=\sum_{j=1}^n(\log\lambda_j+\log\mu_j-\log\nu_j)=-\log\prod_{j=1}^n\frac{\lambda_j\mu_j}{\nu_j}=0,$$

we conclude that (1.11) is equivalent to

$$\sum_{j=1}^n\sum_{k=p}^{\infty}\frac{1}{k}\left\{(1-\lambda_j)^k+(1-\mu_j)^k-(1-\nu_j)^k\right\}\leq 0. \tag{1.12}$$

Now set

$$\Phi(t) = \int_0^t (1-e^{-x})^{p-1}dx, \quad 0 \le t < +\infty,$$

and note that

$$\sum_{k=p}^{\infty} \frac{(1-x)^k}{k} = \int_0^{1-x} \frac{t^{p-1}}{1-t}dt = \int_0^{-\log x} (1-e^{-y})^{p-1}dy = \Phi(-\log x).$$

Thus, denoting $-\log\lambda_j = a_j$, $-\log\mu_j = b_j$ and $-\log\nu_j = c_j$ $(1 \le j \le n)$ one can write (1.12) in the form

$$\sum_{j=1}^{n} \Big\{\Phi(a_j) + \Phi(b_j)\Big\} \le \sum_{j=1}^{n} \Phi(c_j).$$

As $A = I - \lambda P$, where $P$ is a one-dimensional orthogonal projector, we have $a_1 = a_2 = \ldots = a_{n-1} = 0$, $a_n = -\log(1-\lambda)$ and also

$$\sum_{j=1}^{m} b_j = -\log\left(\prod_{j=1}^{m}\mu_j\right) \le -\log\left(\prod_{j=1}^{m}\nu_j\right) = \sum_{j=1}^{m} c_j, \quad 1 \le m \le n-1,$$

and $a_n + \sum_{j=1}^{n} b_j = \sum_{j=1}^{n} c_j$. Consequently

$$\sum_{j=1}^{n}\Big\{\Phi(c_j) - \Phi(b_j)\Big\} = \sum_{j=1}^{n}\int_{b_j}^{c_j}(1-e^{-x})^{p-1}dx,$$

where $c_j \ge b_j$ $(1 \le j \le n)$ and $c_n \ge a_n$ since for each $k(1 \le k \le n)$ we have $\nu_k \le \lambda_k$ and $\nu_k \le \mu_k$ [35, Addition]. Note that at least the first $n-1$ intervals $(b_j, c_j)$ $(1 \le j \le n-1)$ are disjoint. Indeed, let $\mathcal{L}$ be the linear hull of those eigenvectors of $AB$, which correspond to the eigenvalues $\nu_1, \nu_2, \cdots, \nu_{j-1}$, and let $P = (\,\cdot\,, e)e$. Then, using the minimaximal properties of eigenvalues [37, Ch. II] we obtain that

$$\nu_j = \max_{x \perp \mathcal{L}} \frac{(ABx, x)}{(x,x)} \ge \max_{x \perp \mathcal{L}, e} \frac{(Bx, Ax)}{(x,x)} = \max_{x \perp \mathcal{L}, e} \frac{(Bx, x)}{(x,x)} \ge \mu_{j+1}$$

for any $j(1 \le j \le n-1)$. Thus $c_j \le b_{j+1}$. Further,

$$\begin{aligned}\sum_{j=1}^{n}\Big\{\Phi(c_j) - \Phi(b_j)\Big\} &= \sum_{j=1}^{n}\int_{b_j}^{c_j}(1-e^{-x})^{p-1}dx \\ &\ge \int_0^{a_n}(1-e^{-x})^{p-1}dx = \Phi(a_n)\end{aligned}$$

since $(b_j, c_j)$ $(1 \le j \le n)$ are disjoint and the sum of their lengths is $a_n$.

Thus, we proved (1.11) under the assumption that $A$ and $B$ are acting in a finite-dimensional space. It is clear that (1.11) is valid also when $B$ is not invertible. For proving (1.11) in the general case, observe that in virtue of the formula

$$\begin{aligned}\det{}_p AB =& (\det{}_p A)(\det{}_p B)\\ &\times \exp\left\{-\mathrm{Sp}\left(\sum_{k=1}^{p-1}\frac{1}{k}[(I-A)^k+(I-B)^k-(I-AB)^k]\right)\right\}\end{aligned} \quad (1.13)$$

and other simple properties of regularized determinants [37, Ch. IV] the left-hand side of (1.11) depends on the operators $I - A$ and $I - B$ continuously in the $\mathfrak{S}_p$-metric. Consider a monotonely increasing sequence of orthogonal projectors $\{P_n\}_1^\infty$ which strongly tend to $I$ and $P_n A = A P_n$ $(n \ge 1)$. As (1.11) is already proved for $A_n = P_n A P_n$ and $B_n = P_n B P_n$, letting $n \to \infty$ we come to (1.11) in the general case.

*1.5.* **Proof of Theorem 1.2.** As we have proved, $\{z_k\} \cup \{\bar{z}_k\}$ is the zero-set of the function $\mathcal{D}_T$ which belongs to $A^\circ_{p-1+\varepsilon}$ for any $\varepsilon > 0$. In virtue of Proposition 1.1, also the product $\mathcal{D}_0$ constructed by the sequence $\{z_k\}$ belongs to $A^\circ_{p-1+\varepsilon}$. Hence the function $\mathcal{D}_T(z)/\mathcal{D}_0(z)$, which has no zeros in $|z| < 1$, belongs to $A^\circ_{p-1+\varepsilon}$ for any $\varepsilon > 0$, and consequently it allows the representation

$$\mathcal{D}_T(z)/\mathcal{D}_0(z) = e^{i\gamma_\varepsilon} \exp\left\{-\frac{1}{2\pi}\int_{-\pi}^{\pi} S_{p-1+\varepsilon}(e^{-i\vartheta} z)\, d\psi_\varepsilon(\vartheta)\right\}, \quad |z| < 1,$$

where $\operatorname{Im} \gamma_\varepsilon = 0$ and $\psi_\varepsilon$ is a real-valued function of bounded variation in $[-\pi, \pi]$ [19, Ch. IX]. In this formula $\gamma_\varepsilon = 0$ since $\mathcal{D}_T(0), \mathcal{D}_0(0) > 0$. Thus (1.10) is true. For proving that $\psi_\varepsilon$ is continuous, observe that $\log(\mathcal{D}_T/\mathcal{D}_0)$ is holomorphic in $|z| < 1$ and hence it can be expanded in a power series:

$$\log\left(\mathcal{D}_T(z)/\mathcal{D}_0(z)\right) = \sum_{k=0}^{\infty} d_k z^k, \quad |z| < 1.$$

On the other hand, using the power expansion of the kernel $S_{p-1+\varepsilon}$ we find

$$\log\left(\mathcal{D}_T(z)/\mathcal{D}_0(z)\right) = \frac{\Gamma(p+\varepsilon)}{\pi} - \frac{1}{\pi}\sum_{k=1}^{\infty}\frac{\Gamma(p+\varepsilon+k)}{\Gamma(1+k)}\left(\int_{-\pi}^{\pi} e^{-ik\vartheta}\, d\psi_\varepsilon(\vartheta)\right) z^k.$$

Hence

$$-d_k \frac{\Gamma(1+k)}{\Gamma(p+\varepsilon+k)} = \frac{1}{\pi}\int_{-\pi}^{\pi} e^{-ik\vartheta}\, d\psi_\varepsilon(\vartheta), \quad k = 1, 2, \ldots$$

As the same equality is true for any $\varepsilon_1 \in (0,\varepsilon)$, we have

$$|d_k|\frac{\Gamma(1+k)}{\Gamma(p+\varepsilon+k)} = \frac{1}{\pi}\left|\int_{-\pi}^{\pi} e^{-ik\vartheta} d\psi_{\varepsilon_1}(\vartheta)\right| \frac{\Gamma(p+\varepsilon_1+k)}{\Gamma(p+\varepsilon+k)} = o(1)$$

as $k \to \infty$, i.e. the Fourier coefficients of $\psi_\varepsilon$ tend to zero and hence $\psi_\varepsilon$ is continuous on $[-\pi,\pi]$ [122, Ch. III] (note that more might be said on the differential properties of $\psi_\varepsilon$). The uniqueness of the function $\psi_\varepsilon$ follows from the results of [19, Ch. IX]. Moreover, the following inversion formulas are true:

$$\psi_\varepsilon^{(\pm)}(\vartheta_2) - \psi_\varepsilon^{(\pm)}(\vartheta_1) = \lim_{r\to 1-0}\int_{\vartheta_1}^{\vartheta_2}\left[D^{-\alpha}\log\left|\frac{\mathcal{D}_T(re^{i\vartheta})}{\mathcal{D}_0(re^{i\vartheta})}\right|\right]^{\pm} d\vartheta, \tag{1.14}$$

$$\psi_\varepsilon(\vartheta) = \psi_\varepsilon^{(+)}(\vartheta) - \psi_\varepsilon^{(-)}(\vartheta) \ \ (-\pi\le\vartheta_1\le\vartheta_2\le\pi,\ -\pi\le\vartheta\le\pi,\ \alpha=p-1+\varepsilon),$$

where $\psi_\varepsilon^{(\pm)}(\vartheta)$ are nonnegative, continuous, increasing functions defined as the positive and the negative variations of $\psi_\varepsilon$ on $[-\pi,\vartheta]$.

Finally, we shall show that each term of the factorization (1.10) satisfies (1.9). It is well known [12] that to each invariant subspace $\mathfrak{H}_1$ of the operator $T$, on which $T$ is invertible, corresponds a factorization of the characteristic function:

$$W_T(z) = W_1(z)W_2(z), \tag{1.15}$$

where $W_1$ and $W_2$ are holomorphic contractions in $|z|<1$. Now, let $z_1$ be the first eigenvalue of $T$, $Te_1 = z_1e_1$, $\|e_1\| = 1$ and let $\mathfrak{H}_1$ be the one-dimensional invariant subspace born by $e_1$. Then by (1.15) and (1.13)

$$\begin{aligned}\mathcal{D}_T(z) &= \det{}_p\left[W_1(z)W_2(z)W_2^*(\bar z)W_1^*(\bar z)\right] = \det{}_p\left[W_1^*(\bar z)W_1(z)W_2(z)W_2^*(\bar z)\right]\\ &= \det{}_p\left[W_1^*(\bar z)W_1(z)\right]\det{}_p\left[W_2(z)W_2^*(\bar z)\right]\\ &\quad\times\exp\left\{-\mathrm{Sp}\left(\sum_{k=1}^{p-1}\frac{1}{k}\left[(I-A)^k + (I-B)^k - (I-AB)^k\right]\right)\right\},\end{aligned} \tag{1.16}$$

where $A = W_1^*(\bar z)W_1(z)$ and $B = W_2(z)W_2^*(\bar z)$. For finding the elementary factor corresponding to $z_1$, note that

$$\det{}_pW_1^*(\bar z)W_1(z) = \det{}_pW_1(z)W_1^*(\bar z)$$

and use the well known formula [12]

$$W_1(z)W_1^*(\bar z) = I - (1-z^2)D_TP_1(I-zT_1)^{-1}(I-\bar zT_1^*)^{-1}P_1D_T,$$

where $P_1$ is the orthogonal projection on $\mathfrak{H}_1$ and $T_1 = T|\mathfrak{H}_1$. As

$$(1-|z_1|^2)h = (I-T_1^*T_1)h = P_1(I-T^*T)P_1h = P_1D_T^2P_1h,$$
$$(I-zT_1)^{-1}h = (I-zz_1)^{-1}h \quad\text{and}\quad (I-zT_1^*)^{-1}h = (1-z\bar z_1)^{-1}h,$$

for any $h \in \mathfrak{H}_1$, we conclude that the spectrum of the operator $W_1(z)W_1^*(\overline{z})$ coincides with the spectrum of the operator

$$I - (1-z^2)(I - zT_1)^{-1}(I - \overline{z}T_1^*)^{-1}P_1D_T^2P_1$$
$$= I - (1-z^2)(I - zz_1)^{-1}(I - z\overline{z}_1)^{-1}(1-|z_1|^2)P_1$$

having only one eigenvalue $\nu_1(z)$ that differs from 1:

$$\nu_1(z) = 1 - \frac{(1-z^2)(1-|z_1|^2)}{(1-zz_1)(1-z\overline{z}_1)} = \frac{(z-z_1)(z-\overline{z}_1)}{(1-zz_1)(1-z\overline{z}_1)}$$
$$= b_{z_1}(z)b_{\overline{z}_1}(z) = B_1(z).$$

Therefore, obviously

$$\det{}_pW_1(z)W_1^*(\overline{z}) = B_1(z)\exp\left\{\sum_{j=1}^{p-1}\frac{1}{j}[1-B_1(z)]^j\right\}$$

is the desired elementary factor. Further, denoting as $\Phi_k(z)$ the elementary factor corresponding to $z_k$, by (1.16) we obtain

$$\mathcal{D}_T(x)/\Phi_1(x) \leq \det{}_pW_2(x)W_2^*(x), \quad -1 < x < 1. \tag{1.17}$$

Here we used also (1.11) with $A = W_1^*(x)W_1(x)$, $B = W_2(x)W_2^*(x)$, which, as one can easily verify, satisfy the conditions of Lemma 1.2. Since $W_2$ is also a characteristic function of a contraction from $C_p$ [12], the above argument is applicable for $\mathcal{D}_2(z) = \det_pW_2(z)W_2^*(\overline{z})$. By (1.17) this gives

$$0 \leq \mathcal{D}_T(x)/[\Phi_1(x)\Phi_2(x)] \leq \det{}_pW_3(x)W_3^*(x) \leq 1, \quad -1 < x < 1,$$

since $W_3$ is a holomorphic contraction. Taking away the zeros of $\mathcal{D}_T$ in this way, we come to the desired statement, and the proof is complete.

The following question naturally arises: *in which case the exponential factor in (1.10) is absent?* The answer is given by

**Proposition 1.2.** *Let $T$ be a completely non-unitary contraction from $C_p$ ($p \geq 2$). Then its regularized determinant $\mathcal{D}_T$ is exactly the product $\mathcal{D}_0$ of (1.10) if and only if $T$ is complete and normal.*

We omit the proof of this statement since it will not be used in the future.

Note that, as it was stated in the proof of Theorem 1.2, each discrete factor of the factorization (1.10) is a regularized determinant constructed by a divisor of the characteristic operator-function $W_T$, corresponding to an eigenvalue of $T$. So, the considered Blaschke-type factors have a definite spectral interpretation and their application is reasonable in operator theory.

In [19] M.M.Djrbashian used Blaschke-type factors of different form, which permit to obtain inversion formulas (1.14) without division by $\mathcal{D}_0$. Note that products of nearly same type as in (1.10) were introduced in the early works of M.M.Djrbashian [16, 17].

## 2. BOUNDARY BEHAVIOR OF REGULARIZED DETERMINANTS

In this section we find a connection between the behavior of the function $\mathcal{D}_T$ near the boundary point $z = 1$ and some spectral properties of operators $T \in C_p$. Namely, our aim is to prove

**Theorem 2.1.** *Let $p \geq 2$ be an integer, let 1 be not an eigenvalue for an operator $T \in C_p$ and let $\mathcal{D}_T$ be the regularized determinant (1.2). Then the following statements are equivalent:*

(i) $D_T\mathfrak{H} \subset (I - T^*)\mathfrak{H}$ *and* $(I - T^*)^{-1}D_T \in \mathfrak{S}_{2p}$;

(ii) $\sup\limits_{0<r<1} (1 - r)^{-1}\|I - W_T(r)W_T^*(r)\|_p < +\infty$;

(iii) $\sup\limits_{0<r<1} (1 - r)^{-p}[1 - D_T(r)] < +\infty$;

(iv) *if $\{z_k\}$ and $\psi_\varepsilon$ are the parameters of the factorization (1.10), then*

$$\sum_k \left(\frac{1 - |z_k|^2}{|1 - z_k|^2}\right)^p < +\infty \quad \text{and} \quad \int_{-\pi}^{\pi} S_{p-1+\varepsilon}(e^{-i\vartheta}r)d\psi_\varepsilon(\vartheta) = O((1 - r)^p) \tag{2.1}$$

*as $r \to 1 - 0$;*

(v) *the following limits exist*

$$\lim_{r\to1-0}(1 - r)^{-p}\|I - W_T(r)W_T^*(r)\|_p^p, \quad \lim_{r\to1-0}(1 - r)^{-p}[1 - \mathcal{D}_T(r)],$$
$$\lim_{r\to1-0}(1 - r)^{-p}\int_{-\pi}^{\pi} S_{p-1+\varepsilon}(e^{-i\vartheta}r)d\psi_\varepsilon(\vartheta).$$

*Moreover, if any of statements* (i) – (v) *is true, then*

$$\|(I - T^*)^{-1}D_T\|_{2p}^{2p} = 2^{-p}\lim_{r\to1-0}(1 - r)^{-p}\|I - W_T(r)W_T^*(r)\|_p^p$$
$$= p2^{-p}\lim_{r\to1-0}(1 - r)^{-p}[1 - \mathcal{D}_T(r)]$$
$$= \sum_k \left(\frac{1 - |z_k|^2}{|1 - z_k|^2}\right)^p + \lim_{r\to1-0}(1 - r)^{-p}\frac{p}{2^{p+1}\pi}\int_{-\pi}^{\pi} S_{p-1+\varepsilon}(e^{-i\vartheta}r)d\psi_\varepsilon(\vartheta). \tag{2.2}$$

*2.1.* The following lemmas are to be used for proving Theorem 2.1.

**Lemma 2.1.** *If* (ii) *is true, then*

$$|\log \mathcal{D}_T(r)| \leq M_p\|I - W_T(r)W_T^*(r)\|_p^p, \quad 0 < r_0 < r < 1. \tag{2.3}$$

**Proof.** From (1.1) it follows that

$$|\log \mathcal{D}_T(r)| = \sum_k \sum_{j=p}^{\infty} \frac{1}{j}[1-\lambda_k(r)]^j = \sum_k [1-\lambda_k(r)]^p \sum_{j=p}^{\infty} \frac{1}{j}[1-\lambda_k(r)]^{j-p},$$

where $\{\lambda_k(r)\}$ is the set of eigenvalues of the operator $W_T(r)W_T^*(r)$. Further, if $r$ is sufficiently close to 1, then

$$\sum_k [1-\lambda_k(r)]^p = \|I - W_T(r)W_T^*(r)\|_p^p \leq M_p'(1-r)^p < \delta < 1.$$

Thus $1-\lambda_k(r) < \sqrt[p]{\delta}$ for each $k$, and we get

$$|\log \mathcal{D}_T(r)| \leq [p(1-\sqrt[p]{\delta})]^{-1} \sum_k [1-\lambda_k(r)]^p = M_p \|I - W_T(r)W_T^*(r)\|_p^p.$$

**Lemma 2.2.** *The following assertions are equivalent:*

$$\text{(a)} \quad \sup_{0\leq r<1} (1-r)^{-p}[1-\mathcal{D}_0(r)] < +\infty, \qquad \text{(b)} \quad \sum_k \left(\frac{1-|z_k|^2}{|1-z_k|^2}\right)^p < +\infty.$$

**Proof.** If (a) is true and $r$ is sufficiently close to 1, then clearly

$$\sum_k \frac{1}{p}[1-B_k(r)]^p \leq \left|\sum_k \sum_{j=p}^{\infty} \frac{1}{j}[1-B_k(r)]^j\right| = |\log \mathcal{D}_0(r)|$$
$$\leq C[1-\mathcal{D}_0(r)] \leq C_1(1-r)^p.$$

Hence

$$\sum_{k=1}^{n} (1-r)^{-p}[1-B_k(r)]^p \leq C_2$$

for any $n \geq 1$. As

$$\left.\frac{d}{dr}B_k(r)\right|_{r=1} = 2\frac{1-|z_k|^2}{|1-z_k|^2},$$

letting $r \to 1-0$ we get

$$\sum_{k=1}^{n} \left(\frac{1-|z_k|^2}{|1-z_k|^2}\right)^p \leq C_3$$

which implies (b) by arbitrariness of $n \geq 1$.

Conversely, if (b) is true, then observe that

$$1-B_k(r) = \frac{(1-r^2)(1-|z_k|^2)}{|1-z_kr|^2} \quad \text{and} \quad |1-z_kr| \geq 2^{-1}|1-z_k|$$

for $|z_k| < 1$ and $0 \le r < 1$. Hence

$$(1-r)^{-1}[1 - B_k(r)] = (1+r)\frac{1-|z_k|^2}{|1-z_k r|^2} \le 8\frac{1-|z_k|^2}{|1-z_k|^2}$$

and by (b)

$$\sum_k [1 - B_k(r)]^p \le C_4(1-r)^p.$$

Consequently

$$|\log \mathcal{D}_0(r)| \le M_p \sum_k [1 - B_k(r)]^p,$$

and finally we come to $1 - \mathcal{D}_0(r) \le |\log \mathcal{D}_0(r)| \le C_4 M_p (1-r)^p$.

**Lemma 2.3.** *If the statement* (i) *of Theorem 2.1 is true, then*

$$\operatorname*{s-lim}_{r\to 1-0}(I-T)(I-rT)^{-1} = I \quad \textit{and} \quad \operatorname*{s-lim}_{r\to 1-0}(I-T^*)(I-rT^*)^{-1} = I.$$

**Proof.** One can easily verify that the left relation holds if

$$\operatorname*{s-lim}_{r\to 1-0}(1-r)(I-rT)^{-1} = 0. \tag{2.5}$$

To prove (2.5), note that for every $h \in \mathfrak{H}$

$$\begin{aligned}\|(I-rt)^{-1}D_T h - (I-T)^{-1}D_T h\| &\le \|(1-r)T(I-rT)^{-1}(I-T)^{-1}D_T h\| \\ &\le \|(I-T)^{-1}D_T h\|.\end{aligned}$$

Consequently, $\lim_{r\to 1-0}(1-r)(I-rT)^{-1}D_T h = 0$ and $\lim_{r\to 1-0}(1-r)(I-rT)^{-1}f = 0$ for any vector $f$ of the form

$$f = \sum_{k=-n}^{n} T^k D_T h_k, \qquad n \ge 0, \ \{h_k\}_{-n}^{n} \subset \mathfrak{H}.$$

As $\|(1-r)(I-rT)^{-1}\| \le 1$ $(0 < r < 1)$, (2.5) holds if $T$ is completely non-unitary [104, Ch. 1]. The relation (2.5) is true also in the case when $T$ has a unitary component $U$. Indeed, 1 is not an eigenvalue for $U$, as it is not an eigenvalue for $T$. Therefore, by the spectral decomposition of $U$

$$\operatorname*{s-lim}_{r\to 1-0}(1-r)(I-rU)^{-1} = 0.$$

*2.2.* **Proof of Theorem 2.1.** For proving (i) $\Rightarrow$ (ii), we use the following well-known equalities for characteristic operator-functions [104, 12]:

$$\begin{aligned}&W_T(\eta)W_T^*(\xi) = I - (1-\eta\bar{\xi})D_T(I-\eta T)^{-1}(I-\bar{\xi}T^*)^{-1}D_T, \\ &W_T^*(\xi)W_T(\eta) \\ &\quad = I - (1-\eta\bar{\xi})W_T^{-1}(0)D_T T^*(I-\bar{\xi}T^*)^{-1}(I-\eta T)^{-1}T D_T W_T^{-1}(0).\end{aligned} \tag{2.6}$$

Hence $I - W_T(r)W_T^*(r) = (1-r^2)D_T(I-rT)^{-1}(I-rT^*)^{-1}D_T$ $(0 \le r < 1)$ and consequently

$$(1-r^2)^{-1}\|I - W_T(r)W_T^*(r)\|_p = \left\|D_T(I-rT)^{-1}(I-rT^*)^{-1}D_T\right\|_p = \left\|(I-rT^*)^{-1}D_T\right\|_{2p}^2.$$

On the other hand,

$$\left\|(I-rT^*)^{-1}D_T - (I-T^*)^{-1}D_T\right\|_{2p} \le (1-r)\left\|(I-rT^*)^{-1}\right\|\cdot\left\|(I-T^*)^{-1}D_T\right\|_{2p} \le \left\|(I-T^*)^{-1}D_T\right\|_{2p}.$$

Thus, $\|(I-rT^*)^{-1}D_T\|_{2p} < C < +\infty$ $(0 \le r < 1)$, and (ii) holds.

(ii) $\Rightarrow$ (iii). By Lemma 2.1, for $r_0 < r < 1$

$$1 - \mathcal{D}_T(r) \le |\log \mathcal{D}_T(r)| \le M_p\|I - W_T(r)W_T^*(r)\|_p^p \le M_p'(1-r)^p.$$

(iii) $\Rightarrow$ (ii). If $\{\lambda_k(r)\}$ is the sequence of eigenvalues of the nonnegative operator $W_T(r)W_T^*(r)$, then

$$\frac{1}{p}\sum_k [1-\lambda_k(r)]^p < \left|\sum_k \sum_{j=p}^{\infty} \frac{1}{j}[1-\lambda_k(r)]^j\right| = |\log \mathcal{D}_T(r)| \le M[1-\mathcal{D}_T(r)] \le M_1(1-r)^p, \quad 0 \le r < 1,$$

and it suffices to see that $\|I - W_T(r)W_T^*(r)\|_p^p = \sum_k [1-\lambda_k(r)]^p$.

(ii) $\Rightarrow$ (i). From (ii) it follows that for $r_0 \le r < 1$

$$(1-r)^{-p}\|I - W_T(r)W_T^*(r)\|_p^p \le K, \quad (1-r)^{-p}\|I - W_T^*(r)W_T(r)\|_p^p \le K.$$

Therefore, using the formulas (2.6) we obtain

$$\|(I-rT^*)^{-1}D_T\|_{2p} \le K_1, \qquad \|T(I-rT)^{-1}D_T W_T^{-1}(0)\|_{2p} \le K_1$$

and moreover

$$\sup_{0 \le r < 1} \|(I-rT^*)^{-1}D_T h\|_{2p} < +\infty \quad \text{and} \quad \sup_{0 \le r < 1} \|(I-rT)^{-1}D_T h\| < +\infty$$

for any vector $h \in \mathfrak{D}_T$. Since 1 is not an eigenvalue for $T$, we have

$$D_T\mathfrak{H} \subset (I-T^*)\mathfrak{H} \quad \text{and} \quad D_T\mathfrak{H} \subset (I-T)\mathfrak{H} \tag{2.8}$$

by the well-known theorem of B.Sz.-Nagy and C.Foias [104, Ch. IV]. Besides, by the same theorem the following strong limits exist:

$$\operatorname*{s-lim}_{r\to 1-0}(I-rT^*)^{-1}D_T = (I-T^*)^{-1}D_T, \quad \operatorname*{s-lim}_{r\to 1-0}(I-rT)^{-1}D_T = (I-T)^{-1}D_T,$$

and using (2.7) we come to the inclusions [37, Ch. III]

$$(I - T^*)^{-1} D_T \in \mathfrak{S}_{2p} \quad \text{and} \quad (I - T)^{-1} D_T \in \mathfrak{S}_{2p} \tag{2.9}$$

which mean that (i) is true.

For proving the equivalence (iii) $\Leftrightarrow$ (iv), we denote the exponential factor in (1.10) as $G(z)$ and observe that (iii) is equivalent to the following pair of conditions:

$$\sup_{0 \le r < 1} (1 - r)^{-p}[1 - \mathcal{D}_0(r)] < +\infty, \quad \sup_{0 \le r < 1} (1 - r)^{-p}[1 - G(r)] < +\infty. \tag{2.10}$$

The last condition obviously is equivalent to the second condition of (2.1). Therefore, application of Lemma 2.2 proves the equivalence (iii) $\Leftrightarrow$ (iv).

(i) $\Rightarrow$ (v). By Lemma 2.3,

$$\underset{r \to 1-0}{\text{s-lim}} (I - T^*)(I - rT^*)^{-1}(I - T^*)^{-1} D_T = (I - T^*)^{-1} D_T,$$
$$\underset{r \to 1-0}{\text{s-lim}} (I - T)(I - rT)^{-1}(I - T)^{-1} D_T = (I - T)^{-1} D_T.$$

Consequently [37, Ch. III]

$$\lim_{r \to 1-0} \left\| (I - rT)^{-1} D_T - (I - T)^{-1} D_T \right\|_{2p}$$
$$= \lim_{r \to 1-0} \left\| (I - rT^*)^{-1} D_T - (I - T^*)^{-1} D_T \right\|_{2p} = 0,$$

and by the equalities (2.6)

$$\lim_{r \to 1-0} \left\| (I - rT^*)^{-1} D_T \right\|_{2p}^2 = \lim_{r \to 1-0} (1 - r^2)^{-1} \left\| I - W_T(r) W_T^*(r) \right\|_p$$
$$= \left\| (I - T^*)^{-1} D_T \right\|_{2p}^2.$$

Further, if $\{\lambda_k\}$ is the set of eigenvalues of the operator $W_T(t) W_T^*(r)$, then using the already proven equivalence (i) $\Leftrightarrow$ (ii) we obtain

$$\sum_k \sum_{j=p+1}^{\infty} \frac{1}{j} [1 - \lambda_k(r)]^j \le \frac{1}{p+1} \sum_k [\lambda_k(r)]^{-1} [1 - \lambda_k(r)]^{p+1}$$
$$\le C \max_k [1 - \lambda_k(r)] \sum_k [1 - \lambda_k(r)]^p$$
$$= C' \left\| I - W_T(r) W_T^*(r) \right\| \cdot \left\| I - W_T(r) W_T^*(r) \right\|_p^p.$$

Thus, by (ii)

$$(1-r)^{-p}|\log \mathcal{D}_T(r)| = \left|\sum_k (1-r)^{-p}\left\{\log \lambda_k(r) + \sum_{j=1}^{p-1}\frac{1}{j}[1-\lambda_k(r)]^j\right\}\right|$$

$$= \sum_k \sum_{j=p}^{\infty} \frac{1}{j}(1-r)^{-p}[1-\lambda_k(r)]^j = \sum_k \frac{1}{p}(1-r)^{-p}[1-\lambda_k(r)]^p$$

$$+ \sum_k \sum_{j=p+1}^{\infty} \frac{1}{j}(1-r)^{-p}[1-\lambda_k(r)]^j$$

$$= \frac{1}{p}(1-r)^{-p}\|I - W_T(r)W_T^*(r)\|_p^p + o(1) \quad \text{as} \quad r \to 1-0.$$

Now (2.11) implies the existence of the limits

$$p \lim_{r\to 1-0} \frac{1-\mathcal{D}_T(r)}{(1-r)^p} = p \lim_{r\to 1-0} \frac{|\log \mathcal{D}_T(r)|}{(1-r)^p} = \lim_{r\to 1-0} \frac{\|I - W_T(r)W_T^*(r)\|_p^p}{(1-r)^p}.$$

Similar to the above argument (but $B_k(r)$ taken instead of $\lambda_k$), we prove that

$$p \lim_{r\to 1-0} (1-r)^{-p}[1-\mathcal{D}_0(r)] = \lim_{r\to 1-0} \sum_k (1-r)^{-p}[1-B_k(r)]^p,$$

and using the factorization (1.10) and Theorem 2.1 we obtain

$$p \lim_{r\to 1-0} \frac{1-\mathcal{D}_T(r)}{(1-r)^p} = p \lim_{r\to 1-0} \frac{1-\mathcal{D}_0(r)}{(1-r)^p} + p \lim_{r\to 1-0} \frac{1-G(r)}{(1-r)^p}.$$

As the equality (2.4) is true,

$$\lim_{r\to 1-0} \sum_k (1-r)^{-p}[1-B_k(r)]^p = 2^p \sum_k \left(\frac{1-|z_k|^2}{|1-z_k|^2}\right)^p.$$

Consequently

$$\lim_{r\to 1-0} \sum_k (1-r)^{-p}[1-\mathcal{D}_T(r)] = 2^p \sum_k \left(\frac{1-|z_k|^2}{|1-z_k|^2}\right)^p$$

$$+ \lim_{r\to 1-0} \int_{-\pi}^{\pi} \frac{S_{p-1+\varepsilon}(e^{-i\vartheta}r)}{(1-r)^p} d\psi_\varepsilon(\vartheta), \qquad (2.13)$$

and the implication (i) $\Rightarrow$ (v) is proved. The converse implication is an obvious consequence of previous results. Further, formula (2.2) follows from (2.11)–(2.13), and the proof of Theorem 2.1 is complete.

We close this section by two remarks related to Theorem 2.1.

**Remark 2.1.** Theorem 2.1 remains valid if we add to (i) – (v), for instance,

(vi) $\sup_{z\in\Gamma}(1-|z|)^{-1}\|I-W_T(z)W_T^*(z)\|_p<+\infty$,

(vii) $\sup_{z\in\Gamma}(1-|z|)^{-p}|1-\mathcal{D}_T|<+\infty$,

where $\Gamma$ is an angle of opening $<\pi$ in $|z|<1$, symmetric with respect to the real axis and with vertex at $z=1$. Moreover, all the limits in Theorem 2.1 exist as $z\to 1$ non-tangentially. The proof of this extension of Theorem 2.1 is simple and it needs no new idea.

**Remark 2.2.** Formula (2.2) will be discussed also in the next section. One has to note that it is not known weather (2.2) can be observed as a regularized trace formula for operators of $C_p$ ($p\geq 2$). The previous considerations may be used to obtain essentially more general relations of (2.2)-type, which are called trace formulas in the case $p=1$ [75]. The existence of $\varepsilon$ in (2.2) brings some dissatisfaction. The way in which $\varepsilon$ appears was explained in Sec. 1. One may get rid of it by letting $\varepsilon\to 0$ but the properties of the limit function $\psi_0$ (which may be even a distribution) would be hard to analyze. At last, Theorem 2.1 has been proved earlier for the case $p=1$.

## 3. $\mathfrak{S}_p$-PERTURBATIONS OF SELF-ADJOINT OPERATORS

The meaning of the condition (i) of Theorem 2.1 becomes more clear for dissipative unbounded operators $A$ whose Cayley transforms belong to $C_p$. Namely, it appears that (i) is equivalent to the representability of $A$ as a sum of a self-adjoint operator and an operator from $\mathfrak{S}_p$.

*3.1.* Denote by $Q_p$ the set of those operators $A$ whose Cayley transforms $T=(A-iI)(A+iI)^{-1}$ belong to $C_p$. Then $A\in Q_p$ if and only if:

1) $\pm i\notin\sigma(A)$,
2) $\mathrm{Im}\,(Af,f)\geq 0$ for all $f\in D(A)$,
3) The operator $iR_{-i}-iR_{-i}^*-2R_{-i}R_{-i}^*$ (where $R_\lambda=(A-\lambda I)^{-1}$) belongs to $\mathfrak{S}_p$.

Besides, an operator $T\in C_p$ appears to be the Cayley transform of another operator from $Q_p$ if and only if 1 is not an eigenvalue for $T$. In addition, the following statement is true.

**Proposition 3.1.** *Let $A\in Q_p$ be an arbitrary operator. Then its Cayley transform satisfies* (i) *of Theorem 2.1 if and only if*

$$A=A_R+iA_I \tag{3.1}$$

*where* $A_R = A_R^*$ *and* $A_I$ *is nonnegative and belongs to* $\mathfrak{S}_p$.

**Proof.** Let $T$ satisfies (i). Then by (2.8)

$$(I - T^*T)\mathfrak{H} \subset (I - T^*)\mathfrak{H} \quad \text{and} \quad (I - T^*T)\mathfrak{H} \subset (I - T)\mathfrak{H}. \tag{3.2}$$

Since $I - T^*T = 2(iR_{-i} - iR_{-i}^* - 2R_{-i}^*R_{-i})$, $D(A) = (I - T)\mathfrak{H}$ and $D(A^*) = (I - T^*)\mathfrak{H}$, the inclusions (3.2) mean that $(iR_{-i} - iR_{-i}^* - 2R_{-i}^*R_{-i})\mathfrak{H} \subset D(A)$ and $(iR_{-i} - iR_{-i}^* - 2R_{-i}^*R_{-i})\mathfrak{H} \subset D(A^*)$. By the first of these inclusions, $R_{-i}^*Th \in D(A)$ for any $h \in \mathfrak{H}$, and $D(A^*) \subset D(A)$ since $T$ is invertible. Similarly, the second relation gives $D(A) \subset D(A^*)$, and so $D(A) = D(A^*)$. Consequently, we can write

$$i(A^* - iI)R_{-i} - iI - 2R_{-i} = 2^{-1}(A^* - iI)(I - T^*T) = 2^{-1}(A^* - iI)D_T^2,$$

and if we take $h = (A + iI)f$ $(f \in D(A)$, then

$$2^{-1}(A^* - iI)D_T^2(A + iI)f = i(A^* - iI)f - i(A + iI)f - 2f = i(A^* - A)f.$$

Thus

$$A_I = \frac{A - A^*}{2i} = (I - T^*)^{-1}D_T^2(I - T)^{-1}. \tag{3.3}$$

By the left relation in (2.9), $A_I \in \mathfrak{S}_p$. In other words, the closure of $A_I$ (which initially was defined on $D(A)$) is a nonnegative operator from $\mathfrak{S}_p$. On the other hand, $A_R = (A + A^*)/2$ is a self-adjoint operator. Indeed, since $A$ is a closed symmetric operator, it suffices to show that its defect index is $(0, 0)$. If we contrarily suppose that this defect index is $(n, m)$ $(n + m > 0)$, then introducing the operator

$$\mathbb{A} = A \oplus (-A) = \mathbb{A}_R + i\mathbb{A}_I, \quad \mathbb{A}_R = A_R \oplus (-A_R), \quad \mathbb{A}_I = A_I \oplus (-A_I),$$

acting in the space $\mathbb{H} = \mathfrak{H} \oplus \mathfrak{H}$ we conclude that the defect index of $\mathbb{A}_R$ is $(n + m, n + m)$. Thus, $\mathbb{A}_R$ has an extension in $\mathbb{H}$. Since $\mathbb{A}_I$ is bounded, also $\mathbb{A}$ has such an extension. The latter is impossible since $A$ and therefore also $\mathbb{A}$ have $\pm i$ as regular points.

Suppose now that $A$ is representable in the form (3.1). Then similarly we obtain that $A_I f = (I - T^*)^{-1}D_T^2(I - T)^{-1}f$ for any vector $f \in D(A)$ $(= D(A^*))$. Besides, for any $f \in D(A)$

$$\|D_T(I - T)^{-1}f\|^2 = ((I - T^*)^{-1}D_T^2(I - T)^{-1}f, f) = (A_I f, f) \le M\|f\|^2$$

since $A_I$ is bounded. Consequently, the operator $S = D_T((I - T)^{-1}$ can be extended by taking its closure to a continuous operator in $\mathfrak{H}$. As $A_I \in \mathfrak{S}_p$ and $A_I = S^*S$, one can easily deduce that $D_T\mathfrak{H} \subset (I - T^*)\mathfrak{H}$ and $S^* = (I - T^*)-1D_T \in \mathfrak{S}_{2p}$.

*3.2.* Now we shall convert the statement of Theorem 2.1 to an assertion for operators $A \in Q_p$. It is convenient to connect such operators with the functions

$$D_A(w) = D_T\left(\frac{i+w}{i-w}\right), \quad T = (A-iI)(A+iI)^{-1}, \quad \text{Im } w < 0, \qquad (3.4)$$

which are holomorphic in the lower half-plane $G^- = \{w : \text{Im } w < 0\}$ and for which the factorization (1.10) takes the form

$$\mathcal{D}_A(w) = \mathcal{D}_0(w)g(w), \qquad (3.5)$$

where $g$ is a non-vanishing function in $G^-$, corresponding to the exponential factor in (1.10), and $\mathcal{D}_0$ is the product

$$\mathcal{D}_0(w) = \prod_k B_k(w)\exp\left\{\sum_{j=1}^{p-1}\frac{1}{j}[1-B_k(w)]^j\right\}, \quad B_k(w) = \frac{1-\lambda_k w}{1-\overline{\lambda_k} w}\frac{1+\overline{\lambda_k} w}{1+\lambda_k w}$$

($\text{Im } \lambda_k > 0$) constructed by the discrete spectrum $\{\lambda_k\} \in G^-$ of $A$.

Note that a formal representation of the function $g$ can be obtained from the formulas

$$g(w) = G\left(\frac{i+w}{i-w}\right), \quad G(z) = \exp\left\{-\frac{1}{2\pi}\int_{-\pi}^{\pi} S_{p-1+\varepsilon}(e^{-i\vartheta}z)d\psi_\varepsilon(\vartheta)\right\}$$

which are true for $w \in G^-$. Nevertheless, $g(w)$ has another, more natural representation given in the below similarity of Theorem 1.2 for unbounded operators.

**Theorem 3.1.** *Let $p \geq 2$ be an integer, and let $A \in Q_p$ be an arbitrary continuously invertible operator. Then for any $\varepsilon \in (0, 1/2)$*

$$\mathcal{D}_A(w) = \mathcal{D}_0(w)\exp\left\{-\frac{1}{\pi}\int_{-\infty}^{+\infty}\frac{d\mu_\varepsilon(t)}{[i(w-t)]^{p+\varepsilon}}\right\}, \quad \text{Im } w < 0, \qquad (3.6)$$

*where $\mu_\varepsilon(t)$ is a real-valued function of locally bounded variation, satisfying*

$$\int_{-\infty}^{+\infty}\frac{|d\mu_\varepsilon(t)|}{(1+|t|)^{2\varepsilon}} < +\infty.$$

*3.3.* For proving Theorem 3.1, we need the following purely technical lemma which we give without a proof.

**Lemma 3.1.** *Let $G(z) \neq 0$ be a holomorphic function in $|z| < 1$, and let $G(z)$ be holomorphic also in a neighborhood of $z = -1$, where it may be written as*

$$G(z) = 1 + \sum_{k=p}^{\infty} a_k(z+1)^k \quad (p \geq 2, \quad |z+1| < r_0 < 1). \qquad (3.7)$$

*If for some* $\alpha \in (p-1, p)$

$$I = \iint_{|z|<1} (1-|z|^2)^{\alpha-1} |\log|G(z)||d\sigma(z) < +\infty \tag{3.8}$$

*(where* $d\sigma(z)$ *is the area element), then for each* $\gamma \in (\alpha-p+1, 1)$ *the function* $g(w) = G\left(\frac{i+w}{i-w}\right)$, *which is holomorphic in* $G^-$, *satisfies*

$$J = \iint_{G^-} \frac{|\mathrm{Im}\ w|^{\alpha-1}}{1+|\mathrm{Re}\ w|^{\gamma}} |\log|g(w)||d\sigma(w) < +\infty. \tag{3.9}$$

**Proof of Theorem 3.1.** Evidently, $-1 \notin \sigma(T)$ (where $T = (A-iI)(A+iI)^{-1}$) since $A$ is continuously invertible. Using the similarity of Theorem 2.1 for the point $z = -1$ (i.e. for the operator $-T$ for which all remains true with the change $z \to -z$), we obtain

$$\lim_{x \to -1+0} (1+x)^{-p}[1 - D_T(x)] = b_p \quad (0 < b_p < +\infty),$$

and moreover

$$\lim_{x \to -1+0} (1+x)^{-p}[1 - G(x)] = a_p \quad (0 < a_p < +\infty),$$

where $G$ is the exponential factor in (1.10). In our case $G(z)$ is holomorphic in a neighborhood of $z = -1$. Hence, $G(z)$ is representable in the form (3.7). Besides, for any $\alpha > p-1$ and any $r \in (0,1)$

$$\frac{1}{\Gamma(\alpha)} \int_0^r (r-t)^{\alpha-1} \left[\int_{-\pi}^{\pi} \log^+ |G(te^{i\vartheta})| d\vartheta\right] tdt$$
$$\le \int_{-\pi}^{\pi} D^{-\alpha} \log^+ |G(re^{i\vartheta})| d\vartheta < M < +\infty,$$

where $D^{-\alpha}$ is the Riemann–Liouville integro-differentiation and $M$ is a constant independent of $r$ (the latter was seen in Proof of Theorem 1.1). Thus,

$$\sup_{0<r<1} \int_0^r \int_{-\pi}^{\pi} (r-t)^{\alpha-1} \log^+ |G(te^{i\vartheta})| tdtd\vartheta \le M.$$

Further, the equilibrium relation between Nevanlinna characteristics of $G(z)$ leads to (3.11) since $G(z) \neq 0$ in $|z| < 1$. Thus, $G(z)$ satisfies the requirements of Lemma 3.1.

From (3.7) it follows that there exists a neighborhood of $w = \infty$, where $|\log|g(w)|| \le C|w|^{-p}$. Assuming $\alpha = p-1+\varepsilon$ $(0 < \varepsilon < 1/2)$ and using Lemma 1.9 of Ch. 1, we conclude that for $w \in G^-$ with enough great modulus

$$|W^{-\alpha} \log|g(w)|| \le C_1 |w|^{-(1-\varepsilon)}, \tag{3.10}$$

where $W^{-\alpha}$ is Liouville's integro-differentiation and $C_1$ is a constant. Hence by (3.9) (where $\gamma = 2\varepsilon$) we find that for any $v < 0$

$$\int_{-\infty}^{+\infty} |W^{-\alpha} \log |g(u+iv)|| \frac{du}{1+|u|^{2\varepsilon}}$$
$$\leq \sup_{v<0} \frac{1}{\Gamma(\alpha)} \int_{-\infty}^{v} (v-t)^{\alpha-1} dt \int_{-\infty}^{+\infty} |\log |g(u+it)|| \frac{du}{1+|u|^{2\varepsilon}} = \frac{J}{\Gamma(\alpha)} < +\infty.$$

On the other hand, by (3.10)

$$\lim_{R\to+\infty} \frac{1}{R} \int_0^{\pi} |W^{-\alpha} \log |g(Re^{i\vartheta})|| \sin \vartheta d\vartheta = 0.$$

But the function $W^{-\alpha} \log |g(w)|$ is harmonic in the lower half-plane (see Lemmas 1.5 and 1.8 in Ch. 1). Consequently, by the results of Ch. 7

$$W^{-\alpha} \log |g(u+iv)| = \frac{1}{\Gamma(p+\varepsilon)} \frac{v}{\pi} \int_{-\infty}^{+\infty} \frac{d\mu_\varepsilon(t)}{(u-t)^2+v^2}, \quad v < 0, \tag{3.11}$$

where $\mu_\varepsilon$ is a measure with the required properties. Using these properties and the corresponding assertion in Ch. 1, one can verify that

$$W^{-(p-\alpha)} \frac{\partial^p}{\partial v^p} \frac{1}{\Gamma(p+\varepsilon)} \frac{v}{\pi} \int_{-\infty}^{+\infty} \frac{d\mu_\varepsilon(t)}{(u-t)^2+v^2}$$
$$= -\mathrm{Re}\, \frac{1}{\pi} \int_{-\infty}^{+\infty} \frac{d\mu_\varepsilon(t)}{[i(w-t)]^{p+\varepsilon}}. \tag{3.12}$$

On the other hand, the properties of $G(z)$ provide that the holomorphic function $\mathrm{Log}\, g(w)$ is expandable in a neighborhood of $w = \infty$ by a power series:

$$\mathrm{Log}\, g(w) = \sum_{k=p}^{\infty} h_k (iw)^{-k}. \tag{3.13}$$

As

$$W^{-(p-\alpha)} \frac{\partial^p}{\partial v^p} W^{-\alpha} (iw)^{-k} = (iw)^{-k} \qquad (k \geq p),$$

successively applying the operators $W^{-a}$, $\partial^p/\partial v^p$ and $W^{-(p-\alpha)}$ to both sides of (3.13) (the termwise application of these operators does not change the uniform convergence of the sum) and then taking real parts, we get

$$W^{-(p-\alpha)} \frac{\partial^p}{\partial v^p} W^{-\alpha} \log |g(w)| = \log |g(w)|, \quad \mathrm{Im}\, w = v < 0.$$

Thus,

$$g(w) = \exp\left\{-\frac{1}{\pi}\int_{-\infty}^{+\infty} \frac{d\mu_\varepsilon(t)}{[i(w-t)]^{p+\varepsilon}} + iC\right\}, \quad \operatorname{Im} w < 0,$$

where $C$ is a real number. It remains to observe that $C = 0$, since $g(w) \to 1$ as $w \to \infty$ and

$$\lim_{v\to-\infty} \int_{-\infty}^{+\infty} \frac{d\mu_\varepsilon(t)}{(|v|-it)^{p+\varepsilon}} = 0.$$

Hence, our assertion follows from (3.5), and the proof is complete.

*3.4.* The next theorem is a consequence of above results.

**Theorem 3.2.** *If $p \geq 2$ is an integer, then for any operator $A \in Q_p$ the following statements are equivalent:*

(i) *$A$ is representable in the form*

$$A = A_R + iA_I,$$

*where $A_R = A_R^*$ and $A_I \geq 0$ belongs to $\mathfrak{S}_p$;*

(ii) $\sup\limits_{-1<v<0} |v|^{-p}[1 - D_A(iv)] < +\infty$;

(iii) *The limit* $\lim\limits_{v\to-0} |v|^{-p}[1 - D_A(iv)]$ *exists.*

*Besides, if $A$ is continuously invertible, then* (i) – (iii) *are equivalent to*

(iv) $\sum\limits_k (\operatorname{Im} \lambda_k)^p < +\infty$ *and the following limit exists:*

$$\lim_{v\to-0} \frac{p}{4^p\pi} \int_{-\infty}^{+\infty} \frac{d\mu_\varepsilon(t)}{|v|^p(|v|-it)^{p+\varepsilon}},$$

*where $\{\lambda_k\}$ and $\mu_\varepsilon$ are the parameters of the factorization (3.5).*

*In addition, if $A$ is invertible and at least one of the statements* (i) – (iv) *is true, then*

$$\begin{aligned} \|A_I\|_p^p &= \frac{p}{4^p} \lim_{v\to-0} |v|^{-p}[1 - D_A(iv)] \\ &= \sum_k (\operatorname{Im} \lambda_k)^p + \lim_{v\to-0} \frac{p}{4^p\pi} \int_{-\infty}^{+\infty} \frac{d\mu_\varepsilon(t)}{|v|^p(|v|-it)^{p+\varepsilon}}. \end{aligned} \tag{3.14}$$

**Proof** directly follows from Theorem 2.1, Proposition 3.1, and Theorem 3.1, if one uses the formulas

$$(1-x)^{-p}[1 - D_T(x)] = (1-v)^p(2|v|)^{-p}[1 - D_A(iv)], \quad v < 0,$$

$$z_k = \frac{\lambda_k - i}{\lambda_k + i}, \qquad \frac{1-|z_k|^2}{|1-z_k|^2} = \operatorname{Im} \lambda_k$$

and the equality (3.3) by which $\|A_I\|_p^p = \|(I-T^*)^{-1}D_T\|_{2p}^{2p}$.

Note that formula (3.14) is an improvement of the inequality

$$\sum_k (\mathrm{Im}\ \lambda_k)^p \le \|A_I\|_p^p,$$

which is well known at least for bounded dissipative operators. Further, note that for $p = 1$ the quantity Sp $(A_I)$ was calculated by the parameters of the factorization of bounded holomorphic functions in a half-plane in [42].

## 4. COMPLETENESS OF OPERATORS IN $C_p$

The aim of this section is to show how the properties of the regularized determinant $\mathcal{D}_T$ may be applied to the completeness problem of non-weak contractions. Recall that an operator is called complete if the closed linear hull of its root subspaces corresponding to the eigenvalues of that operator coincides with the whole space.

*4.1.* In the case of dissipative operators, the following theorem improves a well-known result due to M.V.Keldisch (see [74] or [37, Ch. V]).

**Theorem 4.1.** *Let $p \ge 1$ be an integer, and let $T \in C_p$ be any operator for which 1 is not an eigenvalue. Further, let the spectrum of $T$ be a sequence $\{z_k\}$ having 1 as its unique limit point, and let*

$$\liminf_{r\to 1-0} \log \frac{\mathcal{D}_T(r)}{\mathcal{D}_0(r)} = 0 \tag{4.1}$$

*for the functions $\mathcal{D}_T(r)$ and $\mathcal{D}_0(r)$ defined by (1.2) and (1.8). Then the operator $T$ is complete.*

Before proving this theorem, we shall show that it really is a generalization of the mentioned result of M.V.Keldisch. Indeed, (4.1) is satisfied "with considerable reserve" if

$$\sup_{0<r<1} (1-r)^{-p}[1-G(r)] < +\infty, \quad \text{where} \quad G(z) = \mathcal{D}_T(z)/\mathcal{D}_0(z).$$

If we additionally demand $\sup_{0<r<1}(1-r)^{-p}[1-\mathcal{D}_0(r)] < +\infty$, then also

$$\sup_{0<r<1} (1-r)^{-p}[1-\mathcal{D}_T(r)] < +\infty,$$

and the requirements of Theorem 2.1 will be fulfilled. Thus, by Theorem 3.2, the corresponding operator $A$ (which is the inverse Cayley transform of $T$) will be representable as $A = A_R + iA_I$, where $A_R = A_R^*$ and $A_I \ge 0$, $A_I \in \mathfrak{S}_p$. Further, $A$ and $A_R$ will have the same continuous spectrum since a compact perturbation does not change that. By the conditions of Theorem 4.1, the continuous spectrum of $A_R$ will be located in $\infty$, i.e. $A_R$ will have only discrete spectrum. In this particular case Theorem 4.1 states: a $\mathfrak{S}_p$-perturbation of a self-adjoint operator with discrete spectrum is a complete operator. Namely this statement was established by M.V.Keldisch [74].

If we put $p = 1$ in the formulas for $\mathcal{D}_T$ and $\mathcal{D}_0$, then $\mathcal{D}_T(z)$ will become the usual determinant of $W_T(z)W_T^*(\overline{z})$ and $\mathcal{D}_0(z)$ will become a Blaschke product. Thus, for $p = 1$ the condition (4.1) is also necessary for the completeness of $T$ if this operator is a weak contraction. The latter fact immediately follows from the well-known completeness criterion for weak contractions [37, Ch. V].

*4.2.* **Proof of Theorem 4.1.** Suppose our statement is not true. Then it is clear that $\mathfrak{H}_1 = \text{Clos span}\{\mathcal{L} : |z_k| < 1\} \neq \mathfrak{H}$, and denoting $\mathfrak{H}_2 = \mathfrak{H} \ominus \mathfrak{H}_1$ we conclude that the following triangulation given by the invariant subspace $\mathfrak{H}_1$ is true:

$$T = \begin{pmatrix} T_1 & \Gamma \\ 0 & T_2 \end{pmatrix}, \qquad T_1 = T|\mathfrak{H}_1, \quad T_2^* = T^*|\mathfrak{H}_2.$$

Here $T_2$ is an operator whose spectrum consists of the single point $z = 1$ which is not an eigenvalue. Thus, the characteristic function $W_T(z)$ admits the factorization [12]

$$W_T(z) = W_1(z)W_2(z) \tag{4.2}$$

the factors of which are defined by the formulas

$$\begin{aligned} W_1(z)W_1^*(0) &= I - D_T P_1 (I - zT_1)^{-1} P_1 D_T, \\ W_2(z)W_2^*(0) &= I - W_1^{-1}(0) D_T P_2 (I - zT)^{-1} P_2 D_T W_T^{*-1}(0), \\ W_2(0) &= W_1^{-1}(0) W_T(0), \end{aligned}$$

where $W_1(0)$ is any invertible solution of the equation $W_1(0)W_1^*(0) = I - D_T P_1 D_T$ [12]. In the above formulas $P_1$ and $P_2$ are the orthogonal projections onto $\mathfrak{H}_1$ and $\mathfrak{H}_2$ correspondingly. As the operator $T_1$ is complete, its characteristic function $W_1$ can be represented by the product

$$W_1(z) = W_{z_1}(z)W_{z_2}(z)\cdots W_{z_n}(z)\cdots = \prod_k^{\rightarrow} W_{z_k}(z),$$

where $W_{z_k}(z)$ are operatorial Blaschke factors. This can be checked by the standard methods introduced into operator theory by M.S.Livšic [78]. Note that this infinite product is convergent in the sense that the operator-function

$$\Lambda_n(z) = \left(\prod_{k=1}^{n} W_{z_k}(z)\right)^{-1} W_1(z)$$

uniformly converges to $I$ inside $|z| < 1$ in $\mathfrak{S}_p$-norm (i.e. $\|I - \Lambda_n(z)\|_p \to 0$ as $n \to \infty$). Thus, from (4.2) it follows that as $n \to \infty$

$$W_T^{(n)}(z) \equiv \left( \prod_{k=1}^{n} W_{z_k}(z) \right)^{-1} W_T(z) = \prod_{k \geq n+1}^{\rightarrow} W_{z_k}(z) W_2(z) \to W_2(z)$$

in $\mathfrak{S}_p$-norm, uniformly in $|z| < 1$. Consequently, by the properties of regularized determinants [37, Ch. IV]

$$\det{}_p W_T^{(n)}(z) W_T^{(n)*}(z) \to \det{}_p W_2(z) W_2^*(z) \quad \text{as} \quad n \to \infty. \tag{4.3}$$

On the other hand,

$$\mathcal{D}_T(x)/\mathcal{D}_0(x) \leq \det{}_p W_2(x) W_2^*(x), \quad -1 < x < 1. \tag{4.4}$$

Indeed, by Lemma 1.2 and relations in the end of the proof of Theorem 1.2

$$\begin{aligned} \mathcal{D}_T(x) &= \det{}_p W_T(x) W_T^*(x) = \det{}_p W_{z_1} W_T^{(1)}(x) W_T^{(1)*}(x) W_{z_1}^*(x) \\ &= \det{}_p W_{z_1}^*(x) W_{z_1}(x) W_T^{(1)}(x) W_T^{(1)*}(x) \leq \Phi_1 \det{}_p W_T^{(1)}(x) W_T^{(1)*}(x), \end{aligned}$$

where $\Phi_1$, as before, is that elementary factor of the product $\mathcal{D}_0$, which corresponds to the eigenvalue $z_1$. Evidently, one can apply the same argument to $\det_p W_T^{(1)}(x) W_T^{(1)*}(x)$ and so on. Thus for any $n \geq 1$

$$\mathcal{D}_T(x) \leq \prod_{k=1}^{n} \Phi_k(x) \det{}_p W_T^{(n)}(x) W_T^{(n)*}(x).$$

Letting here $n \to \infty$ and taking into account (4.3) we come to (4.4). Now denote

$$\mathcal{D}_2(z) = \det{}_p W_2(z) W_2^*(\overline{z}).$$

If $\{\lambda_k\}$ are the eigenvalues of the operator $W_2(x) W_2^*(x)$, then one can see that

$$\begin{aligned} |\log \mathcal{D}_2(x)| &= \sum_{k=1}^{\infty} \sum_{j=p}^{\infty} \frac{1}{j} [1 - \lambda_k(x)]^j \\ &= \sum_{j=p}^{\infty} \frac{1}{j} \left\| (I - W_2(x) W_2^*(x))^j \right\|_1 \geq \sum_{j=p}^{\infty} \frac{1}{j} \|I - W_2(x) W_2^*(x)\|^j. \end{aligned} \tag{4.5}$$

Further, by (4.4) and the second condition of our theorem

$$\lim_{r \to 1-0} |\log \mathcal{D}_2(r)| = \lim_{r \to 1-0} \sum_{j=p}^{\infty} \frac{1}{j} \|I - W_2(r) W_2^*(r)\|^j = 0,$$

and consequently

$$\lim_{r\to 1-0}\left\|I - W_2(r)W_2^*(r)\right\| = 0. \tag{4.6}$$

Let $\Gamma$ be the angle with vertex at $z = 1$, defined by the inequality

$$\frac{|1-z|}{1-|z|} < \alpha \qquad (|z| < 1, \quad \alpha > 1).$$

If $z \in \Gamma$, then by formulas (2.6)

$$((I - W_2(x)W_2^*(x))h, h) = (1 - |z|^2)\left\|(I - \bar{z}T_2^*)^{-1}Q_2h\right\|^2, \quad h \in \mathfrak{D}_T,$$

where $Q_2 = P_2D_TW_1^{*-1}(0)$. Besides, it is obvious that

$$(I - \bar{z}T_2^*)^{-1}Q_2h - (I - rT^*)^{-1}Q_2h = (\bar{z} - r)T_2^*(I - \bar{z}T_2^*)^{-1}(I - rT_2^*)^{-1}Q_2h$$

for $|z| = r$ and

$$\frac{|z-r|}{1-r} \le \frac{1-r}{1-r} + \frac{|1-z|}{1-r} < 1 + \alpha = K$$

for $z \in \Gamma$. Consequently

$$\begin{aligned}\left\|(I - \bar{z}T_2^*)^{-1}Q_2h\right\| &\le \left\|(I - rT_2^*)^{-1}Q_2h\right\| + (1-r)^{-1}|z-r|\left\|(I - rT_2^*)^{-1}Q_2h\right\| \\ &\le K\left\|(I - rT_2^*)^{-1}Q_2h\right\|.\end{aligned}$$

Thus $\big((I - W_2(z)W_2^*(z))h, h\big) \le K\big((I - W_2(r)W_2^*(r))h, h\big)$ for any $h \in \mathfrak{D}_T$, and hence by (4.6)

$$\lim_{z\to 1}\left\|I - W_2(z)W_2^*(z)\right\| = 0, \quad z \in \Gamma.$$

Thus, in any angle $\Gamma$ of the mentioned type

$$\left\|(W_2(z)W_2^*(z))^{-1}\right\| \le \sum_{k=0}^{\infty}\left\|I - W_2(z)W_2^*(z)\right\|^k < C_\alpha < +\infty,$$

where $C_\alpha$ is a constant depending solely on $\alpha$. Henceforth, we shall suppose that $\alpha$ is enough great to provide that the opening of $\Gamma$ is greater than $\pi(1 - 2/p)$. Now we may state that for any $h_1, h_2 \in \mathfrak{D}_T$ the function $f(z) = (W_2^{-1}(z)h_1, h_2)$ is holomorphic in the whole complex plane, except the point $z = 1$, and is bounded in $\Gamma$. Besides, the entire function

$$F(w) = f(z), \quad z = (w - i)/(w + i)$$

is bounded in the upper half-plane, except two angles adjoining the real axis, with openings less than $\pi/p$ (since $W_2(z)$ is a unitary operator when $|z| = 1$

[12]). Aimed to apply the Phragmén–Lindelöf principle in that angles, we shall show that $F$ is of order $p$. To this end, we first note that if $|z| < 1$, then

$$\begin{aligned}\|W_2^{-1}(z)\| &= \|W_2^*(\overline{z}^{-1})\| \le C_1 + C_2\|(I - \overline{z}^{-1}T_2^*)^{-1}\| \\ &= C_1 + C_2|z| \cdot \|(T_2 - zI)^{-1}\|.\end{aligned}$$

Now introduce the operator $A_2 \in Q_p$ which is connected with $T_2$ by the equality $T_2 = (A_2 - iI)(A_2 + iI)^{-1}$. If $z = (w - i)/(w + i)$, then a simple calculation gives

$$(T_2 - zI)^{-1} = (2i)^{-1}(w + i)I + (2i)^{-1}(w + i)(i - w)(A_2 + wI)^{-1}.$$

On the other hand, if $B_2 = A_2^{-1}$, then $B_2 = -i((I - T_2)(I + T_2)^{-1}$ and $B_2 - B_2^* \in \mathfrak{S}_p$ since $B_2 - B_2^* = -2i(I + T_2)^{-1}(I - T_2T_2^*)(I + T_2^*)^{-1}$. As the spectrum of the operator $A_2$ is an empty set, $B_2$ is quasinilpotent and hence it is a compact operator [37, Ch. I]. Besides $B_2 \in \mathfrak{S}_p$ according to a theorem due to V.I.Macaev and, as it is well known, the order of the resolvent $(A_2 - wI)^{-1}$ is equal to $p$ [37, Ch. IV]. Thus, the function $F$ is of order $p$. Therefore, applying the Phragmén–Lindelöf principle we conclude that $F$ is bounded in the whole upper half-plane. By the elementary properties of characteristic operator-functions [12], $F$ is bounded also in the lower half-plane. Consequently, the operator-function $W_2$ is constant in the unit disc, i.e. $T_2$ is a unitary operator. As $\sigma(T_2) = 1$, we come to the equality $T_2 = I$ which contradicts the conditions of our theorem. Thus, Theorem 4.1 is proved.

*4.3.* Examining the proof of Theorem 4.1, one can see that a more general result could be proved. Namely, instead of the condition (4.1) one could require the function $|\log(\mathcal{D}_T(x)/\mathcal{D}_0(x))|$ to be bounded by a constant (which can be calculated) depending solely on $p$. Of course, stronger criterions of completeness given in terms of the regularized determinant $\mathcal{D}_T$ are possible. These can be obtained using more delicate Phragmén–Lindelöf type theorems. For instance, the following result is true for the case $p = 2$ being of special interest.

**Theorem 4.2.** *Let $T$ be a contraction from $C_2$, for which 1 is not an eigenvalue. Further, let the spectrum of $T$ be a sequence $\{z_k\}$ with the unique limit point at 1, and let*

$$\liminf_{r\to 1-0}\,(1 - r)\log\frac{\mathcal{D}_T(r)}{\mathcal{D}_0(x)} = 0. \tag{4.7}$$

*Then $T$ is complete. Moreover, for any given $\alpha < 0$ there exists a non-complete operator in $C_2$, for which the lower limit (4.7) is equal to $\alpha$.*

**Proof.** It follows from (4.4) that

$$\liminf_{r\to 1-0}\,(1 - r)\mathcal{D}_2(r) = 0,$$

where $\mathcal{D}_2$ is the function considered in the proof of Theorem 4.1. Suppose $\mathcal{D}_2(r) = \exp\{-\varphi(r)\}$ ($\varphi(r) > 0$). Then $(1-r)\varphi(r) \to 0$ as $r \to 1-0$, according to Theorem 1.2. Further, (4.5) can be written in the form

$$\varphi(r) > -\log(1 - \|I - W_2(r)W_2^*(r)\|) - \sum_{j=1}^{p-1} \frac{1}{j}\|I - W_2(r)W_2^*(r)\|^j.$$

Using this we get $(1 - \|I - W_2(r)W_2^*(r)\|)^{-1} \leq C\exp\{\varphi(r)\}$ which implies

$$\|(W_2(r)W_2^*(r)^{-1}\| \leq \sum_{j=0}^{\infty} \|I - W_2(r)W_2^*(r)\|^j \leq C\exp\{\varphi(r)\}.$$

Turning as earlier to the entire function $F$, one can see that

$$\limsup_{v\to+\infty} v^{-1}\log|F(iv)| \leq 0 \tag{4.8}$$

and $F$ is of second order and minimal type [37, Ch. V]. Application of the Pragmén–Lindelöf principle to the function $F_1(w) = F(w)\exp\{iw\}$ separately in both quadrants of the upper half-plane gives that $F$ is of first order and, consequently, it belongs to M.Cartwright's class [37, Ch. V]. Thus, by (4.8) the indicator of $F$ is non-positive and hence $F \equiv const.$ The end of the proof of completeness is quite the same as that of Theorem 4.1 and we omit it.

Now introduce an obviously non-complete operator $T \in C_2$ as the orthogonal sum $T = T_1 \oplus T_2$, where $T_1$ is a complete operator from $C_2$, for which (4.7) is true, and $T_2$ is a contraction with one-dimensional defect, the characteristic function of which is of the form

$$W_2(z) = \exp\left\{\frac{a}{2}\frac{1+z}{1-z}\right\}, \quad a < 0.$$

Then $\mathcal{D}_T(z) = \mathcal{D}_{T_1}(z)W_2(z)\exp\{1 - W_2(z)\}$ and consequently

$$\limsup_{r\to1-0} \frac{\mathcal{D}_T(r)}{\mathcal{D}_0(r)} = \lim_{r\to1-0}(1-r)\log W_2(r) = a < 0.$$

NOTES. As it became clear from the later work [5], the measure $d\psi_\varepsilon$ appearing in the similar to (1.10) factorization of the regularized determinant $d_T(z)$, is absolutely continuous and belongs to definite O.V.Besov class.

# REFERENCES


[1] Ahlfors L. V., Heins M. (1949). Questions of regularity connected with the Phragmén–Lindelöf principle. *Ann. of Math.*, 50(2), 341-346.

[2] Akiezer N. I. (1961). *The classical moment problem* [in Russian]. Moscow: FM.

[3] Akutowicz E. J. (1956). A qualitative characterization of Blaschke product in a half-plane. *Amer. J. Math.*, 78, 677-684.

[4] Atkinson F. V. (1964). *Discrete and continuous boundary problems.* New York: Ac. Press.

[5] Avetisyan K. L. (1995). Green type potentials and representability of weighted classes of subharmonic functions. *J. of Contemp. Math. Analysis*, 30, 1-28.

[6] Baghdasaryan D. T., Ohanyan I. V. (1990). Boundary properties of the functions $B_\alpha(z,z_k)$ of M.M. Djrbashian [in Russian]. *DAN of Armenia*, 90, 199-205.

[7] Bergh J., Löfsrtöm J. (1956). *Interpolation spaces. An introduction*, Berlin, Heidelberg, New York: Springer-Verlag.

[8] Bergman S. (1950). *The kernel function and the conformal mapping.* Math. Surveys, 5.

[9] Biberbach L. (1914). Zur theorie und praxis der konformen abbildung. *Palermo Rendiconti*, 38, 98-118.

[10] Boas R. P. (1954). *Entire Functions.* New York: Ac. Press.

[11] Brelot M. (1961). *Éléments de la théorie classique du potential.* Paris: Centre de Doc.

[12] Brodski V. M., Gohberg I. T., Krein M. G. (1971). On characteristic functions of invertible operator [in Russian]. *Acta Sci. Math. (Szeged).* 32, 141-164.

[13] Carleman T. (1922). Über die approximation analytischer funktionen durch lineare aggregate von vorgegebenen potenzen. *Arkhiv för Mat., Astr. och Fys.*, 17.

[14] De'Branges L. (1968). *Hilbert spaces of entire functions.* Englewood Cliffs, N. Y., Prentice-Hall Inc.

[15] Djrbashian A. E., Shamoian F. A. (1988). *Topics in the theory of $A^p_\alpha$ spaces*, Leipzig, Teubner Verlag.

[16] Djrbashian M. M. (1945). On canonical representation of functions meromorphic in the unit disc [in Russian]. *DAN of Armenia*, 3, 3-9.

[17] Djrbashian M. M. (1948). On the representability problem of analytic functions [in Russian]. *Soobsch. Inst. Math. and Mech. AN of Armenia*, 2, 3-40.

[18] Djrbashian M. M. (1966). Classes of functions and their parametric representations [in Russian]. In *Contemporary problems in theory of analytic functions, International Conference on the Theory of Analytic Functions, Yerevan, 1965* (pp. 118-137). Moscow, Nauka.

[19] Djrbashian M. M. (1966). *Integral transforms and representations of functions in the complex domain* [in Russian]. Moscow: Nauka.

[20] Djrbashian M. M. (1968) A generalized Riemann–Liouville operator and some of its applications. *Math. USSR Izv.*, 2, 1027-1065.

[21] Djrbashian M. M. (1969). Theory of factorization of functions meromorphic in the unit disc. *Math. USSR Sbornik*, 8, 493-591.

[22] Djrbashian M. M. (1975). Theory of factorization and boundary properties of functions meromorphic in the disc. *Proceedings of the International Congress of Mathematicians, Vancouver 1974, V. 2*, (pp. 197-202), Canadian Mathematical Congress, Vancouver.

[23] Djrbashian M. M. (1984). Some open problems in the theory of representations of analytic functions. In *Lecture Notes in Mathematics, Vol. 1043* (pp. 522-526). Berlin, Springer-Verlag.

[24] Djrbashian M. M., Djrbashian A. E. (1985). Integral representations of some classes of functions analytic in the half-plane [in Russian]. *DAN SSSR*, 285, 547-550.

[25] Djrbashian M. M., Zakarian V. S. (1967). Boundary properties of meromorphic functions of the class $N_\alpha$ [in Russian]. *Izv. Ac. of Sci. of Armenia, Matematika*, 2, 275-294.

[26] Djrbashian M. M., Zakarian V. S. (1968). On factorization of the function $B_\alpha$ [in Russian]. *Mat. Zametki*, 4, 3-10.

[27] Djrbashian M. M., Zakarian V. S. (1969). On boundary properties of meromorphic functions of the class $N_\alpha$ [in Russian]. *DAN SSSR*, 173(121), 1247-1250.

[28] Djrbashian M. M., Zakarian V. S. (1970). Boundary properties of some subclasses of meromorphic functions of bounded type. *Math. USSR Izv.*, 4, 1273-1354.

[29] Djrbashian M. M., Zakarian V. S. (1971). Boundary properties of some subclasses of meromorphic functions of bounded type [in Russian]. *Izv. Ac. of Sci. of Armenia, Matematika*, 6, 182-194.

[30] Djrbashian M. M., Zakarian V. S. (1993). *Classes and boundary properties of functions meromorphic in the disc* [in Russian], Moscow, Nauka.

[31] Dunford N., Schwarz J. I. (1971). *Linear operators*. Vol. II. New York: Interscience Publishers.

[32] Duren P. L. (1970). *Theory of $H^p$ spaces*. New York: Ac. Press.

[33] Frostman O. (1939). Sur les produits de Blaschke. *Fysiogr. Söldsk. Land. föhr.*, 12, 1-14.

[34] Gakhov F. D. (1958) *Boundary problems* [in Russian]. Moscow: GIFML.

[35] Gantmaher F. R. (1959). *Theory of matrices* [in Russian]. Moscow, Nauka. English Transl. (1959), Vols. 1, 2, New York, Chelesa.

[36] Garnett J. B. (1981). *Bounded analytic functions*. New York, Ac. Press, 1981.

[37] Gohberg I. C., Krein M. G. (1969). *Introduction to the theory of linear nonselfadjoint operators*. Providence, RI, AMS.

[38] Goldberg A. A., Ostrovskii I. V. (1970). *Value distribution of meromorphic functions* [in Russian]. Moscow, Nauka.

[39] Govorov N. V. (1965). On the indicator of functions of non-integer order, analytic and of completely regular growth in the half-plane [in Russian]. *DAN SSSR*, 162, 495-498.

[40] Gubreev G. M., Jerbashian A. M. (1991). Functions of generalized bounded type in spectral theory of non-weak contractions. *Rev. Romanie de Math. Pures et Appl.*, 36, 147-154.

[41] Gubreev G. M., Jerbashian A. M. (1991). Functions of generalized bounded type in spectral theory of non-weak contractions. *Journal of Operator Theory*, 26, 155-190.

[42] Gubreev G. M., Kovalenko A. I. (1983). On a new property of the perturbation determinant of non-weak contractions [in Russian]. *Journal of Operator Theory*, 10, 39-49.

[43] Hardy, Gh., Littlewood, J. E. (1932). Some properties of functional integrals II. *Mathematische Zeitschrift*, 34, 403-439.

[44] Hayman W. K., Kennedy P. B. (1976). *Subharmonic functions.* London, Ac. Press.

[45] Hedenmalm H., Korenblum, B., Zhu, K. (2000). *Theory of Bergman Spaces.* Graduate Texts in Mathematics, Berlin, Springer Verlag.

[46] Herglotz G. (1911) Über Potenzreihen mit positiven reelen Teil im Einheitskreise. *Ber. Verh. Sächs Acad. Wiss. Leipzig*, Bd. 63, 501-511.

[47] Hille E., Tamarkin J. D. (1935). On the absolute integrability of Fourier transforms. *Fund. Math.*, 25, 325-352.

[48] Hoffman K. (1962). *Banach spaces of analytic functions.* Englewood Cliffs, N. J., Prentice-Hall Inc.

[49] Jerbashian A. M. (1979). Blaschke type functions for the half–plane. *Soviet Math. Dokl.*, 20, 607-610.

[50] Jerbashian A. M. (1980). Generalization of formulas of R. and F. Nevanlinna and Carleman. *Soviet Math. Dokl.* 21, 248-251.

[51] Jerbashian A. M. Factorization of some general classes of functions meromorphic in a half-plane. In *Colloquia Mathematica Soc. Janos Bolyai, no. 35, Functions, Series, Operators* (pp. 423-429), Budapest.

[52] Jerbashian A. M. (1981). Factorization of some general classes of functions meromorphic in a half-plane. *Soviet Math. Dokl.*, 257, 21-25.

[53] Jerbashian A. M. (1981). Factorization of analytic functions of any finite order in a half-plane. *Soviet Math. Dokl.*, 23, 289-292.

[54] Jerbashian A. M. (1983). Blaschke type functions for the half-plane. *Sov. J. of Contemp. Math. Analysis*, 18, 1-36.

[55] Jerbashian A. M. (1986). Equilibrium relations and factorization theorems for functions meromorphic in a half-plane. *Sov. J. of Contemp. Math. Analysis*, 21, 213-279.

[56] Jerbashian A. M. (1987). Parametric representations of classes of meromorphic functions with unbounded Tsuji characteristic. *Sov. J. of Contemp. Math. Analysis*, 22, 34-60.

[57] Jerbashian A. M. (1988). Classes of functions of bounded type in the half-plane, a uniqueness theorem and a Phragmén–Lindelöf type theorem. *Sov. J. of Contemp. Math. Analysis*, 22, 398-401.

[58] Jerbashian A. M. (1989). Herglotz-Riesz type theorems [in Russian]. *Matem. Zametki*, 45, 19-26.

[59] Jerbashian A. M. (1989). A representation of the Blaschke product for a half-plane. *Sov. J. of Contemp. Math. Analysis*, 24, 46-53.

[60] Jerbashian A. M. (1990). Uniform approximation of functions of generalized bounded type in the half-plane and analog of Akutowicz theorem. *J. of Contemp. Math. Analysis*, 25, 75-83.

[61] Jerbashian A. M. (1991). On boundary behavior of functions of generalized bounded type. *J. of Contemp. Math. Analysis*, 26, 187-209.

[63] Jerbashian A. M. (1991). On functions of bounded and generalized bounded type in the half-plane. *Rev. Roum. Math. Pur. et Appl.*, 36, 425-432.

[63] Jerbashian A. M. (1993). On some classes of subharmonic functions which have nonnegative harmonic majorants in a half-plane. *J. of Contemp. Math. Analysis*, 28, 42-61.

[64] Jerbashian A. M. (1994). On some general classes of subharmonic functions. *J. of Contemp. Math. Analysis*, 29, 29-41.

[65] Jerbashian A. M. (1995). An extension of the factorization theory of M.M.Djrbashian. In *Theory of Functions and Applications, Collection of works dedicated to the memory of M.M.Djrbashian* (pp. 65-74), Yerevan, Louys.

[66] Jerbashian A. M. (1995). An extension of the factorization theory of M. M. Djrbashian. *J. of Contemp. Math. Analysis*, 30, 39-61.

[67] Jerbashian A. M. (1999). On the embedding of Nevanlinna type classes $N\{\omega\}$ [in Russian]. *Izv. of Russian Ac. of Sci.*, 63, 687-705.

[68] Jerbashian A. M. (2002). *Weighted classes of regular functions area integrable over the disc.* Preprint 2002 - 01, Institute of Mathematics, National Ac. of Sci. of Armenia.

[69] Jerbashian A. M., Mikaelyan G. V. (1980). Construction and main properties of a family of Blaschke type functions for the half-plane. *Sov. J. of Contemp. Math. Analysis*, 15, 461-474.

[70] Jerbashian A. M., Mikaelyan G. V. (1986). The asymptotic behavior of Blaschke-type products for the half-plane. *Sov. J. of Contemp. Math. Analysis*, 21, 38-58.

[71] Jerbashian A. M., Mikaelyan G. V. (1991). On the boundary behavior of Blaschke type functions. *J. of Contemp. Math. Analysis*, 26, 435-442.

[72] Kats I. S. (1956). On integral representations of analytic functions mapping the upper half-plane into its part [in Russian]. *UMN*, 11, 139-144.

[73] Keldysch M. V. (1941). Sur les conditions pour qu'un systeme de polynomes orthogonaux avec un poids soit ferme. *Coptes Rendus (Doklady) Academie des Sciences USSR (ns)*, 30, 778-780.

[74] Keldysch M. V. (1951). On eigenvalues and eigenfunctions of some classes of nonselfadjoint equations [in Russian]. *Soviet Math. Dokl.*, 77, 11-14.

[75] Krein M. G. (1987). On perturbation determinants and trace formula for some classes of pairs of operators [in Russian]. *J. of Operator Theory*, 17, 129-187.

[76] Krilov V. I. (1939). On functions regular in a half-plane [in Russian]. *Mat. Sbornik*, 6(48), 95-138.

[77] Levin B. Ja. (1964). *Distribution of roots of entire functions.* Transl. Math. Monographs, Vol. 5, Providence, RI, Amer. Math. Soc.

[78] Livšic M. S. (1954). On spectral resolution of linear nonselfadjoint operators [in Russian]. *Mat. Sbornik*, 3, 144-199. English Transl. (1960). Amer. Math. Soc. Transl. (2), 13, 61-63.

[79] Livšic M. S. (1974). Linear discrete systems and there connection with the theory of factorization of meromorphic functions of M.M.Djrbashian [in Russian]. *DAN SSSR*, 219, 793-796.

[80] Nevanlinna F., Nevanlinna R. (1922). Über die Eigenschaften analytischer Funktionen in der Umgebung einer singulär en Stelle oder Linie. *Acta Soc. Sci. Fennicae*, 50.

[81] Nevanlinna R. (1922). Asymptotische Entwickelungen beschränkter Funktionen und das Steiltjessche Momentenproblem. *Ann. Acad. Sci. Fenn. Ser. AI*, 18.

[82] Nevanlinna R. (1924). Über eine Klasse von meromorphen Funktionen. *Math. Ann.*, 92.

[83] Nevanlinna R. (1925). Über die Eigenschaften meromorpher Funktionen in einem Winkelraum. *Acta Soc. Sci. Fennicae*, 50, 3-45.

[84] Nevanlinna R. (1936). *Eindeutige Analytische Funktionen.* Berlin: Springer.

[85] Paley R.E.A.C., Wiener R.E.A.C. (1934). *Fourier transforms in the complex domain.* New York, Amer. Math. Soc. Coll. Publ., 19.

[86] Privalov I. I. (1937). *Subharmonic functions* [in Russian]. Moscow-Leningrad, GITTL.

[87] Privalov I. I. (1941). *Boundary properties of univalent analytic functions* [in Russian]. Moscow, MGU.

[88] Riesz F. (1911). Sur certaines systemes singulieres d'equations integrales. *Ann. Ec. Norm.*, 28, 33-62.

[89] Rudin W. (1980). *Function theory in the unit ball of* $\mathbb{C}^n$. Berlin, Springer-Verlag.

[90] Saitoh S. (1988). *Theory of reproducing kernels and its applications.* Pitman Res. Notes in Math. Ser., 189, England, Longman Sci.& Tech.

[91] Saitoh S. (1997). *Integral transforms, reproducing kernels and their applications.* Pitman Res. Notes in Math. Ser., 369, England: Addison Wesley Longman.

[92] Samko S. G., Kilbas A. A., Marichev O. I. (1993). *Fractional integrals and derivatives, theory and applications.* Gordon & Breach Sci. Publishers.

[93] Schur J. (1916). Über Potenzreihen, die im Inneren des Einheitskreises beschräuktsind. *J. Reine und Angew. Math.*, 147, 205-232 and 148, 122-145.

[94] Shamoian F. A. (1978). The factorization theorem of M.M.Djrbashian and a characterization of zeros of analytic in the disc functions with majorants of finite order. *Sov. J. of Contemp. Math. Analysis*, 13, 405-422.

[95] Shamoian F. A. (1990). Diagonal mapping and representation questions in anisothropic spaces of functions holomorphic in the polydisc [in Russian]. *Sib. Mat. J.*, 31, 197-216.

[96] Shamoian F. A. (1993). Some remarks on the parametric representation of Nevanlinna–Djrbashian classes. *Math. Notes*, 52, 727-737.

[97] Shamoian F. A. (1999). Parametrical representation and description of the root sets of classes of functions holomorphic in in the disc [in Russian]. *Sib. Mat. J.*, 40, 1420-1440.

[98] Shamoian F. A., Chasova N. A. (2003). Diagonal mappings in Hardy mixed-norm spaces [in Russian]. *Trudi Mat. Senter of N.I.Lobachevski, Kazan* 19, 242-256.

[99] Shamoian F. A., Shubabko E. N. (2000). Parametrical reprezentation of some classes of holomorphic function in the disc. *Operator Thory; Advances and Application,* Basel, Birkhauser Verlag, 113, 332-339.

[100] Shamoian F. A., Shubabko E. N. (2201). On a class of functions holomorphic in the disc [in Russian]. *Investigations on linear operators and function theory, St. Petersburg, POMI*, 29, 242-256.

[101] Shamoian F. A., Yaroslavtseva O. V. (1997). Continuous projectors, duality and diagonal mappings in some mixed-norm spaces of holomorphic functions [in Russian]. *Zapiski Nauch. Sem. POMI*, 247, 268-275.

[102] Shohat J. A., Tamarkin J. D. (1943). *The problem of moments.* New York: AMS.

[103] Solomentsev E. D. (1959). On classes of functions subharmonic in a half-space [in Russian]. *Vestnik MGU, Ser. Math.*, 5, 73-91.

[104] Sz.-Nagy B., Foiaş B. (1970). *Harmonic analysis of operators in Hilbert space.* Budapest, Akademiai Kiado.

[105] Tsuji M. (1950). On Borel's directions of meromorphic functions of finite order. *Tohoku Math. J.*, 2, 97-112.

[106] Tsuji M. (1975). *Potential theory in modern function theory.* Tokyo, Maruzen Co. Ltd.

[107] Walsh J. L. (1935). *Interpolation and approximation by rational functions.* New York, Amer. Math. Soc. Publ.

[108] Weyl H. (1917). Bemerkungen zum Begriff des Differentialquotienten gebrochener Ordnung. *Vierteljahrschrift der Naturforschenden Geselschaft in Zurich*, 62, 296-302.

[109] Wirtinger W. (1932). Über eine minimumaufgabe im gebiet der analytischen funktionen. *Motatshefte für Math. und Phys.*, 39, 377-384.

[110] Zakarian V. S. (1963). On radial boundary values of a class of functions meromorphic in the disc [in Russian]. *Izv. AN SSSR, Ser. Mat.*, 27, 801-818.

[111] Zakarian V. S. (1963). Uniqueness theorems for some classes of functions meromorphic in the disc [in Russian]. *DAN of Armenia*, 36, 3-9.

[112] Zakarian V. S. (1963). Uniqueness theorems for some classes of functions holomorphic in the disc [in Russian]. *Mat. Sbornik*, 10, 3-22.

[113] Zakarian V. S. (1967). On segmental variation of a class of analytic functions [in Russian]. *Izv. Ac. of Sci. of Armenia, Matematika*, 2, 117-122.

[114] Zakarian V. S. (1968). Radial limits and radial variations of the product $B_\alpha$ [in Russian]. *Izv. Ac. of Sci. of Armenia, Matematika*, 3, 38-51.

[115] Zakarian V. S. (1968). On radial boundary values of the function $B_\alpha$ [in Russian]. *Izv. Ac. of Sci. of Armenia, Matematika*, 3, 287-300.

[116] Zakarian V. S. (1974). A growth estimate for meromorphic functions of the class $N_\alpha$ [in Russian]. *Izv. Ac. of Sci. of Armenia, Matematika*, 9, 85-106.

[117] Zakarian V. S. (1978). Segmental variations of the Green type potential [in Russian]. *DAN of Armenia*, 66, 212-215.

[118] Zakarian V. S., Sekhposyan L. A. (1979). A growth estimate for meromorphic functions of the class $N\{\omega\}$ [in Russian]. *Izv. Ac. of Sci. of Armenia, Matematika*, 15, 348-368.

[119] Zakarian V. S., Hunanyan A. G. (1981). Variations of Green type potentials on a segment [in Russian]. *DAN of Armenia*, 73, 3-8.

[120] Zakarian V. S., Madoyan S. V. (1984). On boundary values of functions from classes $N_\alpha$ $(-1 < \alpha < 0)$ [in Russian]. *DAN of Armenia*, 39, 348-368.

[121] Zakarian V. S. (1988). On a growth estimate for the products of M.M.Djrbashian [in Russian]. *Izv. Ac. of Sci. of Armenia, Matematika*, 23, 189-192.

[122] Zygmund A. (1959). *Trigonometric series. Vol. I.* Cambridge, University Press.

# INDEX